NUTRITION and MENTAL HEALTH

NUTRITION and MENTAL HEALTH

RUTH LEYSE-WALLACE, PhD

CRC Press is an imprint of the
Taylor & Francis Group, an **informa** business

CRC Press
Taylor & Francis Group
6000 Broken Sound Parkway NW, Suite 300
Boca Raton, FL 33487-2742

© 2013 by Taylor & Francis Group, LLC
CRC Press is an imprint of Taylor & Francis Group, an Informa business

No claim to original U.S. Government works

Printed in the United States of America on acid-free paper
Version Date: 20121115

International Standard Book Number: 978-1-4398-6335-0 (Hardback)

This book contains information obtained from authentic and highly regarded sources. Reasonable efforts have been made to publish reliable data and information, but the author and publisher cannot assume responsibility for the validity of all materials or the consequences of their use. The authors and publishers have attempted to trace the copyright holders of all material reproduced in this publication and apologize to copyright holders if permission to publish in this form has not been obtained. If any copyright material has not been acknowledged please write and let us know so we may rectify in any future reprint.

Except as permitted under U.S. Copyright Law, no part of this book may be reprinted, reproduced, transmitted, or utilized in any form by any electronic, mechanical, or other means, now known or hereafter invented, including photocopying, microfilming, and recording, or in any information storage or retrieval system, without written permission from the publishers.

For permission to photocopy or use material electronically from this work, please access www.copyright.com (http://www.copyright.com/) or contact the Copyright Clearance Center, Inc. (CCC), 222 Rosewood Drive, Danvers, MA 01923, 978-750-8400. CCC is a not-for-profit organization that provides licenses and registration for a variety of users. For organizations that have been granted a photocopy license by the CCC, a separate system of payment has been arranged.

Trademark Notice: Product or corporate names may be trademarks or registered trademarks, and are used only for identification and explanation without intent to infringe.

Visit the Taylor & Francis Web site at
http://www.taylorandfrancis.com

and the CRC Press Web site at
http://www.crcpress.com

Contents

List of tables xix
Preface... xxi
Introduction xxiii
    How Nutrients Affect Mental Health... xxiv
    Levels of Evidence... xxviii
Author biography... xxxi

Chapter 1   Historical perspective 1
Introduction 1
Scientific contributions.. 1
    Psychological consequences of starvation 1
    Biological psychiatry.. 2
    An experimental self-induced folic acid deficiency 3
    Orthomolecular medicine.. 3
    Other pioneers 5
    Nutrition for the brain 5
    Content of meals affect the brain 6
    Nutrition in psychiatry.. 6
    Functional medicine and biological individuality 7
    Nutritional biochemistry 8
    Nutrigenomics and nutrigenetics 9
Public action in mental health in the United States 9
    National symposium 9
    Surgeon General's Report on Mental Health 10
    New Freedom Commission on Mental Health 10
    Mental Health Parity and Addiction Act... 10
References 11

Chapter 2   Addiction—food, alcohol, and caffeine 13
Introduction 13
Food addiction... 13
    Food as drug model: Neurochemistry and genetics......................... 13
    Malnutrition in food addiction 14
    Assessment for food addiction... 15

      Yale Food Addiction Scale 15
      Self-assessment tools for food addiction 15
Alcohol addiction 15
  Metabolism of alcohol.. 15
  Appetite and nutrient intake 16
  Alcohol and pregnancy 16
  Lipids... 17
      Essential fatty acids, impulsive behavior and alcoholics............. 17
  Vitamins 17
      Interaction of vitamins and alcohol 17
      Nutritional status: Mildly and heavily dependent alcoholics..... 18
      Vitamin relationships... 18
      Predicting thiamin deficiency 18
      Alcoholic neuropathy distinct from thiamin-deficiency
      neuropathy... 19
      Recovery from Wernicke–Korsakoff syndrome............................ 19
      Vitamin A deficiency: A case study 19
      Vitamin C deficiency: A case study 19
      Vitamin K lowered risk of alcoholism: A 30-year follow-up 19
      Moderate level supplements during alcohol rehabilitation......... 19
  Minerals.. 20
      Zinc, copper, and withdrawal from alcohol 20
  Other 20
      Alcohol use and cognitive decline 20
      Alcohol and psychiatric comorbidity 21
      Alcohol and psychiatric comorbidity: Eating disorders 21
      Screening for alcohol use... 21
      Discontinuing use of alcohol—withdrawal and depression........ 22
      Nutrition education and recovery 22
      Alcohol, diabetes, and Antabuse/disulfiram 22
Caffeine addiction... 22
  Caffeine-induced psychosis: A case study 23
  Caffeine tolerance... 24
  Noncoffee caffeine.. 24
  Drinks combining alcohol and caffeine—an FTC warning 24
  Journal of Caffeine Research 25
Conclusions... 26
References 26

Chapter 3   Aggression, anger, hostility, and violence........................... 31
Introduction 31
Lipids 32
  Essential fatty acids, anger, and anxiety... 32
  Violent behavior and essential fatty acids.. 32

Intake of fish, omega-3, and omega-6 fatty acids, and hostility in
young adults 32
Low cholesterol and violent crime 32
Cholesterol-lowering drugs 33
Vitamins 33
Aggression, empty calories, and thiamin 33
Minerals 34
Copper-to-zinc ratios in assaultive young males 34
Selenium (Se): Environmental exposure 34
Supplements 34
Criminal behavior: Lessons from the past 34
Disciplinary infractions in prison and nutrition supplements 35
Reported incidences reduced, but not aggressiveness 36
Juvenile delinquency and vitamin–mineral supplementation 36
Rage and labile mood in two children 37
Other 37
Hostility, BMI, waist–hip ratio, calorie intake, and lipids 37
Anger and metabolic syndrome 37
Cholesterol, violent behavior, and genes 38
Conclusions 38
References 38

**Chapter 4   Autism spectrum disorders (ASDs) and attention
deficit hyperactivity disorder (ADHD) 41**
Introduction 41
Autism spectrum disorders 41
Nutrients and autism 42
Consensus report and evaluation guidelines for pediatric ASD
digestive problems 43
Environmental and genetic factors in autism 44
Autism and environmental factors 45
Autism and nongenetic risk factors 45
Lipids 46
Fatty acid: Mental retardation vs. autism 46
Fatty acids in erythrocytes and plasma lipids 46
Carbohydrates 46
Gene expression of enzymes involved in carbohydrate
metabolism 47
Protein 47
Gluten- and casein-free diet intervention 47
Casein- and gluten-free diets: A Cochrane review 47
Vitamins 47
Vitamin D and autism: Hypothesis 47

       Vitamin B_2, B_6, and magnesium supplements and
       dicarboxylic acids in children with autism 48
  Minerals 48
       Fatal overdose of magnesium 48
       Magnesium and vitamin B_6: A Cochrane review 48
  Supplements 49
       Vitamin/mineral supplement 49
       Lower baseline values normalized with supplements 50
       A systematic review: Folate metabolites, interventions, and
       genes 50
  Other 50
       Research on aspartame 50
       Common diets reported 51
       Digestive enzyme supplementation 51
       Substitutive and dietary approaches 51
       Management using micronutrients and medication 52
       Parental perceptions and treatment choices 52
Attention deficit hyperactivity disorder (ADHD) 53
  Lipids 53
       Effect of high-dose olive, flax, or fish oil on phospholipids 53
       Omega-3 assessment and effect of supplement on behavior 53
  Carbohydrates 54
       Sugar and hyperactivity 54
  Vitamins 55
       Vitamin D and ADHD 55
  Minerals 55
       Zinc and ADHD 55
  Supplements 55
       A review of nutritional supplements by
       the Canadian Pediatric Society 55
  Other 56
       Western diet patterns, adolescents, and ADHD 56
       Self-medication with nicotine 56
Conclusions 56
References 57

**Chapter 5 Genetics, nutrition, and inherited disorders
               of metabolism 61**
Nutrient–gene interaction 61
  Vitamins and genetic stability 61
  Genetic polymorphism and vitamin-dependent enzymes 62
  Biomarker for genome stability influenced by vitamin and
  mineral intake 63
  Alcoholism and genetics: Polymorphism 63

Folate and genetic testing... 64
Inherited disorders of metabolism 64
  Phenylketonuria (PKU) 65
    Phenylalanine requirement as indicated by
    amino acid oxidation 65
    Screening, diagnosis, signs, and symptoms 65
    Treatment 66
    Assessment of response to BH4 therapy 67
    BH4 and patients with psychiatric illness 68
  Lipids 68
    Supplementation with DHA 68
    Plasma DHA and EPA associated with bone mineral density 68
  Minerals 68
    Zinc, selenium, and copper not correlated with
    dietary formula 68
  Other 68
  Galactosemia 69
  Maple syrup urine disease (MSUD) 70
  Homocystinuria 70
Additional information on inherited disorders of metabolism 70
Conclusions 72
References 72

Chapter 6 Intellect, cognition, and dementia 75
Introduction 75
Intellect 75
  Lipids 75
    Infants and essential fatty acids 75
  Vitamins 76
    Vitamin B_{12} and infants 76
    Fluid intelligence: Vegan diets, B_{12}, and adolescents 76
  Minerals 77
    Lead toxicity in adults 77
    Lead toxicity in children 77
    Lead levels in children and effect on children's
    achievement in school 78
    Lower threshold for lead exposure 78
    Lead in spices from India 78
    Lead in Mexican pottery in Oklahoma 79
    Lead, arsenic, and mercury in Ayurvedic medications 79
    Iron deficiency 79
    Iodine and selenium 80
    Zinc 80
  Supplements 81

　　　　Vitamin–mineral supplementation and intelligence in
　　　　school children 81
　　　　Supplements and academic performance in schoolchildren 81
Cognition 82
　　Lipids 82
　　　　Cognition and essential fatty acids: DHA, EPA, and AA 82
　　　　Aging, cognition, and fish oil supplements: Assessment at
　　　　11 and 64 years of age 82
　　　　DHA and cognition in midlife adults free of
　　　　neuropsychiatric disorders 83
　　Carbohydrate 83
　　　　Observation of hypoglycemia and hyperglycemia using
　　　　magnetic resonance imaging 83
　　Vitamins 84
　　　　Homocysteine, B vitamins, brain atrophy, and
　　　　cognitive impairment 84
　　　　Cognitive impairment and vitamin B_{12} 84
　　　　Folic acid, B_{12} supplements, and cognitive decline 84
　　　　Vitamin D and cognition 84
　　　　Evidence report by Agency for Healthcare Research and
　　　　Quality (AHRQ) 85
　　Minerals 85
　　　　Boron, brain function, and cognitive performance 85
　　　　Free plasma copper vs. bound copper 85
　　　　Risks of copper toxicity 86
　　　　High copper, high saturated fat, and trans fats 86
　　　　Effect of vitamin C, E, beta-carotene, zinc, and copper on
　　　　cognition 86
　　Other 86
　　　　Resveratrol and cognition 86
　　　　Cognition and diabetes 87
　　　　Cognitive impairment in diabetes 87
Dementia 87
　Preventing Alzheimer's disease and cognitive decline:
　2010 NIH conference statement 87
　　Lipids 88
　　　　Oral doses of DHA and changes in synaptic characteristics
　　　　in animals 88
　　Vitamins 89
　　　　Alzheimer's disease and niacin 89
　　　　Alzheimer's disease and vitamin E 89
　　　　Brain atrophy and folate 90
　　Minerals 90
　　　　Dementia and minerals/metals 90

      Amyloid aggregation and toxicity 90
      Dementia and copper 90
      Copper and cognition in Alzheimer's disease 91
      Plasma levels of Cu, Fe, and Zn and cognitive function
      differ in men and women 91
      Iron and dementia 91
      Dementia and mercury toxicity 92
      Mercury in older adults 92
Conclusions 92
References 93

Chapter 7   Mood disorders: Depression, bipolar disorder, and suicide 97

Depression 97
    Introduction 97
    Lipids 97
      EPA vs. DHA 98
      Omega-3 fatty acids and depression 98
      Essential fatty acids, depressive symptoms, and neuroticism 99
      No effect of EPA and DHA on depression 99
      Fatty acids in plasma and erythrocytes, and enzyme
      activity in depression 99
      EPA supplements along with standard drugs 100
      EPA treatment for depression 100
      Lipids, zinc, albumin, T cells, and depression 100
      Essential fatty acids and depression: The Rotterdam study 101
      Community-living adults and essential fatty acids 101
      Elderly depressed women, omega-3 fatty acid supplements,
      and quality of life 102
      Cholesterol and the brain 102
      Depression and low cholesterol 102
    Proteins—amino acids 103
      Amino acids, immune-inflammatory response, and
      depression 103
      Tryptophan and chronic insomnia 103
      Tryptophan metabolism and dieting 103
    Carbohydrates 104
      Mixed meals and effect on mood 104
    Vitamins 104
      Ascorbic acid deficiency 104
      Ascorbic acid, depression, and personality changes 105
      Ascorbic acid: Depression, self-induced starvation, and
      scurvy—a case study 105
      Vitamin C and mood 106

Biotin: Sequence of deficiency symptoms 106
B_{12} and depression 107
Folate and depression 107
Folic acid: A Cochrane review 108
Folic acid: A population study 108
Folic acid, homocysteine, depression, and MRI scans 108
Niacin deficiency 109
Riboflavin deficiency 109
Vitamin D and depression 109
Depression and vitamin D status in the elderly 109
Tocopherol and depression 110
Minerals 110
Chromium and depression 110
Chromium supplementation for depression: Case studies 110
Chromium supplements: Toxic or not? 111
Electrolytes and mood 111
Magnesium: Deficiency or excess 111
Selenium 112
Selenium, mood, and quality of life 112
Selenium, depression, and nursing home residents 112
Depression and zinc 113
Supplements 113
Depression and 5-HTP (5-hyroxytryptophan) supplement:
A Cochrane review 113
Caution regarding tryptophan supplements and
drug interactions 114
Serotonin syndrome: Too much of a good thing? 114
Other 115
Vegetarian diets and mood states 115
Dieting and depression: 5-HTP supplements 115
Depression and folic acid fortification 116
Depression and anxiety in adolescents with diabetes 116
Depression and osteoporosis 117
Depression and eating patterns (Healthy Eating
Index scores) 118
Traditional diets vs. Western diets associated with
depression and anxiety 118
Bipolar disorder 118
Introduction 118
Bipolar disorder and comorbidities 119
Bipolar disorder and nutrition-related behavior 120
Genetics 120
Lipids 121
Omega-3 fatty acids and psychiatry 121

　　　　Bipolar disorder and DHA supplement 121
　Amines... 121
　　　　Choline and rapid-cycling bipolar disorder 121
　Vitamins 122
　　　　Folic acid–sensitive genetic sites.................................. 122
　Minerals... 122
　　　　Magnesium/lithium binding sites................................ 122
　Supplements.. 122
　　　　Micronutrient supplementation: A case report........... 122
　　　　ABAB treatment with micronutrient formula 123
　　　　Self-reported multivitamin/mineral supplementation 123
　Other 123
　　　　Inositol and bipolar disorder 123
　　　　Inositol supplementation with medication 123
　　　　Low-inositol diet for low-response patients 124
　　　　Inositol in bipolar disorder patient with psoriasis:
　　　　A case study... 124
　　　　Myo-inositol in food 124
　　　　Complementary and alternative treatments................ 124
　　　　Bipolar disorder and metabolic syndrome 124
　　　　Obesity as bipolar disorder? 125
　　　　Weight gain and lithium.. 125
Suicide.. 125
　Introduction 125
　　　　Diet and mental health in the arctic............................. 126
　Lipids.. 126
　　　　Cholesterol and suicide in anorexia nervosa............... 126
　　　　Cholesterol and suicide: A prospective study 126
　　　　HDL and suicide attempt in healthy women 126
　　　　Leptin, cholesterol, and suicide attempters 127
　　　　Postmortem fatty acids in brain following suicide..... 127
　　　　5-HIAA, cholesterol, and suicide attempts 127
　　　　Review of evidence.. 127
　Other 128
　　　　Food insufficiency and suicidal ideation 128
Conclusions... 128
References... 129

Chapter 8　Schizophrenia.. 137
Introduction 137
　Niacin skin flush test for diagnosis of schizophrenia 138
　Nutrition-related comorbidities 138
　Schizophrenia and oxidative stress................................... 139
Lipids 139

AA and DHA with different diets 139
EPA = placebo 139
Psychopathology and EPA/DHA, vitamins E and C 139
Omega-3 fatty acids and antipsychotic properties:
A Cochrane review 140
Omega-3 fatty acids decreased progression to psychosis 140
Remission of symptoms with EPA: Case study 140
Variation in study results 140
Phospholipids 141
Amino acids and protein 141
One-carbon metabolism and gluten sensitivity 141
Vitamins 141
Folate and negative symptoms 141
Low folate levels associated with fourfold to sevenfold risk of
schizophrenia 142
Vitamin D and mental illness 142
Minerals 142
Zinc and copper in schizophrenic males 142
Magnesium deficiency 143
Minerals (Mg, Zn, Cu) and antipsychotic medications 143
Supplements 143
Vitamin C, oxidative stress, and outcome of schizophrenia 143
Nutritional status and response to supplements 144
Megavitamin therapy 144
Other 145
Tetrahydrobiopterin (BH4) and schizophrenia 145
Water intoxication/dilutional hyponatremia 146
Schizophrenia and calorie needs 146
Schizophrenia and diet history 147
Schizophrenia and diet pattern 147
Prevention of weight gain following start of
antipsychotic medications 148
Programs for management of weight gain induced by
antipsychotic medications 148
Conclusions 149
References 150

**Chapter 9 Starvation, eating disorders, craving, dieting,
and bariatric surgery 153**
Introduction 153
Starvation 154
Mental changes during starvation: The historic work of
Ancel Keys, Josef Brozek, and Austin Henschel 154
Emotional and personality changes 155

Partial completions 156
No change in intellect 157
Behavioral changes related to stress of starvation 157
Eating disorders 158
    Comparison of starvation of conscientious objectors to patients with eating disorders 158
    Eating disorders: Nutritional consequences 160
        Observations of plasma, enzyme functions, and skin 161
        Pellagra in anorexia nervosa 161
        Neuroimaging indicates change in brain 161
        Binge-eating and genes 162
        Twin study suggesting genetic link for eating disorders 162
    Lipids 162
        Cholesterol and suicide in anorexia nervosa 162
        Anorexia nervosa and polyunsaturated fatty acids 163
    Vitamins 163
        Scurvy in anorexia nervosa: A case study 163
    Minerals 164
        Magnesium and anorexia nervosa 164
    Other 164
        BMI: Hospitalization and death 164
Food craving 164
    Craving for chocolate 165
    Craving carbohydrate 165
    Craving with calorie restriction 165
    Brain function and craving 166
Dieting: Fasting and restricting food intake 166
    Physical effects of dieting 166
    Psychological effects of dieting 166
    Health at Every Size (HAES) 167
    A measure of intuitive eating 167
Bariatric surgery 167
    Psychological aspects of severe obesity: Early reports 168
    Candidates for bariatric surgery and psychiatric diagnoses 168
    Previous maltreatment and bariatric surgery candidates 169
    Bariatric surgery and eating disorders 169
    Psychopathology pre- and postbariatric surgery 169
    Long-term effects 170
    Binge eating before and after bariatric surgery 170
    Postsurgical avoidance of food/eating 171
    Anxiety, depression, and quality of life following bariatric surgery ... 171
    Impact of bariatric surgery on health status 172
    Reappearance of symptoms after nine years 172
    Factors influencing unfavorable outcomes of bariatric surgery 173

Psychological support 173
Cognitive behavioral therapy following bariatric surgery 173
Bariatric Quality of Life Index 174
Nutritional effects of bariatric surgery with
psychological implications 174
Pre- and postoperative nutrition 174
Vitamins 174
    Vitamin B_{12} 174
    Thiamin: A case study 175
    Thiamin levels before bariatric surgery 175
Minerals 176
    Calcium 176
    Iron 176
Supplements 176
    Standard supplementation may not be adequate 176
Other 176
    Nutritional monitoring following bariatric surgery 176
Managing the diet following bariatric surgery 177
    Stressful consequences of bariatric surgery 177
Conclusions 179
References 180

Chapter 10 Quality of life, well-being, and stress 185
Quality of life 185
Nutritional quality of life (NQOL) 185
Barriers to using knowledge 186
Well-being and mental energy 186
Vitamins 187
    Thiamin supplement 187
    Annual large-dose vitamin D and mental well-being 187
Supplements 187
    Healthy male volunteers 187
    Supplements, cognition, and multitasking 188
Other 188
    Foods, nutrients, and alcohol related to anxiety, depression,
    and insomnia 188
Stress 189
Oxidative stress 190
Ethane as a biomarker for lipid peroxidation 190
Oxidative stress in psychiatric disorders 190
Amino acids 191
    Lysine and anxiety 191
Carbohydrate 191
    Awareness of hypoglycemia and stress 191

Contents   xvii

        Vitamins 191
                B_6 and bereavement stress 191
        Other 192
                Stress and gastrointestinal symptoms 192
                Psychological distress and inadequate dietary intake in
                immigrant women 192
                Vitamin D hypovitaminosis and chronic pain 193
                Vitamin D and mental well-being 193
Conclusions 193
References 194

**Chapter 11 Additional links between mental status and
                     nutritional status 197**
Introduction 197
Nutrients 197
        Vitamins 197
                Agoraphobia and vitamin status 197
                Niacin for treatment of migraines 199
                Riboflavin and migraine headaches 199
                Symptoms of vitamin B_{12} deficiency 200
                Inositol: Review of effects 200
        Minerals 201
                Copper and Wilson's disease 201
                Mercury toxicity 201
                Mercury in fish 203
                Mercury in consumers of large amounts of fish 203
                Selenium 204
                Selenium levels and enzyme activity in epileptic and
                healthy children 205
                Selenium and seizures: a case study 205
                Selenium and quality of life 205
                Magnesium 206
                Magnesium and migraine headache 206
                Manganese 206
        Other 206
                Choline, cognition, and dementia 206
                Obsessive compulsive disorder and micronutrients
                following cognitive behavioral therapy 207
                Low fat vs. low carbohydrate diet: Effect on mood and
                cognitive function 207
                Unexplained medical symptoms not necessarily psychiatric
                problems 207
                Gluten sensitivity and psychiatric presentation:
                A case study 208

Another dimension 208
  Spiritual traditions and diet 208
  Biofeedback 209
  Biology of Belief 210
Conclusions 212
References 212

Chapter 12 Conclusions and recommendations 215
Introduction 215
  Nutrition education for mental health professionals 216
Nutritional assessment 216
  Assessment for nutritional status 217
    Diet assessment 217
    Nutrition-focused physical examination 217
    Laboratory assessment 218
Suggestions for research in nutrition and mental health 218
Selected topics of interest 220
Notes regarding specific nutrients 221
  Lipids: EPA, DHA, essential fatty acids, and omega-3 fatty acids 221
    Precautions regarding essential fatty acids 222
  Protein and amino acids 222
    Protein 222
    Amino acids 224
  Carbohydrates 224
    Blood glucose levels 225
  Vitamins 225
    Vitamin B_{12} assessment 225
    Vitamin B_{12} treatment 226
    Low-dose vitamin B_{12} supplementation 226
    Thiamin 226
  Minerals 227
    Chromium 227
    Magnesium 227
    Mercury 227
    Mercury in fish 228
    Selenium 228
Concluding statement 228
References 229

Appendix A: Tools for nutritional assessment 231
    Stages of nutritional injury 255
Appendix B: Mental health concerns: Nutritional concerns matrix 257
Glossary 265
Index 279

List of tables

Table I.1	How Nutrients Affect Mental Health	xxiv
Table I.2	Levels of Evidence	xxviii
Table 1.1	A Timeline: Nutrition and Mental Health	2
Table 1.2	Vitamin B_{12} Status in Patients Admitted to a Psychiatric Hospital	4
Table 1.3	Excretion Rates of Ascorbic Acid in Patients with Schizophrenia and Controls	5
Table 1.4	Links between Nutrients and Psychiatric Concerns	7
Table 2.1	Calories in Alcoholic Beverages	16
Table 2.2	Common Foods and Beverage Sources and Approximate Caffeine Content	25
Table 3.1	Cu:Zn Ratios Found in Assaultive Youth	34
Table 4.1	Changes Following Vitamin/Mineral Supplementation of Children with Autism	49
Table 5.1	Classifications of Altered Phenylalanine Metabolism	66
Table 6.1	Low, Intermediate, and High Mineral Levels in Plasma	91
Table 7.1	Prevalence of Associated Symptoms in Major Depression	98
Table 7.2	Fasting Serum Phospholipids as Percentage of Total Lipids	99
Table 7.3	Changes Observed with Vitamin C Deficiency	105
Table 7.4	Associations of Modified Diet with Mood Changes	112
Table 7.5	Comparison of Vegetarians and Omnivores	116
Table 8.1	Ratios of Zinc and Copper in Criminals with Schizophrenia and Noncriminals with Schizophrenia	143

Table 8.2	Changes Observed in Subjects Who Took Recommended Supplements	145
Table 8.3	Changes Observed in Control Subjects	145
Table 9.1	Selected Lesions and Frequencies Observed in Patients with Eating Disorders	160
Table 9.2	Evidence of Genetic Links in Eating Disorders	163
Table 9.3	Suicide Rates per 10,000 Individuals	172
Table 9.4	Nutritional Status Prior to Bariatric Surgery	175
Table 9.5	Thiamin Levels before Bariatric Surgery	176
Table 9.6	Recommendations for Monitoring Nutrient Deficiencies Following Bariatric Surgery	178
Table 11.1	Abnormal Nutritional Status in Agoraphobic Patients ($N = 12$)	198
Table 11.2	Copper: Facts, Signs, and Symptoms, and Treatments for Wilson's Disease	202
Table 11.3	Screening of Patients for Blood Mercury Levels	204
Table 11.4	Selenium and Epilepsy	205
Table 12.1	Recommendations for Intake of Essential Fatty Acids	222
Table 12.2	Food Sources of Essential Fatty Acids	223
Table 12.3	Approximate Protein Obtained from a Variety of Food Groups	224
Table 12.4	Recommended Intakes of Essential Amino Acids for Adults	224
Table 12.5	Comparison of Measures of Blood Glucose Levels	225
Table A.1	Selected Standards for Assessment of Nutritional Status	241
Table A.2	Clinical Signs Potentially Related to Nutritional Deficiencies	246
Table A.3	Selected Drug: Nutrient/Food Interactions	250
Table A.4	Selected Food Sources of EPA (20:5n-3) and DHA (22:6n-3)	252
Table A.5	Brief Assessment of Dietary History	253
Table B.1	Mental Health Concerns: Nutritional Concerns Matrix	258
Table B.2	Health Conditions or Health Histories That Warrant Evaluation for Nutritional Status	262
Table B.3	Level I Evidence Reports	263

Preface

I have had a great interest in physical and mental health as well as other areas of science for over 40 years, beginning with my first job as clinical dietitian at Osawatomie State Hospital in Osawatomie, Kansas. This interest expanded during my work at The Menninger Foundation in Topeka, Kansas, and at Mesa Vista Hospital in San Diego, California, as well as in other positions. I have a deep interest in the study of the person as a whole, integrated being; the physical, intellectual, emotional, and spiritual aspects interacting and influencing each other. The growth of knowledge and capability of discovery in these areas is awesome, in the true sense of the word.

Acknowledging that nutrition and biochemistry are not the entire answer to all health problems (mental or physical), I believe consideration of nutritional status is a vital part of preventing, diagnosing, and treating many health conditions. If for no other reason, it should be ruled out as an influence, if not a cause, for a health condition.

This book is a selection from the huge body of historical and current literature published in scientific journals that drew my attention relevant to the proposal that nutrients and nutritional status influence mental health. I hope this selection will stimulate the interest of readers about the subject, illustrate the diverse research published, and give a glimpse of complementary or supplementary views. Some selections are included as reports of research; others are intended as resources for readers wanting to delve further into a subject. As a scientific community, we are just beginning to be *able* to study the combination of nutrition and mental health in the depth necessary for meaningful conclusions. As fundamental as they are to a healthy physical life, I believe we will ultimately realize how deeply nutrients and nutritional status affect mental life as well. I can hardly wait to see what happens next.

I propose a new term for conceptualizing this bidirectional influence: the PsychoNutriologic Person (PNP). I define a PNP as the individual who has both nutritional and mental health concerns that interact and influence quality of life. This term parallels the terms Pharmacologic Person,

Functionologic Person, and Psychodermatologic Person, the focus of pharmacy, nursing, and dermatology, respectively.

If all research on mental disorders included an assessment of the nutritional biochemistry and genetics of each participant, with scientists and clinicians establishing and reporting this aspect of equality between experimental groups, I suspect new knowledge would explain some discrepancies present at this time.

Recognizing that some readers are trained in a field other than nutrition, I include some explanations that may be familiar to nutritionists. Likewise, some explanations for nutritionists who are not trained in mental health are included. Interested nonprofessionals are also accommodated with additional explanations.

I believe the content of this book demonstrates that nutritional status influences mental status, in a variety of ways, at different levels of severity. My hope is that in-depth assessment of nutritional status and research on outcomes of nutritional interventions for improved mental status will be funded more liberally and that the evidence gained will be used clinically on an individually focused, routine basis. What a waste it is to *not* use an inexpensive and intuitively accepted modality such as good nutrition in support of mental health.

I commend you, the reader, for your interest and thoughtful consideration of these complex aspects of the human condition.

Introduction

More than twice as many Americans (62%) fear losing their mental capacity as they age than fear a diminished physical capacity (29%) (Parade 2006). This finding from a poll of 1000 Americans regarding attitudes about aging may be partly captured in the quip, "Of all the things I've lost, I miss my mind the most." During all stages of life, ordinary people want to have more energy, be able to function well at work or school, and cope effectively with the stresses of life, all of which may be influenced by nutrition.

Vitamin and mineral deficiencies are thought to be rare, and an advanced deficiency necessary, before health is affected. However, many common health conditions deplete nutritional status, including suboptimal dietary intake, frequent alcohol intake, changes in appetite due to aging, food allergies or sensitivities, special diets, and eating disorders. Additionally, inadequate income for food will affect dietary choices and intake. All may contribute to a deficient state of nutrition with potential to alter mental status. Inherited genes and genes altered by the environment are now known to be influenced by nutrients and influence mental status as well.

People express an intuitive sense that "of course mental health is influenced by nutrition." However, the intuitive sense does not satisfy the scientist's or clinician's desire to understand when, how, and why this might be so. Nutrition and mental health are each difficult to describe and measure precisely. Combined, the study of the interactions between aspects of life is even more complex. Both areas, involving human subjects, contain fundamental ethical issues plus issues of the adequacy of science to explain and predict. Reductionist research does not ultimately tell the story of an entire, living being.

We eat whole foods rather than capsules of nutrients and real food does not lend itself to double-blind experimental conditions: taste, aroma, and mouth feel are giveaways. Mental status is inner and personal, and therefore described differently by each experimental subject or investigator. Physical measures such as MRIs, blood levels of one molecule, or an active enzyme can be recorded at one, or several, moments but rarely capture the connectivity of life. Researchers do not take many brain samples or require lengthy periods of eating a single food or food combination. Those who volunteer for experiments may have a built-in bias that makes results invalid for application to others.

Despite the inherent complications, it is the pull of wanting to understand, predict, and influence that drives theoretical scientists and clinical professionals to observe, experiment, and apply new knowledge for the betterment of the human condition.

The how and why of interactions between nutrition and mental health are a two-way street. It is readily understood that a variety of mental health conditions may affect ability and interest in getting and eating a well-rounded selection of foods. In addition to knowing that nutrition affects mental status, it is important to understand *how* nutrition affects mental status. Since nutrients affect mental health at the metabolic and genetic level, effects are more difficult to discover and not easily observable. The accompanying summary in Table I.1 describes how nutrients potentially affect mental health. The body of this book provides selected reports of historical and recent science that explain and add to this understanding.

Table I.1 How Nutrients Affect Mental Health

1. Nutrients and other elements may support or interfere with normal development and maintenance of the brain and central nervous system of the fetus, growing child, or adult. (Essential fatty acids are essential to myelin sheath formation. Altered metabolism and excretion of amino acids or carbohydrates, such as phenylalanine or galactose, may result in accumulation of levels toxic to the brain.)
2. Nutrients may serve as precursors to the manufacture of neurotransmitters. They may contribute skeletons of a molecule or a required component of the neurotransmitter molecule. (Folic acid is a source of methyl groups.)
3. Nutrients are needed to supply the brain with an energy source and the ability to use the energy. (Carbohydrate is needed for glucose, which is the brain's primary energy source.)
4. Nutrients may influence genetic transcription. (Nutrients influence metabolic signals for stimulating or failing to signal for transcription of genes.)
5. Nutrients may have pharmacologic functions at doses higher than nutriologic requirements—either to accommodate altered genetic transcription or as toxic elements. (Nutrients may function as pro-oxidants as well as anti-oxidants.)
6. Nutrients and food contribute to mood, sense of well-being, and psychological function, perhaps related to changes in nutritional status. (Mood may change because of food restriction, dieting, and starvation.)
7. There may be changes in the entrance or exit of nutrients through receptors on cell surfaces related to thoughts and emotions. (Stress-related dysfunction of cells and systems may affect nutrient use.)

Four similar explanatory models were proposed by Kaplan et al. (2007) to serve as possible explanations for why micronutrient supplementation could ameliorate some mental symptoms.

- Expressions of inborn errors of metabolism
- Manifestations of deficient methylation reactions
- Alterations of gene expression by nutrient deficiency
- Long latency deficiency diseases

The defining terms: mental health, mental status, mental illness, mental disorder, well-being, mental energy, and the "worried well," are important in the understanding the status of a person's psychological well-being.

In his book, *The Human Mind* (1930), Dr. Karl Menninger of The Menninger Foundation summarized mental health:

> Mental health is the adjustment of human beings to the world and to each other with a maximum of effectiveness and happiness, not just efficiency or contentment, or the grace of obeying the rules of the game cheerfully. It is all of these together. It is the ability to maintain an even temper, an alert intelligence, socially considerate behavior and a happy disposition. This, I think, is a healthy mind.

The World Health Organization (2010) describes mental health as "not just the absence of mental disorder. It is defined as a state of well-being in which every individual realizes his or her own potential, can cope with the normal stresses of life, can work productively and fruitfully, and is able to make a contribution to her or his community".

Mental health is defined as a state of successful performance of mental function, resulting in productive activities, fulfilling relationships with other people, and the ability to adapt to change and to cope with adversity. Mental health is indispensable to personal well-being, healthy family and interpersonal relationships, and positive contribution to community or society (HealthyPeople.gov.2010).

Mental energy is defined as the ability to perform mental tasks, the intensity of feelings of energy/fatigue, and the motivation to accomplish mental and physical tasks (International Life Sciences Institute Workshop 2006).

Nutritional health and nutritional status

Nutrients at inadequate, excessive, or imbalanced levels may produce a variety of effects on the brain and central nervous system. These effects may range from preventable mental retardation to transient discomfort resulting from low blood glucose or caffeine cravings. Food insecurity can influence the quality of life or even the will to live. The will, beliefs, emotions, and decisions made by the mind, as well as environmental influences such as culture and food availability, ultimately result in our food and nutrient intake. The emotional influence of food on the experience of comfort and security begins in infancy and continues throughout life.

Nutrition has been defined as the sum of the process concerned with growth, maintenance, and repair of the living body as a whole, or studied *in vitro* (in test tubes), within cells, organs, entire animals, or humans.

Nutritional status refers to the health of an individual as it is affected by intake, storage, and use of nutrients and nutritional support of normal body functions.

Clinical nutrition generally refers to the application of nutritional knowledge to the individual for growth, maintenance of health, or prevention of disease.

Mental status, nutritional status, and quality of life

At any given time, an individual falls somewhere on a continuum of mental health and mental illness, although mental illness and mental health are not quite "opposites." The point on the continuum at a given time may be described as one's mental status. A diagnosis of a mental illness "medicalizes" mental status in a sense, while mental health and mental status encompass everyday symptoms and states of mind. The totality of one's situation may be thought of as the "quality of life." Quality of life (QOL) has been described as "enjoying life," "being happy and satisfied with life," and "being able to do what you want to do when you want to do it."

At any given time, an individual also falls somewhere on a continuum of nutritional health. The point on the continuum at a specified time may be described as one's nutritional status. Nutritional status may also be defined in terms of nutritional injury. The continuum of nutritional status/nutritional injury ranges from (1) healthy or possible risk of nutritional injury to (2) diminishment of body stores to (3) biochemical alterations to (4) nonspecific, physically observable, but

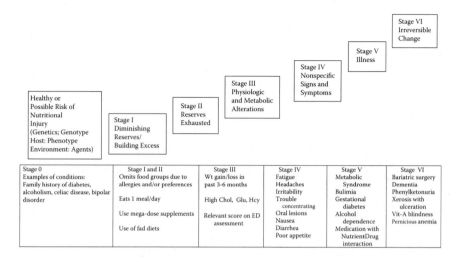

Figure I.1 Stages of nutritional injury. (Adapted from Arroyave, G., Genetic and biologic variability in human nutrient requirements, *Am J Clin Nutr* 1979; 32:486–500. With permission.)

reversible signs and symptoms of nutritional injury to (5) development of a defined psychiatric or medical condition to (6) irreversible signs and symptoms of nutritional injury (Arroyave 1979) (see Figure I.1). An individual's nutritional quality of life (NQOL) may include "being able to eat what I want, when I want." The taste, pleasure, enjoyment, comfort, or socialization that we experience from food and mealtimes are part of NQOL (Winkler 2010).

Research in mental health and nutrition

Scientists from neurosciences, biochemistry, biology, medicine (allopathic, complementary, and functional medicine), psychiatry, psychoneuroimmunology and psychology, nursing, behavioral science, genetics, metabolomics, and nutrition all research the influence of diet and supplements on behavior and mental functioning.

This book gathers and reports on selected reports of research from scientific literature linking nutrition and nutrients to mental health, mental illness, and mental functioning. Chapters address psychiatric diagnoses or general mental concerns along with reports of knowledge and research regarding the associated nutrients or diet. An indication of the level of evidence offered is included.

The following adapted levels of evidence will be used in describing the literature reported (Table I.2).

Table I.2 Levels of Evidence

Level I	Experimental, prospective, randomized, double-blind, placebo-controlled, clinical trials; not all characteristics present in all reported studies
Level II	Observational, analytical, descriptive, case control, cohort, cross-sectional, epidemiological analysis of data collected for study of multiple outcomes
Level III	Meta-analyses and reviews; reviews may be formal, systematic reviews; less formal reviews may be included in the category "expert opinion"
Level IV	Individual case studies, expert opinion
Level V	Other: hypotheses, theory, instrument validation

Sources: U.S. Department of Health and Human Services; Therapeutic Goods Administration; Rechkemmer 2005.

Prospective, randomized, double-blind, placebo-controlled clinical trials may allow a conclusion of cause and effect. They are especially valuable if validated by various researchers in different populations.

Observational studies provide evidence that two conditions may be related in some way. Correlations may be positive or negative and generate hypotheses for further study. Confounding factors not accounted for may introduce bias. Accounting for 100% of circumstances that influence a human body or real life is impossible due to inner, personal experiences and perceptions.

Meta-analyses allow statistical combining of data from a number of studies with perhaps smaller numbers of participants to see if the combined outcomes give a more meaningful picture. Studies combined need to have similar procedures, standards, and definitions. Food for thought includes a view of meta-analyses succinctly described by Owens (1994): When no other answers are available, the results of pooled analyses are better than nothing, and contrarily, no answer is better than one that is probably misleading.

Case studies reflect observations of one subject by a limited number of observers, and may raise questions from which additional hypotheses are defined. Research from all levels of the literature is reported, illustrating where we have been, where we are, and suggesting where we need to go.

References

Arroyave, G., 1979. Sequence of events in the development of clinically evident nutritional disease. Figure 2 in Young, V. R. and Scrimshaw, N. S. Genetic and Biologic Variability in Human Nutrient Requirement. *Am J Clin Nutr* 1979; 32:486–500.

World Health Organization, 2007. *What Is Mental Health*. http://www.who.int/features/qa/62/en/index.html (accessed 8/3/2010).

HealthyPeople.gov.2010. Mental Health and Mental Disorders. www.healthypeople.gov/2020/topicsobjectives2020/overview.aspx?topicid=28, (accessed 11/29/12).

International Life Sciences Institute Workshop. Mental Energy: Defining the science. From the International Life Sciences Institute Workshop. Washington DC: *Special Supplement to Nutr Rev* 64(7) (2006):(Part II) (S1–S16).

Kaplan, BJ, SG Crawford, CJ Field, and JS Simpson. Vitamins, minerals, and mood. *Psychol Bull*. 2007; 133(5):747–760.

Menninger, K. *The Human Mind*. Alfred A. Knopf. New York and London, 1930. p. 447.

Owens, D. Cholesterol and violent death. Editorial. *BMJ* 1994; 309:1228.

Parade Publications. What Americans think about aging and health. *Par Mag*. February 5, 2006: 11.

Rechkemmer, G. Existing approaches for evaluating the health effects of functional foods, PASSCLAIM. International Life Sciences Institute-Europe. See http://ods.od.nih.gov/pubs/conferences/bfc2005/Approaches%20for%20Evaluating%20Health%20Effects%20of%20Functional%20Foods%20G-Rechkemmer.pdf

Therapeutic Goods Administration. Glossary of Terms, Guidelines for levels and kinds of evidence to support claims. 2001. http://www.tga.gov.au/docs/pdf/tgaccevi.pdf (accessed 8/16/2010).

U.S. Department of Health and Human Services. Systems to Rate the Strength of Scientific Evidence: Evidence Report/Technology Assessment Number 47. Prepared for the Agency for Healthcare Research and Quality (AHRQ). http://www.ncbi.nlm.nih.gov/bookshelf/br.fcgi?book=hserta&part=A73054 (accessed 8/16/2010).

Winkler, MF. 2009 Lenna Frances Cooper Memorial Lecture: Living with enteral and parenteral nutrition: How food and eating contribute to quality of life. *J Amer Diet Assoc* 2010; 110 (2):169–177.

World Health Organization. 2010. Mental health: A state of well-being. http://www.who.int/features/factfiles/mental_health/en/(accessed 11/29/12).

Author biography

Ruth Leyse-Wallace, PhD, RD, received her BS degree from the University of California at Davis as a member of Phi Kappa Phi, earned her MS degree while completing her dietetic internship at the University of Kansas Medical Center in Kansas City, and in 1998 was awarded her PhD from the University of Arizona in Tucson, Arizona.

She began practicing clinical dietetics at Osawatomie (Kansas) State Hospital, then at the Menninger Foundation in Topeka, Kansas, from 1977 to 1984. She was later employed at Mesa Vista Hospital (now Sharp-Mesa Vista) in San Diego, California, and HCA Willow Park Hospital in Plano, Texas. Her practice included providing nutritional care for patients of all ages hospitalized for eating disorders, alcohol and drug abuse, and general psychiatric diagnoses. While attending graduate school in Tucson, Dr. Leyse-Wallace served at Sierra Tucson and Hospice Family Care, as well as Group Health Medical Associates, a general medicine clinic, on a part-time basis. She has served as an adjunct faculty member at Pima County College in Tucson and Mesa College in San Diego. A long-term member of the American Dietetic Association (ADA), she has been an active contributor to the Behavioral Health Nutrition dietetic practice group in the ADA (now the Academy of Nutrition and Dietetics).

Dr. Leyse-Wallace retired from clinical practice and engages in professional writing and speaking.

chapter one

Historical perspective

Introduction

New terms and concepts such as behavioral phenotypes, biotransformation, epigenetics, biological psychiatry, functional medicine, mental energy, metabolomics, nutraceuticals, nutritional neuroscience, psychobiology, PsychoNutriologic Person, transgenic, and systems biology are being coined in psychiatry, medicine, and nutrition, supplementing the older concepts of biochemical individuality and orthomolecular medicine. Perhaps as the science and methodologies of nutrition and mental health continue to evolve, a new integration will emerge, although the ethics of clinical studies makes the study of these two complex areas especially difficult.

> The *PsychoNutriologic Person* is defined by the author as an individual who has both nutritional and mental health concerns that interact and influence the quality of life.

Scientific contributions

See Table 1.1 for a timeline of contributions to the study of nutrition and mental health.

Psychological consequences of starvation

In the 1940s, Ancel Keys, Josef Brozek, and Austin Henschel conducted a scientific study of starvation with volunteer conscientious objectors at the University of Minnesota, in a setting for optimal monitoring and observation by a multidisciplinary team. The twenty-four male participants were provided approximately two-thirds of their needed calories. This semistarvation period lasted six months during which volunteers lost approximately 25% of their body weight. Rehabilitation was monitored after the period of semistarvation, and some experimental groups received a multivitamin supplement.

Participants were monitored for physiological, radiological, biochemical, and psychological changes (behavior, personality, and intelligence). The authors commented that "the bond between the physiological status of the organism and the 'psyche' is closer than is sometimes realized.

Table 1.1 A Timeline: Nutrition and Mental Health

1940s: Ancel Keys, Josef Brozek, and Austin Henschel conducted a scientific study of starvation with volunteer conscientious objectors at the University of Minnesota.
1945: The Society of Biological Psychiatry was founded.
1950s: Roger Williams, professor at the University of Texas, theorized biochemical individuality linked nutrition with genetic variation.
1954: Linus Pauling, chemist and molecular biologist, began work on the molecular basis of mental disease; over time formulating a general theory of the dependence of function on molecular structure; coining the term "orthomolecular."
1962: Victor Herbert, professor of Medicine, Mount Sinai-New York University, reported on symptoms observed while on self-induced folate-deficient diet.
1970: Michael Crawford, research scientist at London Metropolitan University, Institute of Brain Chemistry and Human Nutrition, predicted that the same dietary causes of poor heart health would inevitably lead to increased problems in brain health and mental health.
1970: John Fernstrom and Richard Wurtman, at the Massachusetts Institute of Technology, demonstrated that carbohydrate and protein content of a meal influenced neurotransmitter levels in the brain.
1982: One of the first conferences addressing the interaction of nutrients, food, and human behavior was organized by Richard Wurtman.
1983: Peter Roy Byrne, MD, reported a case study description of scurvy in a psychiatric patient, commenting, "psychopathology and nutritional status are interdependent."
1984: "Diet and Behavior: A Multi-Disciplinary Evaluation Symposium" was held, cosponsored by the American Medical Association, The International Life Sciences Institute, and the Nutrition Foundation, Inc.
1999: U.S. Surgeon General David Satcher released the first ever *Surgeon General's Report on Mental Health.*
2002: "Disorders of the Brain: Emerging Therapies in Complex Neurologic and Psychiatric Conditions," the Ninth Symposium of the Institute for Functional Medicine was held.
2002: David Horrobin, a neuroendocrinologist from Scotland, wrote, "Nutritional biochemistry is soon going to be essential for anyone working with mentally ill patients."
2002: President George W. Bush established the New Freedom Commission on Mental Health.
2010: The federal Mental Health Parity and Addiction Act mandated that insurance coverage for mental health.

The dominance of the 'body' becomes prominent under severe physical stress" (Keys et al. 1950). (See also Chapter 9 "Starvation, Eating Disorders, Craving, Dieting, and Bariatric Surgery.")

Biological psychiatry

The Society of Biological Psychiatry was founded in 1945 to emphasize the medical and scientific study and treatment of mental disorders. Its purpose is to foster scientific research and education and to increase knowledge in the field of psychiatry.

Major aspects of biological psychiatry include modern psychiatric medicine, humanitarian aspects, psychological treatment, and socio-cultural orientation. Biological psychiatry uses the traditional medical model that incorporates psychology, social setting, and emotions into patient care.

The society publishes the journal *Biological Psychiatry*, which includes the whole range of psychiatric neuroscience, and supports a Web site located at http://www.sobp.org. The concept of biological psychiatry includes the fields of nutritional neuroscience, and genetic and other environmental factors.

An experimental self-induced folic acid deficiency

In 1962, Herbert reported that four months on a self-induced folate-deficient diet (5 µg/day intake) resulted in insomnia, irritability, fatigue, and forgetfulness. Serum folate fell from 7 ngs/ml to <3 ngs/ml after three weeks of deprivation. The onset of anemia and red cell macrocytosis as well as the mental changes occurred in the eighteenth week. The symptoms abated following oral folate replacement of 250 µg (Herbert 1962; Thornton and Thornton 1978).

Orthomolecular medicine

The term *orthomolecular* was first used by Linus Pauling in a paper published in *Science* in 1968. He wrote,

> I might have described this therapy as the provision of the optimum molecular composition of the brain. The brain provides the molecular environment of the mind. I use the word mind as a convenient synonym for the functioning of the brain. The word orthomolecular may be criticized as a Greek-Latin hybrid. I have not, however, found any other word that expresses as well the idea of the right molecules in the right amounts. (Pauling 1968)

The key ideas in orthomolecular medicine are that genetic factors are central not only to the physical characteristics of individuals, but also to their biochemical milieu. Biochemical pathways of the body have significant genetic variability in function and concentration of molecules. Variation has been found in transcriptional potential and individual enzyme concentrations, receptor-ligand affinities, and protein transporter efficiency. Diseases such as atherosclerosis, cancer, schizophrenia, or depression are associated with specific biochemical abnormalities that are either causal or aggravating factors of the illness. In the orthomolecular view, it is possible that the provision of vitamins, amino acids, trace elements, or fatty acids in amounts sufficient to correct biochemical abnormalities will be therapeutic in preventing or treating such diseases.

In 1954, Pauling decided to work on the molecular basis of mental disorders and felt there were a number of plausible mechanisms by which the concentration of a vitamin could affect the functioning of the brain. One mechanism involved the occurrence of a genetic mutation and an altered apoenzyme, which greatly reduced affinity for the coenzyme. Increased concentration of the coenzyme could counteract this effect and lead to the formation of enough of the active enzyme to catalyze the reaction effectively (Pauling 1968). With more advanced scientific methods available, mechanisms similar to those hypothesized by Pauling have been described by Ames and colleagues at The University of California at Berkeley in 2002 (Ames 2002).

In an article on orthomolecular psychiatry, Pauling reported that Edwin et al. determined the amount of B_{12} in the serum of every patient over 30 years old admitted to a mental hospital in Norway during a period of 1 year (Table 1.2). The estimate for B_{12} levels was based on the reported frequency of pernicious anemia in the area: 9.3 per 100,000 persons per year (Pauling et al. 1973) (See Table 1.2).

Pauling is well known for his recommendations related to high doses of ascorbic acid (vitamin C). He reports a study of 106 schizophrenic patients who had recently been hospitalized in a private hospital, a county-university hospital, or a state hospital who were compared with 89 control subjects (hospital staff). A large dose of ascorbic acid (~1.5 to 2.0 grams based on body weight) was administered followed by a 6-hour excretion test, which indicated how much ascorbic acid was reserved by the body to reach storage capacity (Table 1.3).

Pauling stated in schizophrenia there was apparently an adequate vitamin intake for growth and development until the illness became manifest in the teens or early adult life (Pauling 1973).

Table 1.2 Vitamin B_{12} Status in Patients Admitted to a Psychiatric Hospital

Category re B_{12} Status	Blood Levels	Frequency Observed
Normal	150–1300 pg/ml	
Subnormal	101–150 pg/ml	38/396 = 0.095%
Pathologically low	<101 pg/ml	23/396 = 0.058%
Subtotal of subnormal + pathologically low in patients		15%
Subnormal + pathologically low in general population		5%

Source: From Pauling, L. et al., Results of a loading test of ascorbic acid, niacinamide and pyridoxine in schizophrenic subjects and controls. In: *Orthomolecular Psychiatry: Treatment of Schizophrenia*, Hawkins, D. and Pauling, L., Eds. San Francisco, CA: W.H. Freeman and Co., 1973, pp. 18–34. With permission.

Table 1.3 Excretion Rates of Ascorbic Acid in Patients with Schizophrenia and Controls

Category re Ascorbic Acid Status	6-h Excretion of Loading Dose: 1.5–2.0 g Vitamin C	Patients with Schizophrenia (N = 106)	Control Group Hospital Staff (N = 89)
Deficient	Excretion of <17% of loading dose	81 (76%)	27 (30%)
Great deficiency	Excretion of <4% of loading dose	24 (22%)	1 (<1%)

Source: From Pauling, L. et al., Results of a loading test of ascorbic acid, niacinamide and pyridoxine in schizophrenic subjects and controls. In: *Orthomolecular Psychiatry: Treatment of Schizophrenia*, Hawkins, D. and Pauling, L., Eds. San Francisco, CA: W.H. Freeman and Co., 1973, pp. 18–34. With permission.

Other pioneers

Carl C. Pfeiffer, PhD, MD, founder of the Brain Bio Center in Princeton, New Jersey, and Abram Hoffer, a Canadian psychiatrist, were early proponents of nutritional therapy for mental disorders. They advocated megadose levels of some vitamins or other nutrients for treating mental disorders, and supported the orthomolecular theories of Pauling. This is not to be confused with the Brain Bio Centre in Richmond, London, founded in 2003 by Patrick Holford, a psychologist and student of Pfeiffer (Pfeiffer 1975). (For other pioneers, see http://orthomolecular.org/history/index.shtml.)

William Walsh, a scientist with more than 25 years experience in biochemistry research, was a former research collaborator with the late Dr. Carl Pfeiffer. In the 1980s, Walsh founded the nonprofit Health Research Institute and its clinical arm, the Pfeiffer Treatment Center in Warrenville, Illinois.

Orthomolecular methods and treatments have not gained general acceptance in the psychiatric community. Perhaps the development and precision of new scientific methods of measurement, diagnosis, and assessment of nutritional status and the effects on mental status will produce evidence demonstrating what will and will not prove clinically useful in times to come. It took many small steps of researchers in biology, biochemistry, neurology, and immunology to demonstrate the hows of the links between emotions, stress, and general health (Sternberg 2001). Science is not as far along the path linking nutrition and mental health.

Nutrition for the brain

Reasoning over 30 years ago that the nutrients necessary for keeping a healthy heart were the same as those needed for a healthy brain, Michael

Crawford, research scientist at London Metropolitan University, Institute of Brain Chemistry and Human Nutrition, warned that the explosion in the prevalence of heart disease seen in the West during the 20th century would soon be followed by a rise in mental health problems. His predictions are coming true. Depression is now the leading cause of disability worldwide and mental health problems presently cost £100 billion each year in the UK alone (Lang 2005). In the U.S., the cost of diagnosing and treating mental disorders totaled approximately $69 billion in 1996.

Content of meals affect the brain

In the 1970s and 1980s John Fernstrom and Richard Wurtman, scientists at Massachusetts Institute of Technology (MIT), illustrated that previous beliefs that the brain was impervious to variations in short-term food intake due to protection by the blood–brain barrier were not warranted. Their work demonstrated that the carbohydrate and protein content of a meal influenced neurotransmitter levels in the brain by way of the amino acids tryptophan and tyrosine, the neurotransmitter pre-cursors (Fernstrom and Wurtman 1971).

In 1982, Richard Wurtman at MIT organized one of the first conferences addressing the interaction of nutrients, food, and human behavior, reported by Gina Kolata (1982) in *Science*. Wurtman and colleagues reported tryptophan doses of at least 1 g helped mildly insomniac patients fall asleep more quickly and wake less frequently. The amino acid tyrosine is the precursor to the neurotransmitters dopamine and norepinephrine, which play a role in regulating mood as well as in motor activity. One early study suggested that tyrosine relieved depression in some patients. After a high-protein meal, tyrosine increases in the blood and in the brain. Wurtman concluded that researchers need to inquire about the carbohydrate and protein in the meals of their subjects and to pay attention to blood levels of amino acids and other precursors of neurotransmitters (Kolata 1982).

> Mild insomnia is defined as taking about 30 min to fall asleep and then waking during the night.

Nutrition in psychiatry

Peter Roy-Byrne, in conjunction with a case study description in 1983, wrote,

> We suggest that, in general, psychopathology and nutritional status are interdependent. A patient's mental state may contribute to unusual eating habits leading to nutritional deficiencies. These deficiencies may in turn lead to biochemical changes which

could exacerbate psychopathology already present... Therefore, nutritional status should be an important consideration in evaluating psychiatric patients and attention should be paid to the eating habits and dietary intake of all patients. (Roy-Byrne, Gorelick, and Marder 1983. See also Chapter 7–Case Study)

Functional medicine and biological individuality

The Institute for Functional Medicine (IFM) founded in Gig Harbor, Washington, in 1991, bases its work on the biochemical and physiological functions of the body and the effects of genetics and changes in metabolism, diet, and lifestyle. Founded by Jeffrey Bland, nutritional biochemist, IFM held The Ninth Symposium of The Institute for Functional Medicine, "Disorders of the Brain: Emerging Therapies in Complex Neurologic and Psychiatric Conditions," in May 2002.

"Diet and nutrition are major factors influencing the translation of genotype to the phenotype," stated Bland (2002). This reflects the principles and application of systems biology, which sees health as the reflection of the genetic, molecular, and biochemical interactions between bodily systems, a network of relationships that has the capacity for self-regulation (Table 1.4). (For more information on systems biology, visit http://www.systemsbiology.org.)

Nutritional signals modulate human genes, thereby influencing a person's health. Functional medicine involves treating genomic variability (personalized medicine). (See also the IFM Web site at http://www.functionalmedicine.org.)

Table 1.4 Links between Nutrients and Psychiatric Concerns

1. Iron overload and psychiatric symptoms
2. Zinc, brain, and behavior
3. Thyroid and psychiatric symptoms
4. Gastroenterology and psychiatry
5. Omega-3 fatty acids and bipolar disorder
6. Omega-3 fatty acids and red cell membranes in depression
7. Psychobiological effect of carbohydrates
8. Tryptophan and mood and psychiatric patients
9. Diabetes in patients with manic depression
10. Cyanocobalamin (vitamin B_{12}) deficiency without anemia

Source: From Hedaya, R.J., The role of functional medicine in the treatment of depressive disorders: Research-based methods of maximizing patient outcomes. In: *Disorders of the Brain: Emerging Therapies in Complex Neurologic and Psychiatric Conditions. Proceedings of the Ninth Symposium of the Institute for Functional Medicine*, Fort Lauderdale, FL, 67–127, 2002.

Roger Williams, whose nutritional research at the University of Texas in the late 1950s through the 1990s, is part of the theoretical support for functional medicine. Williams (1958) stated, "Nutrition applied with due concern for individual genetic variations, which may be large, offers the solution to many baffling health problems." Significant differences in enzyme activity in different individuals were reported by Williams (1956), who called this concept "biochemical individuality."

Robert J. Hedaya describes the guiding principles of what he terms "whole psychiatry" as an integration of the principles of functional medicine, traditional medicine, general systems theory, environmental medicine, and the psycho-social-spiritual and pharmacological methods of traditional psychiatry.

This integration recognizes and utilizes scientifically established interactions between the brain and cognition; perception; culture, spirituality, and lifestyle factors (including exercise, sexual function, and relationships); physiological body systems, nutrition, and toxicology; and genetic vulnerabilities. Functional medicine addresses disorders via reduction of inflammation and oxidative stress, detoxification, hyperinsulinism, and hormonal modulation. Because functional or absolute deficiencies of B vitamins (B_1, B_2, B_6, B_{12}, folic acid, and biotin) can be associated with depression, Hedaya recommended functional testing of B-vitamin status in every patient with affective disorders. Functional testing (enzyme-dependent activity) is necessary for detection of suboptimal levels (Hedaya 2002).

Richard J. Wurtman and Judith J. Wurtman described the process that links the consumption of specific foods to the production of selected brain constituents and the physiologic consequences of such links, such as (1) the relationship between carbohydrate craving and depression; (2) reports of research subjects who eat, not from hunger, but to combat tension, anxiety, or mental fatigue; and (3) levels of precursors uridine, cytidine, choline, and fatty acids in the control of phosphatide synthesis rates of neurons and other cells (Wurtman and Wurtman 2002).

Nutritional biochemistry

In 2002, David Horrobin, MD, a neuroendocrinologist from Scotland, wrote,

> Nutritional biochemistry is soon going to be essential for anyone working with mentally ill patients. ... Without an adequate intake of all the required essential nutrients, the brain simply cannot function normally. Trying to apply any treatment modality, whether psychological, pharmacological or social,

to a brain that cannot function normally because of lack of an essential nutrient is like trying to run a 220-volt electrical appliance on a 120-volt system. Yet anyone who works with mentally ill patients ... knows that diets are often appalling and micronutrient deficits are common and serious.

Nutrigenomics and nutrigenetics

Following the sequencing of the human genome in 2003, J. Bruce German, Steven M. Watkins, and Fay Laurent-Bernard (2005) as well as other authors have described how metabolomics and nutrigenomics may influence an individual's health assessment, treatment, and outcomes. Nutrients may be used as "pseudo-drugs" or nutraceuticals. For example, in treating depression and bipolar disorders, doses of omega-3 fatty acids are higher than the probable intake from fish for most individuals.

> Nutrigenetics refers to the influence of the genes; gene alteration is the problem, foods and nutrients are the solution.
>
> Nutrigenomics refers to the influence of nutrients (an environmental influence) on genes and gene expression.

Public action in mental health in the United States

National symposium

The Diet and Behavior: A Multi-Disciplinary Evaluation Symposium, co-sponsored by the American Medical Association, The International Life Sciences Institute, and The Nutrition Foundation, Inc., was held in 1984 in Arlington, Virginia. In preparation for the symposium, researchers attended a conference and published a statement that could be echoed today:

> The panel recommends that scientifically sound research in these [diet, behavior and multi-disciplinary evaluation] and other areas be encouraged. Furthermore, this research requires an interdisciplinary approach involving input from fields such as nutrition, neuroscience, endocrinology and the behavioral sciences. In addition, the continued development and evaluation of methodology, including protocols, is essential to studying the effect of diet on behavior. Finally, the panel warns against changes in public policy that outpace the acquisition of scientific information. (American Medical Association 1984)

Surgeon General's Report on Mental Health

In early December 1999, U.S. Surgeon General David Satcher released the first ever *Surgeon General's Report on Mental Health*. This landmark, science-based report calls attention to the fact that mental illness is an "urgent health concern" that the United States must address. The report states that mental illness is the second leading cause of disability and premature mortality. In established market economies such as the United States, mental illness is comparable to heart disease and cancer as a cause of disability. The single explicit recommendation of the report is: "Seek help if you have a mental health problem or think you have symptoms of a mental disorder" (U.S. Department Health and Human Services 1999). The consequences of untreated mental illness produce negative results for the individuals, their families, communities, and society (National Association of Social Workers 1999).

New Freedom Commission on Mental Health

Between 1991–1992 and 2001–2002 the prevalence of major depression among U.S. adults increased from 3.33% to 7.06% (Compton et al. 2006). In 2002, President George W. Bush established the New Freedom Commission on Mental Health. The commission's recommendations included addressing mental health with the same urgency as physical health, using research to develop new evidence-based practices, and putting discoveries into practice immediately so that Americans would fully benefit from the enormous increases in the scientific knowledge base (The Carter Center 2003).

Mental Health Parity and Addiction Act

It was feared that (1) estimates of the prevalence of mental disorders—based on expansive definitions—would lead to uncontrolled spending increases and (2) if lawmakers insist on laboratory evidence to validate every claim for a behavioral health benefit (a standard that is not applied to medical benefits), then adequate coverage will be impossible to achieve. In fact, states that have passed strong parity laws have not seen rapid spending increases because managed behavioral health care has reduced the use of costly inpatient care more than enough to offset increased use of outpatient services (Mechanic 2003).

As of January 1, 2010, the federal Mental Health Parity and Addiction Act mandated that insurance coverage for mental health be comparable to that for other medical interventions for group plans of companies employing 50 or more. What it means for those who are covered is that reimbursements, number of visits, annual and lifetime caps, copayments, and any out-of-pocket costs for psychotherapy and related medications should be in line with, for example, treatment for diabetes (Department of Health and Human Services 2011).

References

The American Medical Association, International Life Sciences Institute, and The Nutrition Foundation, Inc. 1984. Statement of the Resource Conference on Diet and Behavior. Diet and Behavior: A Multi-Disciplinary Evaluation Symposium Panel Statement. *Nutr Rev* 42(5):200–201.

Ames, B.N., I.E. Schwan, and E.A. Silver. 2002. High-dose vitamin therapy stimulates variant enzymes with decreased coenzyme binding affinity (increased Km): Relevance to genetic disease and polymorphisms. *Am J Clin Nutr* 75(4):616–658.

Bland, J. 2002. Functional approaches to psychiatric disorders. In: *Disorders of the Brain: Emerging Therapies in Complex Neurologic and Psychiatric Conditions. Proceedings of the Ninth Symposium of the Institute for Functional Medicine* 19–44 (31–34), Fort Lauderdale, FL.

The Carter Center. 2003. *The President's New Freedom Commission on Mental Health: Transforming the Vision.* The Nineteenth Annual Rosalynn Carter Symposium on Mental Health Policy. November 5–6, 2003. http://www.cartercenter.org/documents/1701.pdf. Accessed October 1, 2010.

Compton, W.M., K.P. Conway, F.S. Stinson, and B.F. Grant. 2006. Changes in the prevalence of major depression and comorbid substance use disorders in the United States between 1991–1992 and 2001–2002. *Am J Psyc* 163:2141–2147.

Department of Health and Human Services (DHHS). 1999. Report of the Surgeon General Executive Summary: Mental Health, Rockville, MD, http://www.surgeongeneral.gov/library/mentalhealth/(accessed October 1, 2010.)

Department of Health and Human Services. Centers for Medicaid and Medicare Services. The Mental Health Parity and Addiction Equity Act, http://www.cms.hhs.gov/healthinsreformforconsume/04_thementalhealthparityact.asp. Accessed September 1, 2011.

Edwin, I., K. Holten, K.R. Norum, et al. 1965. Vitamin B12 hypovitaminosis in mental diseases. *Acta Med Scand* 177:689–699.

Fernstrom, J.D. and R.J. Wurtman. 1971. Brain serotonin content: Physiological dependence on plasma tryptophan levels. *Science* 173:149–152.

German, J.B., S.M. Watkins, and F. Laurent-Bernard. 2005. Metabolomics in practice: Emerging knowledge to guide future dietetic advice toward individualized health. *J Am Diet Assoc* 105(1):1424–1432.

Hedaya, R.J. 2002. The role of functional medicine in the treatment of depressive disorders: Research-based methods of maximizing patient outcomes. In: *Disorders of the Brain: Emerging Therapies in Complex Neurologic and Psychiatric Conditions. Proceedings of the Ninth Symposium of the Institute for Functional Medicine* 67–127, Fort Lauderdale, FL.

Herbert, V. 1962. Experimental nutritional folate deficiency in man. *Translation Assoc Amer Phys* 75:307–320.

Horrobin, D.F. 2002. Food, micronutrients, and psychiatry. Guest editorial. *Intl Psychoger* 14(4):331–334.

Keys, A., J. Brozek, A. Henschel, et al. 1950. *The Biology of Human Starvation*, Minneapolis: University of Minnesota Press, pp. 767–905.

Kolata, G. 1982. Food affects human behavior. *Science* 1209–1210.

Lang, T. 2005. Forward. In: *Changing Diets, Changing Minds: How Food Affects Mental Health and Behavior*, Van de Weyer, C. United Kingdom: Sustain: The Alliance for Better Food and Farming. See also www.sustainweb.org or www.mentalhealth.org.uk. Accessed October 1, 2010.

Mechanic, D. 2003. Is the prevalence of mental disorders a good measure of the need for services? *HealthAffairs* 22(5):8–20. http://content.healthaffairs.org/cgi/content/full/22/5/8. Accessed October 1, 2010.

National Association of Social Workers. 1999. Bulletin: Groundbreaking Surgeon General's Report on Mental Health. http://www.socialworkers.org/practice/behavioral_health/surgeon_gen.asp. Accessed September 1, 2011.

Pauling, L, et al. 1968. Orthomolecular psychiatry. Varying the concentrations of substances normally present in the human body may control mental disease. *Science* 160(825): 265–271.

Pauling, L., A.B. Robinson, S.S. Oxley, et al. 1973. Results of a loading test of ascorbic acid, niacinamide and pyridoxine in schizophrenic subjects and controls. In: *Orthomolecular Psychiatry: Treatment of Schizophrenia*. D. Hawkins and L. Pauling, Eds. San Francisco: W H Freeman and Co., pp. 18–34.

Pfeiffer, C.C. 1975. *Mental and Elemental Nutrients*. New Canaan, CT: Keats Publishing, Inc.

Roy-Byrne, P., D.A. Gorelick, and S.R. Marder. 1983. Unusual dietary habits in a patient with schizotypal personality disorder: Interaction of nutritional status and psychopathology. *J Psych Treat and Eval* 5:67–69.

Sternberg, E.M. 2001. *The Balance Within*. New York: W H Freeman and Co.

Thornton, W.E. and B.P. Thornton. 1978. Folic acid, mental function and dietary habits. *J Clin Psychiatry* 39(4):315–322.

Williams, R.J. 1956. *Biochemical Individuality: The Basis for the Genetotrophic Concept*. New York: John Wiley & Sons. 1998 ed., New Canaan, CT: Keats Publishing.

Wurtman, R.J. and J. Wurtman. 2002. Foods that affect brain biochemistry and behavior. In: *Disorders of the Brain: Emerging Therapies in Complex Neurologic and Psychiatric Conditions. Proceedings of the Ninth Symposium of the Institute for Functional Medicine* 223–232, Fort Lauderdale, FL.

chapter two

Addiction—food, alcohol, and caffeine

Introduction

The basic aspects of addiction often include abuse, tolerance, and withdrawal. Abuse is defined as a psychological feeling of wishing to get or take more and more of something, regardless of painful consequences, with or without dependence. Tolerance and withdrawal reflect dependence. Sex addiction, eating or food addiction, or Internet addiction are seen as psychological abuse (Ghaemi 2010).

Diagnosis of a physiological dependence is connected to tolerance and withdrawal. Addiction or dependence can be diagnosed using entirely behavioral criteria (persistent desire, time spent seeking, interference with important activities, use despite adverse consequences). Addiction has been implicated in craving, bingeing, and obesity. Corwin and Grigson suggest that even highly palatable food is not addictive in and of itself, but the manner the food is presented and consumed results in the addiction-like process. Research regarding loss of control appears relevant (Corwin 2009).

Food addiction

From an evolutionary perspective, it would be adaptive for food to be rewarding, especially foods that are high in fat and sugar. In today's environment, food differs from other addictive substances because it is legal, necessary, cheap, and widely marketed. The concept of addiction does not negate the role of free will and personal choice, but variability in genetics and personality traits are thought to increase vulnerability or susceptibility to an individual's overuse of reward pathways (Taylor 2009).

Food as drug model: Neurochemistry and genetics

Self-reported craving and food "addictions" are thought by some to be a combination of (1) ambivalence, (2) normal mechanisms of appetite, (3) the hedonistic effects of certain foods, and (4) social or cultural ideas of appropriate intakes and uses of foods, rather than true addictions (Rogers 2000).

The food-as-drug model proposes that chronic overeating can be a form of substance abuse. In comparing individuals who abuse alcohol and those who abuse food, their brain responses are not very different. Food may be competing and substituting in the brain's reward pathways (Gold 2004). Support for the food addiction hypothesis comes from alterations in neurochemistry (dopamine, endogenous opioids) and neuroanatomy (limbic system). Over 100 peer-reviewed articles showed that humans produce opioids (the chemically active ingredient in heroin, cocaine, and other narcotics) as a derivative of the digestion of excess sugars and fats. Genetic research, brain-imaging research, and research on endogenous opioids converge at the pleasure centers of the brain, the D2 dopamine receptors. Intense sweetness—not just refined sugar, but also artificial sweeteners—surpasses cocaine as a reward in laboratory animals. Some obese adults who binge on carbohydrates and who were addicted to neither alcohol nor drugs were shown to have the same D2 dopamine gene marker present as found in alcoholism and other drug addictions.

Research on out-of-control food consumption related to pain reduction focuses on the serotonin mechanisms in the brain. Research shows malfunctions in the serotonin pathway correlate with an addiction to sugars and flours. Other biochemical explanations for out-of-control eating include low leptin levels, especially those with Prader-Willi syndrome who exhibit behavior similar to those with very strong biochemical urges to eat and binge on all foods. Self-assessed food addicts call this "volume addiction."

Laboratory-based scientific research corresponds closely with clinical observations by professionals working with food addicts. Foods said to have potential addictive properties include (1) sweets, (2) artificial sweeteners, (3) carbohydrates, (4) fats, (5) sweet/fat combinations, (6) caffeine, and (6) possibly processed or high-salt foods. Self-assessed food addicts in Overeaters Anonymous are successful in weight loss by dealing first with physical craving and abstaining completely from their major binge foods. These findings provide a strong argument for the existence of both physical craving and food addiction (Cheren 2009).

Chronic food cravings, compulsive overeating, and binge eating may represent a phenotype of obesity. Future research needs to include a focus on human food addiction, the impact of treatment on underlying neurochemistry, and prevention or reversal of food addiction in humans (Corsica 2010) (See also "Craving for Chocolate" section in Chapter 9).

Malnutrition in food addiction

Although it is sometimes assumed that high calorie intake results in an adequate intake of vitamins and minerals, it is conceivable that high calorie malnutrition could occur in abusers of food or alcohol. Excess intake of unfortified foods high in sugar may not be balanced with the nutrients needed to metabolize the refined carbohydrate: vitamins B_1, E, A, C, magnesium, and others. The 2003–2006 National Health and Nutrition Examination Survey (NHANES) data analysis indicated that above 5 to 10% of calorie intake, nutrient intake was less with each 5% increase in added sugars. The 2005 U.S. Dietary Guidelines recommend limiting discretionary calories (which includes both added sugar and solid fat) to 13% of energy requirement (U.S. Department of Health & Human Services 2005). In their Diet and Lifestyle Recommendations Revision 2006, the American Heart Association Nutrition Committee also stated that the primary reasons for reducing the intake of beverages and foods containing added sugars are to lower total calorie intake and promote nutrient adequacy (Lichtenstein 2006). Thirteen percent of the population has added sugars intake greater than 25% of calorie intake (Marriott 2010). This suggests that nutritional assessment and treatment of those who eat compulsively could decrease or eliminate the mental consequences of nutritional inadequacy.

Assessment for food addiction

Yale Food Addiction Scale

A food addiction scale for professional use has been initially validated and shows promise in identifying food addiction (see Yale Food Addiction Scale at http://www.yaleruddcenter.org/resources/upload/docs/what/addiction/FoodAddictionScale09.pdf) (Gearhardt, Corbin, and Brownell 2009).

Self-assessment tools for food addiction

Several self-assessment tools for food addiction are available on the Internet. Assessment topics include experiencing lack of control of eating, shame or embarrassment around eating, using food for comfort or relief from stress, eating differently when alone, eating when not hungry, and spending a lot of time and thought on food. These characteristics parallel issues addressed in the identification and treatment of those with eating disorders. Users of online assessment tools are advised to seek professional help for in-depth assessment and diagnosis.

Alcohol addiction

Metabolism of alcohol

Ethyl alcohol is quickly absorbed and is circulated to the body and brain within minutes of consumption. Change of mood and decrease of inhibition occurs quickly, which is one appeal of drinking alcohol.

Table 2.1 Calories in Alcoholic Beverages

Beverage	~Calories
Regular beer: 12 oz.	150
Regular beer: 6 pack	900
Table wine: 7 oz.	145
90-proof liquor: 1.5 oz. jigger	110
Liqueur (e.g., crème de menthe): 1.5 oz.	175
Four Loko® energy drink: 23.5 oz. @12% alcohol: 2.8 oz. alcohol is equal to 5 cans of beer	660[a]

Source: Pennington et al. 2005.

[a] http://dailyburn.com/nutrition/four_loko_approximation_calories.

Alcohol sedates the inhibitory nerves, ultimately acting as a depressant affecting all nerve cells. Because it is a toxin, ethanol is given top priority for processing by the liver. The liver can detoxify ½ oz. of ethanol per hour. A "drink" is defined as the amount of a beverage that contains ½ oz. of *pure* ethanol. "Proof" is twice the percent of alcohol (80-proof is 40% ethanol). Alcohol is a two-carbon molecule, and it enters the energy metabolism system via the same route as fatty acids. Metabolism of fat in the liver is delayed so the liver can metabolize ethanol. This competition for metabolic systems contributes to the development of "fatty liver."

Appetite and nutrient intake

Ethanol and drugs (illegal, legal, over-the-counter, or prescribed) often alter an individual's appetite, choice of foods, digestion and nutrient needs, metabolism, and excretion. The excretion of water-soluble vitamins may increase due to the increased urination associated with alcohol intake. With nutrient intake down, needs unchanged or increased, and excretion up, the risk is high for malnutrition in those with substantial alcohol intake.

For moderate drinkers, alcohol may stimulate the appetite and increase total calorie intake, especially if fruit, cream, or carbonated mixers are used. Awareness of calories causes some individuals to consume alcohol as substituted energy rather than additional energy, which creates greater imbalance between nutrient need and nutrient intake (Table 2.1).

Alcohol and pregnancy

Women who consume alcohol during pregnancy increase the possibility of fetal alcohol effects (FAE) or fetal alcohol syndrome (FAS) in their child. In addition to physical malformations, the child may develop varying degrees of mental slowness or retardation, hyperactivity, mood disorders, and aggressive behavior. Since alcohol readily crosses the placenta, the percentage of alcohol in the blood is higher in the fetus due to its small body size. Since the detoxification system in the fetus is not mature, there is no "safe" level of alcohol intake defined for a pregnant woman. Critical periods for fetal development that are affected by alcohol occur before a woman realizes she is pregnant. Therefore, it is wise for a woman who may conceive to eliminate alcohol intake.

Paternal alcohol consumption before pregnancy has been studied in animals. One acute dose of alcohol before insemination caused early developmental delays in one study and found that alcohol-sired animals are less fearful and more aggressive as adults (Meek 2007). The equivalent of two or more alcoholic drinks per day before conception may impair learning ability, as well as survival, of the offspring. Effects appear linked to alteration in paternal DNA transcription (Whitney 2005).

Lipids

Essential fatty acids, impulsive behavior and alcoholics —level II evidence

Impulsive violence, suicide, and depression are strongly associated with low concentrations of 5-hydroxyindoleacetic acid (5-HIAA) and homovanillic acid (HVA) in cerebrospinal fluid (CSF), which are metabolites of neurotransmitters. The relationship of DHA (an O-3 fatty acid) to CSF 5-HIAA and HVA in healthy volunteers is significantly different from the alcoholic groups. Studies of oral intake (diet and supplements) may help determine if essential fatty acids can influence neurotransmitters and modify impulsive behaviors (Hibbeln 1998).

Vitamins

Interaction of vitamins and alcohol

Alcohol makes a minimal contribution to vitamins and mineral intake although the metabolism of alcohol requires the same vitamins and minerals needed for metabolism of carbohydrates, proteins, and fats. Alcohol interferes with the absorption of vitamins as well as with their function. Regular consumption of a well-formulated vitamin-mineral supplement may be helpful in some cases but will not prevent the health and nutritional consequences of excessive alcohol consumption. Deficiency of some vitamins affects mental status in addition to the direct effects of alcohol.

- Niacin (as NAD and NADH) is part of the initial processing of ethanol. Excess buildup of NADH results in buildup of lactic acid, changing the acid/base balance and suppressing nervous system activity.

- Alcohol toxicity decreases the intestine's normal retrieval of folic acid, excretion of folic acid by the kidney increases, and folic acid is not retained as efficiently by the liver. Combined, these factors lead to increased risk of folic acid deficiency.
- Thiamin is required for alcohol metabolism. Chronic alcoholics often have a marked reduction in blood transketolase, an enzyme involved in thiamin and alcohol metabolism. A deficiency of thiamin can result in development of Wernicke–Korsakoff syndrome (WKS) or Wernicke's encephalopathy (WE). In addition to physical symptoms, about 90% of patients with WKS show a derangement of mental function manifesting as confusion, ataxia, apathy, listlessness, inattentiveness, indifference to surroundings, disorientation, minimal spontaneous speech, disorientation in time and place, misidentification of those nearby, inability to grasp the meaning of the illness or the immediate situation, and irrational inconsistent remarks. Confusion, ataxia, and nystagmus are present in only 10% of patients and 80% of patients with WE are not diagnosed before postmortem (Lingford-Hughes 2004).
- Alcohol dislodges vitamin B_6 from its associated protein, inactivating it and causing a vitamin B_6 deficiency.
- The metabolism of vitamins A and D is also affected by heavy alcohol intake, but has not been shown to effect mental status.

Nutritional status: Mildly and heavily dependent alcoholics—level II evidence

Both mildly dependent and heavily dependent alcoholics who had consumed at least 100 g (~3.5 oz.) of ethyl alcohol per day for five years have been found to have low intakes of vitamins A, B_1, B_2, B_6, C, D, E, and folate, and the minerals calcium, selenium, and zinc (Mariari 2003).

Vitamin relationships—level II evidenc

Alcoholics may exhibit (1) peripheral neuropathy (thiamin, pyridoxine, vitamin B_{12} deficiency), (2) darkened pigmentation associated with pellagra (niacin deficiency), (3) cutaneous lesions of the mucous membranes (folic acid, riboflavin, pyridoxine deficiency), and (4) decreased activity of vitamin-dependent enzymes (erythrocyte transketolase activity [ETA]). CSF levels reflect neurological status more accurately than do blood concentrations (Dastur 1976).

Predicting thiamin deficiency—level II evidence

Two highly specific items predicting thiamin deficiency on a screening instrument are (1) patients who indicated they missed meals due to lack of funds and (2) the clinical co-occurrence of medical conditions potentially related to poor nutrition (Sgouros 2004).

Alcoholic neuropathy distinct from thiamin-deficiency neuropathy—level II evidence

Neuropathy not related to thiamin deficiency may also occur in chronic alcoholics, related to the direct toxic effect of ethanol and has distinct differences shown in electrophysiological examination and loss of axonal fibers. Treatment with thiamin improves only thiamin-deficiency neuropathy (Mellion 2011; Koike 2003).

Recovery from Wernicke–Korsakoff syndrome—level IV evidence

Even with adequate thiamin, recovery is slow. An injection or intravenous (IV) dose of 100 mg of thiamin followed by 100 mg given intramuscularly daily for five days and oral doses at maintenance level plus balanced meals over time are often enough to normalize biochemical indicators of the deficiency. This compares to the Dietary Reference Intake (DRI) of 1.1 to 1.4 mg/day for adults. Complete recovery occurs in less than 20% of patients, suggesting that this symptom is due to irreversible structural changes. Examinations after death show lesions in brain tissue including destruction of nerve cells and fibers (Roman 2006).

Vitamin A deficiency: A case study—level IV evidence

A case of depression-related, alcohol-induced malnutrition resulted in vitamin A deficiency. The patient complained of dry eyes and difficulty seeing at night and had alcoholic cirrhosis, protein deficiency, and megaloblastic anemia. Treatment included vitamin supplementation and artificial tears (Roncone 2006).

Vitamin C deficiency: A case study—level IV evidence

A case of scurvy in a 65-year-old male who reported heavy alcohol abuse for years and a diet of only cheese pizza resulted in serum ascorbic acid of <0.12 mg/dL (N = 0.20–1.9 mg/dL). The following vitamin deficiency symptoms were observed: ecchymosis of lower limbs, poor dentition, and corkscrew hairs with perifollicular petechia (Wang 2007).

Vitamin K lowered risk of alcoholism: A 30-year follow-up—level II evidence

A comparison of 234 individuals at infancy and at 30 years of age supported the hypothesis that vitamin K supplementation at birth is associated with lower rates of problem drinking and alcohol dependence (Manzardo 2005).

Moderate level supplements during alcohol rehabilitation—level I evidence

A randomized, double-blind, placebo-controlled study of the effect of 21 days of supplements (6 mg beta-carotene, 120 mg vitamin C, 30 mg vitamin E, 20 mg zinc, and 100 mcg selenium) found that serum indicators were significantly improved for all five nutrients. The placebo group showed mixed changes for the time period (Guequen 2003).

Minerals

Inadequate intake of calcium, magnesium, and zinc occur frequently with chronic alcohol overconsumption. Serum magnesium may be normal in spite of total body deficit. Inadequate zinc is linked to a less efficient immune system, and a low intake of calcium is believed to contribute to occurrence of fractures and osteoporosis in chronic alcoholics.

Bone loss is more intense in alcoholics than in controls with similar BMI, especially in those with irregular eating habits. Cross-sectional data suggests that alcoholic osteopenia may be interpreted as a form of nutritional osteoporosis (Santolaria 2000).

Zinc, copper, and withdrawal from alcohol—level II evidence

Thirty patients hospitalized for seizures, delirium tremens, or hallucinatory states during alcohol withdrawal had significantly higher copper and lower zinc concentrations. Normal range for the copper-to-zinc ratio was determined to be 0.8–2.0 µg/L. Patients who developed delirium tremens had elevated Cu:Zn ratios: 2.14, 3.17, 3.44, and 4.75 µg/L. Patients who experienced only seizures had a Cu:Zn of 1.72 ± 0.52 µg/L. Plasma copper-to-zinc ratios could be of value in predicting which patients might have metabolic imbalance and might develop these symptoms (Bogden 1978).

Other

Alcohol use and cognitive decline—level II evidence

A study of a dementia-free cohort examined cognitive decline 2.5 years and 4.5 years after baseline. Severe cognitive decline was defined as loss of ≥1 point per year on the Mini-Mental State Examination (MMSE) or a diagnosis of incident dementia. An intake of ½ to 1 oz. of alcohol per day did not appear to protect against cognitive decline. The risk of severe cognitive decline with intake of ½ to 1 oz. alcohol per day in women appears to be 2½ times that of those who abstain from alcohol in this analysis (Lobo 2010). In contrast, using data from the large English Longitudinal Study of Ageing, it was concluded that better cognition, subjective well-being, and fewer depressive symptoms were associated with moderate levels of alcohol consumption (not over 1 drink per day) compared to lifetime abstinence from alcohol (Lang 2007).

> One Drink, Defined*
> 12 fluid ounces regular beer (5% alcohol)
> 5 fluid ounces wine (12% alcohol)
> 1.5 fluid ounces 80-proof distilled spirits (40% alcohol)
> *U.S. Department of Agriculture, U.S. Department of Health and Human Services 2010 Dietary Guidelines for Americans, www.dietaryguidelines.gov.

Alcohol and psychiatric comorbidity—level IV evidence

Clinical observations indicate alcohol use/abuse often coexist in individuals with anxiety, obsessive-compulsive, depressive, and paranoid symptoms. Those with severe alcohol dependency demonstrate more psychiatric symptoms (Baigent 2005; Lima 2005). Both alcohol misuse and a number of psychiatric disorders are influenced by genetic factors. A large twin registry study demonstrated that genetics accounted for 49% of the variance for the age of onset of drinking and 25% of variance for the time to progress to alcohol dependence (Liu 2004).

Alcohol and psychiatric comorbidity: Eating disorders—level II evidence

Alcohol-use disorders rise significantly in patients with a history of bulimia, anorexia nervosa, or both. Alcohol-use disorders are associated with more depression and anxiety disorders as well as being psychologically more complex (Bulik 2004). The full syndrome of alcohol use disorder may occur with subthreshold eating disorders and anxiety disorders (Lewinsohn 2004; Pincus 1999).

> Subthreshold conditions are those that do not meet the full diagnostic criteria for a syndrome, not meeting the diagnostic cutoff (Pincus 1999).

Screening for alcohol use—level III evidence

Some recommend that all psychiatric inpatients be screened for alcohol use disorders as part of their risk assessment (Baigent 2005). Tools for assessment include the Psychiatric Disorders Screening Questionnaire, the Alcohol Use Disorders Identification Test (AUDIT), and the CAGE test. CAGE screening consists of four questions regarding the following: having tried repeatedly without success to cut down on drinking, being annoyed by criticism about drinking habits, feeling guilty about drinking, and needing an eye opener drink in the morning. Answering yes to any of these questions signifies possible hazardous drinking.

Discontinuing use of alcohol—withdrawal and depression—level II evidence

Hangovers are a mild form of alcohol withdrawal. One cause is congeners, the co-occurring substances in distilled alcohol. Gin, vodka, and bourbon contain different congeners. Another cause of a hangover is dehydration. Alcohol reduces the water content of brain cells, and when the cells rehydrate later, nerve pain accompanies the swelling involved in returning to their normal size.

Discontinuing alcohol consumption after dependency develops is often followed by depression. The risk of major depression is 4.2 times higher in those with a history of alcohol dependence (Hasin 2002). Anxiety and depression are much worse for problem drinkers compared with moderate drinkers and abstainers. Alcohol abstainers (fewer than 12 drinks in their lifetime) have a much lower likelihood of major depression than moderate drinkers do (Sareen 2004).

Nutrition education and recovery—level II evidence

Improved treatment outcomes relate significantly to provision of nutrition education during substance abuse treatment programs. Nutrition education improved scores on the Addiction Severity Index (ASI) in the psychological domain by 68% and in the medical domain by 56%. Individual nutrition/substance abuse education predicted improvements in ASI scores of 99% in the family and social domains (Grant 2004).

Alcohol, diabetes, and Antabuse/disulfiram—level III evidence

Diabetics who take insulin and drink alcohol may induce hypoglycemia by inhibiting glucose release by the liver. The symptoms of hypoglycemia closely resemble alcohol intoxication, so they may not be recognized. Persons with diabetes who take sulfonylurea drugs such as Glipizide or Glucotrol may have a reaction to alcohol similar to that from taking Antabuse (flushing, nausea, palpitations, and dyspnea) (Alpers 1995).

Caffeine addiction

Caffeine belongs to the methylxanthine chemical family along with theophylline and theobromine. These and other alkaloids are present in plants worldwide.

Comparative potency of alkaloids	
Highest:	Theophylline
2nd:	Caffeine
3rd:	Theobromine

More than 99% of orally ingested caffeine is absorbed. The maximal concentration of caffeine in plasma is reached after 60 minutes, but individual differences are large. Oral intake of 250 mg of caffeine, comparable to 2 to 3 cups of coffee, results in plasma concentrations between 5 and 25 µg/mL, 15% being bound to plasma proteins. The plasma half-life of caffeine ranges between 3 to 7.5 hours. Various drugs, smoking, and pregnancy can affect caffeine clearance.

Intake of beverages containing caffeine can activate the sympathetic nervous system in nonhabitual coffee drinkers, but, interestingly, this is not observed in habitual coffee drinkers.

Selected effects of drinking coffee include the following:

- Does not reverse the sedating effect of alcohol.
- Reduces tiredness and somnolence, decreases response time, improves concentration, and extends intellectual efficiency.
- Causes a small tremor and a reduction in control of fine movement, affecting tasks requiring fast response times or optimal muscle coordination.
- Extends sleep latency, shortens total sleep duration, and worsens subjective sleep quality when consumed 30 to 60 minutes before going to sleep.
- Increases respiratory rate up to 20%.

Recent research indicates that habitual moderate coffee intake does not represent a health hazard and may even be associated with beneficial effects on cardiovascular health (Sudano 2005).

Caffeine-induced psychosis: A case study—level IV evidence

Excess caffeine can cause anxiety, agitation, and aggression, acting as a competitive adenosine antagonist and affecting dopamine transmission. A 47-year-old male had increased his coffee consumption from 10 to 12 cups per day to ~36 cups per day, drinking more than a gallon of coffee per day. Within the following 7 years, he developed depression, explosive anger, anhedonia, and feelings of worthlessness. He slept as little as 4 hours a night and experienced decreased concentration and appetite. According to his wife, he interpreted many everyday occurrences as a plot.

When seeking care, he reported a life-long difficulty sustaining attention, excessive talking, disorganization, distraction, and forgetfulness, but had no history of psychosis or hospitalization. After beginning care, he reduced his intake by 50% and was less paranoid. Two months later, he was drinking 1 to 2 cups of coffee per day, his paroxetine medication had been tapered and discontinued, and he had no evidence of paranoia or other psychosis. Later, he increased his caffeine intake for short periods with a return of paranoia. Reducing caffeine intake again resolved his symptoms (Hedges 2009).

Caffeine tolerance—level II evidence

Although modest effects may occur following initial use, tolerance to these effects appears to develop with habitual use. Controlled studies show that the positive effects of caffeine on performance and mood are due to reversal of adverse withdrawal effects associated with short periods of abstinence.

Acute withdrawal included impairment of cognitive performance, increased headache, and reduced alertness and clear-headedness. Cognitive performance was affected adversely by acute caffeine withdrawal, even in the context of alertness. Caffeine (vs. placebo) did not significantly improve cognitive performance following long-term withdrawal. These findings support the hypothesis that the effects of caffeine intake are related to reversal of withdrawal from previous caffeine intake (Rogers 2005).

In a randomized, placebo-controlled study, measurements of performance and mood confirmed a psycho-stimulant action of caffeine (1.2 mg/kg vs. placebo), but only after 8 hours of caffeine abstinence. Moderate to high caffeine consumption was defined as an average of 370 mg caffeine per day (Heatherley 2005) (See Table 2.2).

Noncoffee caffeine—level IV evidence

Red Bull®, a canned beverage containing sugar, taurine, glucuronolactone, and caffeine, is said to enhance endurance, concentration, and reaction speed. The effects are largely from the caffeine (Woojae 2003). One can of Red Bull provides approximately 80 mg caffeine, just slightly less than one cup of coffee, but if consumed more quickly than a cup of hot coffee, plasma caffeine would rise quickly.

Drinks combining alcohol and caffeine—an FTC warning

Suggesting that alcoholic beverages containing added caffeine present unusual risks to health and safety, in November 2010 the Federal Trade Commission (FTC) sent letters to producers warning that marketing of such beverages may constitute an unfair or deceptive practice that violates the FTC Act. Comments include "Consumers—particularly young, inexperienced drinkers—may not realize how much alcohol they have consumed because caffeine can mask the sense of intoxication." Several brands of these beverages, with 23.5-oz cans, provide alcohol content

Table 2.2 Common Foods and Beverage Sources and Approximate Caffeine Content

Beverage	Quantity	~Caffeine
Coffee – Regular	6 oz	100 mg
Coffee – Espresso	6 oz	376 mg
Coffee – Instant	6 oz	60 mg
Coffee – Decaffeinated	6 oz	2 mg
Tea – Regular	6 oz	40 mg
Tea – Instant	6 oz.	20 mg
Hot chocolate	6 oz.	5 mg
Colas – Regular and diet	12 oz.	35–50 mg
Red Bull	1 can	80 mg
Vitamin Water	8 oz.	35–45 mg
Chocolate ice cream	1 cup	5 mg
Chocolate topping	1 tablespoon	1–3 mg
Coffee yogurt	8 oz.	45 mg
Milk chocolate	8 oz.	5–10 mg
Semisweet chocolate	1 oz.	1–15 mg
Baking chocolate	1 oz.	55–60 mg
NoDoz maximum strength	1 tablet	200 mg
Other sources: Green tea, guarana, mate, cola nut		Variable

Source: Organic Compounds of Interest in Selected Foods. Table A-24. 2005. In: *Modern Nutrition in Health & Disease*. M. E. Shils, M. Shike, A. C. Ross, B. Caballero, and J. Cousins, Eds. New York: Lippincott Williams & Wilkins, p. 1989.
Pronsky, Z. M. 2004. *Food Medications Interactions*, 15th ed. Birchrunville, PA: pp. 65, 350, 355.

equal to 3 to 7 light beers as well as caffeine, taurine, guarana, and ginseng. (See also http://www.ftc.gov/opa/2010/11/alcohol.shtm.)

> "Light beer" refers to comparative calorie intake. Light beers provide from 55 to 125 calories per 12 oz.
> Regular beer provides 150 calories per 12 oz.

Journal of Caffeine Research

A new publication, the *Journal of Caffeine Research: The International Multidisciplinary Journal of Caffeine Science* was launched in April 2011. It reported that caffeine is the most widely consumed psychoactive substance in history with its near-universal use in beverages, food, and medicines, and it exceeds usage of any other psychoactive substance, including nicotine, alcohol, and all illicit drugs. The journal expects to cover all aspects of caffeine, including pharmacology and toxicology, psychopharmacology, neurological and endocrine function, cognitive performance and mood, dependence and withdrawal, psychopathology, gastrointestinal function, headache, sleep, and others (see www.liebertpub.com).

Conclusions

1. Individuals with food and eating addictions have symptoms and neurological similarities to those addicted to alcohol, caffeine, and drugs. Patterns of eating rather than individual foods may be the cause or source of addiction-like behaviors.
2. Addictions affect food and nutrient intake and availability of nutrients. Malnutrition may be related to intake of low-nutrient calories, decreased overall intake of food, or alteration in absorption or metabolism of nutrients. Deficiency of individual nutrients and the effects of multiple deficiencies can affect mental and physical health. Nutritional assessment is recommended to eliminate the possible consequences of malnutrition in those with food or alcohol addictions.
3. Altered ratios of minerals and fatty acids may be cause for concern and an indication for further assessment.
4. Caffeine intake and withdrawal may affect physical and mental status.
5. In addition to environmental influences, genetics may influence development of an addiction.

References

Alpers, D, WF Stenson and DM Bier. *Manual of Nutritional Therapeutics.* 449–450. New York: Little, Brown & Company, 1995.
Baigent, MF. Understanding alcohol misuse and comorbid psychiatric disorders. *Curr Opinion Psychiatry* 2005; 18(3):223–228.
Bogden, JD and RA Troiano. Plasma calcium, copper, magnesium, and zinc concentrations in patients with the alcohol withdrawal syndrome. *Clin Chem* 1978; 24 (9):1553–1556.
Bulik, CM, KL Klump, L Thornton, et al. Alcohol use disorder comorbidity in eating disorders: A multi-center study. *J Clin Psychiatry* 2004; 65:1000–1006.
Cheren, M, M Foushi, E Helga, et al. Chapter 2 in *Binge Eating and the Addictive Model Revisited. Physical Craving and FoodAddiction.* The Food Addiction Institute, Sarasota, FL. 2009.
Corsica, JA and ML Pelchat, Food addiction: true or false? *Curr Opin Gastroenterology* 2010; 26 (2):165–169. http://journals.lww.com/co-gastroenterology/Abstract/2010/03000/Food_addiction__true_or_false_.16.aspx
Corwin, RL and PS Grigson. Symposium overview-food addiction: Fact or fiction? *J Nutrition* Suppl. 2009; 139:617–619.
Dastur, DK, N Santhadevi, EV Quadros, et al. The B-vitamins in malnutrition with alcoholism. A model of intervitamin relationships. *Br J Nutr* 1976; 36(2):143–159.

Gearhardt, AN, WR Corbin, and KD Brownell. Preliminary validation of the Yale Food Addiction Scale. *Appetite* 2009; 52:430–436.

Ghaemi, Nassir. What Do We Mean by Addiction? http://boards.medscape.com/forums?128@826.8eIfaM8Hn5K@.29fa7171!comment=1 Posted Feb 16, 2010.

Gold, Mark. Link between obesity and addiction. Research Brief. *Today's Dietitian* 2004; 6(9):30–31.

Grant, LP, B Haughton, and DS. Sachan. Nutrition education is positively associated with substance abuse treatment program outcomes. *J Amer Diet Assoc* 2004; 104(4):604–610.

Guequen, S, P Pirollet, P Leroy, et al. Changes in serum retinol, alpha-tocopherol, vitamin C, carotenoids, zinc and selenium after micronutrient supplementation during alcohol rehabilitation. *J Am Coll Nutr* 2003; 22(4):303–310.

Hasin, DS, and BF Grant. Major depression in 6050 former drinkers: Association with past alcohol dependence. *Arch Gen Psychiatry* 2002; 59(9):794–800.

Heatherley, SV, RC Hayward, HE Seers, and PJ Rogers. Cognitive and psychomotor performance, mood and pressor effects of caffeine after 4,6 and 8 h caffeine abstinence. *Psychopharmacology (Berlin)*, 2005; 178(4):461–470.

Hedges, DW, FL Woon, and SP Hoopes. Caffeine-induced psychosis. *CNS Spectr* 2009; 24(3):127–129.

Hibbeln, JR, M Linnoila, JC Umhau, et al. Essential fatty acids predict metabolites of serotonin and dopamine in cerebrospinal fluid among healthy control subjects, and early and late-onset alcoholics. *Biolog Psychiatry* 1998; 44(4):235–242.

Journal of Caffeine Research: The International Multidisciplinary Journal of Caffeine Science. http://www.liebertpub.com/products/product.aspx?pid=388

Koike, H, M Lijima, M Sugiura, et al. Alcoholic neuropathy is clinicopathologically distinct from thiamin-distinct neuropathy. *Ann Neurol* 2003; 54(1):19–29.

Lang, I, RB Wallace, FA Huppert and D Melzer. Moderate alcohol consumption in older adults is associated with better cognition and well-being than abstinence. *Age and Ageing* 2007; 36:256–261.

Lewinsohn, PM., SA Shankman, J M Gau, and DN Klein. The prevalence and co-morbidity of subthreshold psychiatric conditions. *Psychological Med* 2004; 34:613–622.

Lichtenstein, AH, LJ Appel, M Brands, et al. Diet and lifestyle recommendations revision 2006: A scientific statement from the American Heart Association Nutrition Committee. *Circulation*. 2006; 114:82–96. see http://circ.ahajournals.org/content/114/1/82.long

Lima, AF, F Pechansky, MP Fleck, and R De Boni. Association between psychiatric symptoms and severity of alcohol dependence in a sample of Brazilian men. *J Nerv Ment Dis* 2005; 193(2):126–130.

Lingford-Hughes, AR, S Welch, and DJ Nutt. Evidence-based guidelines for the pharmacological management of substance misuse, addition and comorbidity: Recommendations for the British Association for Psychopharamacology. *J Psychopharmacology* 2004; 18(3):293–335.

Liu, I-Chou, DL Blacker, RP. Xu, et al. Genetic and environmental contributions to the development of alcohol dependence in male twins. *Arch Gen Psychiatry* 2004; 6:897–903.

Lobo, E, C Dufouil, G Marcos, et al. Is there an association between low-to-moderate alcohol consumption and risk of cognitive decline? *Am J Epidemiol* 2010; 172(6):708–716.

Manari, AP, VR Preedy, and TJ Peters. Nutritional intake of hazardous drinkers and dependent alcoholics in the UK. *Addict Biol* 2003; 8(2):201–210.

Manzardo, AM, EC Penick, J Knop, et al. Neonatal vitamin K might reduce vulnerability to alcohol dependence in Danish men. *J Study Alcohol* 2005; 66(5):586–592.

Marriott, BP, L Olsho, L Hadden, and P Connor. Intake of added sugars and selected nutrients in the United States:

National Health and Nutrition Examination Survey (NHANES) 2003–2006. *Crit Rev Food Sci Nutr.* 2010; 50(3):228–58.

Meek, LR, K Myren, J Sturm, D Burau. Acute paternal alcohol use affects offspring development and adult behavior. *Physiol Behav.* 2007; 91(1):154–160.

Mellion, M, JM Gilchrist, and S de la Monte. Alcohol-related peripheral neuropathy: Nutritional, toxic, or both? *Muscle Nerv* 2011; 43(3) 309–316.

Pennington, Jean A. T. and Judith S. Douglass. *Bowes and Church's Food Values of Portions Commonly Used.* Philadelphia: Lippincott Williams & Wilkins. 2005: xxxii.

Pincus, HA, WW Davis, and LE McQueen. Subthreshold mental disorders. A review and synthesis of studies on minor depression and other brand names. *BJP* 1999; 174:288–296.

Rogers, PJ, S Heatherley, RC Hayward, et al. Effects of caffeine and caffeine withdrawal on mood and cognitive performance degraded by sleep restriction. *Psychopharmacology (Berl)* 2005; 179(4):742–52.

Rogers, PJ and HJ Smit. Food craving and food "addiction": A critical review of the evidence from a biopsychosocial perspective. *Pharmacology Biochem Behav* 2000; 66(1):3–14.

Roman, G C. Nutritional Disorders of the Nervous System. In *Modern Nutrition in Health and Disease* 1362–1380. Ed. ME. Shils, M Shike, AC Ross, B Caballero, and R J Cousins. Lippincott Williams & Wilkins, Philadelphia. 2006.

Roncone, DP. Xerophthalmia secondary to alcohol-induced malnutrition. *Optometry* 2006; 77(3):124–133.

Santolaria, F, E Gonzalez-Reimers, JL Perez-Manzano, et al. Osteopenia assessed by body composition analysis is related to malnutrition in alcoholic patients. *Alcohol* 2000; 22(3):147–157.

Sareen, J, L McWilliams, B Cox, and MB Stein. Does a U-shaped relationship exist between alcohol use and DSM-III-R mood and anxiety disorders? *J Aff Dis* 2004; 82:113–118.

Sgouros, X., M Baines, and R N Bloor, et al. Evaluation of a clinical screening instrument to identify states of thiamine deficiency in in-patients with severe alcohol dependence syndrome. *Alcohol and Alcoholism* 2004; 39(3):227–232.

Sudano, I, L Spieker, and C Binggeli, et al. Coffee blunts mental stress-induced blood pressure increase in habitual but not in nonhabitual coffee drinkers. *Hypertension* 2005; 46:521–526.

Taylor, VH, CM Curtis and C Davis. The obesity epidemic: The role of addiction. *Can Med Assoc J.* 2010; 182(4):327–328.

U.S. Department of Health and Human Services, U.S.D.A. Dietary Guidelines for Americans 2005. see http://www.health.gov/dietaryguidelines/dga2005/document/pdf/dga2005.pdf

Wang, AH and C Still. Old world meets modern: A case report of scurvy. *Nutr Clin Pract* 2007; 22(4):445–448.

Whitney, EN and SR. Rolfes. *Understanding Nutrition.* 540–541. Thomson Wadsworth. Belmont California, 2005.

Woojae, K. Debunking the effects of taurine in Red Bull energy drink. *Nutrition Bytes* 9(1) (2003): article 6 in the Scholarship Repository of UCLA's Department of Biological Chemistry, located at http://repositories.cdlib.org/ Then follow links to ucla biol chem and nutrition bytes http://www.ftc.gov/opa/2010/11/alcohol.shtm

chapter three

Aggression, anger, hostility, and violence

Introduction

Neurotransmitters influence mental status in a variety of ways. Nutritional intake of essential amino acids and essential fatty acids has been linked to generation and function of neurotransmitters.

> Neurotransmitters include acetylcholine, dopamine, histamine, norepinephrine, and serotonin.

Impulsive behaviors such as criminal and aggressive acts, suicide attempts, and apparent genetic loading for early onset alcoholism appear to be linked by reduced functions of serotonin. The effects of normal levels of serotonin include better impulse control, decreased irritability, and decreased tendency toward aggression, as well as decreased depression, increased relaxation, ability to go to sleep, and less cravings for sweets.

Tryptophan, an essential amino acid, is a precursor to the neurotransmitter serotonin. With a higher level of tryptophan present in the brain, more serotonin can be produced. Dietary intake of tryptophan and the body's response to it influence how much is taken up by the brain and is available for producing serotonin (Fernstrom 1971). When carbohydrate is eaten, a healthy body secretes insulin to facilitate the transfer of glucose into the cells. Insulin also causes amino acids except tryptophan to be taken up by cells all over the body, leaving tryptophan in a comparatively higher concentration in the blood and being carried into the brain. A higher level of glucose, a higher level of insulin, and less competition from other amino acids results in more amino acid tryptophan entering the brain, available for serotonin synthesis.

Levels and ratios of essential fatty acids appear to be linked to anger, violence, hostility, and aggressive behavior. These include (1) highly unsaturated fatty acids (HUFA), (2) eicosapentaenoic acid (EPA), a 22-carbon, omega-3 fatty acid important in cell membrane function, (3) docosahexaenoic acid (DHA), a 22-carbon, omega-3 fatty acid important in cell membrane structure, (4) alpha-linoleic acid (ALA) and omega-6 fatty acids, (5) arachidonic acid (AA), and (6) gamma-linolenic acid (GLA).

Lipids

Essential fatty acids, anger, and anxiety—level I evidence

In a 3-month, double-blind, placebo-controlled, randomized trial of supplementation with essential fatty acids, 22 patients being treated in a substance abuse outpatient clinic were supplemented with 2.25 g of EPA, 500 mg DHA, and 250 mg omega-3 PUFA per day in 5 capsules. Before and after supplementation dietary intake, blood levels of EPA, DHA, and other lipids were measured and assessed in relationship to scores of anxiety and anger. Diet records of usual intake indicated the treatment group consumed 4.2% of the recommended intake of omega-3 fatty acids (3.5g/2000 kcal) and placebo groups consumed 4.7% of recommended levels. Intake of omega-6 fatty acids was eleven times greater than their intake of omega-3 fatty acids.

Anger scores declined in the supplemented group but not the placebo group, related to increased blood levels of DHA and DPA (docosapentaenoic acid). Anxiety scores declined in the supplemented group, but not in the placebo group, related to an increase of EPA (Buydens-Branchey 2008).

Violent behavior and essential fatty acids—level II evidence

Serotonin turnover rate in the central nervous system may be modulated by DHA. In violent subjects, but not in nonviolent subjects, plasma DHA was negatively correlated with the main metabolite of serotonin in the human body: CSF 5-HIAA (cerebrospinal fluid 5-Hydroxyindoleacetic acid). Violent subjects had lower concentrations of CSF 5-HIAA. It is suggested that well-controlled dietary studies are needed to determine whether *dietary intake* of DHA will increase or decrease either CSF 5-HIAA or the impulsive violent behavior (Hibbeln 1998).

Intake of fish, omega-3, and omega-6 fatty acids, and hostility in young adults—level II evidence

Analysis of data from a cross-sectional observational cohort study of 3581 young adults showed that a higher intake of DHA and fish might be related to lower likelihood of hostility in young adulthood (Iribarren 2006).

Low cholesterol and violent crime—level II evidence

In a community cohort (N = 79,777) of enrollees in a health screening project, cholesterol levels of those who had police records of having committed violent crimes were compared to nonoffenders. Individuals were matched on age, gender, education, and alcohol use. Violent criminals

(N = 100) had significantly lower cholesterol than did nonoffenders. Low cholesterol was described as "below the median." Age was a strong confounder (Golomb 2000).

In a study of children, serum total cholesterol below the 25th percentile (145 mg/dl) was significantly associated with aggressive behaviors in children (Mascitelli 2008).

Cholesterol-lowering drugs—level IV evidence

Six case studies reported individuals who experienced severe irritability (defined as homicidal impulses, threats to others, road rage, threats to family members, and damage to property) while on statin drugs for lowering cholesterol levels. The personality changes disappeared after drug cessation and reappeared with use of other similar drugs (Golomb 2004). While case studies cannot be interpreted as evidence of cause and effect, they add one perspective to the literature on aggression linked to low cholesterol levels.

Vitamins

Aggression, empty calories, and thiamin—level II evidence

In an early study of functional thiamin deficiency (which measures thiamin-dependent enzyme activity), it was noted that 20 juvenile subjects reported symptoms generally considered to be (1) neurotic dysfunction, (2) conversion reactions, or (3) hysterical, and also reported diets high in refined carbohydrates (described as "high sugar," "junk foods," or "empty-calorie foods"). Subjects' parents described sleep disturbance, poor impulse control, personality changes, hostility, anger, irritability, and aggressiveness.

The study compared transketolase activity (TKA) and thiamin percentage effect/uptake (TPPE) as well as levels of folate and vitamin B_{12}. TKA and TPPE reliably detect thiamin deficiency. TPPE is most sensitive to thiamin levels and is thought to become abnormal before TKA levels fall. All 20 participants had elevated TPPE initially. They responded to thiamin supplementation with increased uptake activity. Eight had subnormal TKA levels. All noticed marked symptomatic improvement or loss of symptoms after thiamin supplementation.

The researchers commented that the early symptoms of beriberi (vitamin B1 deficiency) are nonspecific and "functional" in nature. Beriberi has been described as the prototype of autonomic dysfunction. Symptoms are of such a subjective nature that "they would represent a trap for the unwary physician since he would be unable to find any objective physical sign other than variations of normal which would be easily classed as the effects of a chronic state of anxiety" (Lonsdale 1980).

Table 3.1 Cu:Zn Ratios Found in Assaultive Youth

	Normal	Control Group	Assaultive Group
Copper (serum)	Range 70–155 mcg/dl		
Zinc (plasma)	Range 60–130 mcg/dl		
Cu:Zn ratio	Calculated from above Range lows 70:60 = 1.16 Range highs 155:130 = 1.19	1.02	1.4

Source: Walsh et al. 1997.

Minerals

Copper-to-zinc ratios in assaultive young males—level II evidence

Males 3 to 20 years old who had a history of assaultive behavior were compared with control subjects who did not have such a history. Significant differences in the levels and ratios of serum copper and plasma zinc were found (Table 3.1). Assaultive behavior included verbal assault, destructive rage, physical assault, and aggravated assault. Frequencies reported ranged from monthly to many times per day (Walsh 1997).

Selenium (Se): Environmental exposure—level IV evidence

Signs and symptoms of excess Se may include impaired learning, memory loss, hyperactivity, aggressiveness, and defensiveness, and may occur at low doses. Symptoms from environmental contamination with selenium include: (1) neurological disturbances found in individuals who live in North Dakota, China, and Venezuela (areas with high SE in the soil and resulting in increased levels in the food supply), and (2) gradual central nervous system neurotoxicity with behavior and emotional changes related to environmental exposure rather than intake of foods with pesticide residues (Hatchcock 1994).

Supplements

Criminal behavior: Lessons from the past—level III evidence

In 1982, Stasiak reviewed publications involving nutrition, criminal behavior, hypoglycemia, allergies, and vitamins. He concluded that while nutritional deficiencies and metabolic deficits may be associated with maladaptive behavior, the relationship had not been demonstrated to a sufficient degree to advocate widespread use of special diets and vitamin supplementation in large and possibly harmful pharmacological doses with criminal offenders. At that time, he recommended areas of future research establishing the nutritional status of the offender population and the efficacy of dietary manipulation in improving behavior and recidivism.

Stasiak recommended that in addition to overt deficiencies, the incidence of metabolic disorders, allergies, and metal toxicity should be examined and that well-controlled, scientifically rigorous experiments be conducted and analyzed with statistical precision. He further concluded that biological treatments needed to be adjunctive to traditional rehabilitative techniques. He states that "even if correcting nutritional factors does result in a more emotionally and psychologically stable individual with the capacity to cope more effectively with his or her environment, years of behaving produce a habit of thought and set of skills that will continue to guide behavior. Intensive interventions for establishing strong competing ways of thinking and behaving will be needed" (Stasiak 1982).

Disciplinary infractions in prison and nutrition supplements—level I evidence

Twenty years later a randomized, double-blind study of young adult prisoners who were provided moderate-level nutritional supplements of vitamins, minerals, and essential fatty acids was conducted. The null hypothesis was that the supplements would make no difference in the number of minor or major infraction reports. The placebo and treatment groups were matched regarding diet, behavior, and psychological test results as well as history of disciplinary infractions. Participants were in the study for a minimum of two weeks, many much longer.

Statistical analysis of 314 reports showed that the supplemented group had a reduction of minor disciplinary incidents of 33.3%, from 16 to 10.4 incidents per thousand-person-days. The group that received no supplements reduced their incidents by 6.5%. The number of more serious incidents involving violence (338 reports) showed the supplemented group reduced infractions by 37% ($p = 0.005$). The placebo group reduced incidents by 10.1%.

The investigators suggested that future studies should include biochemical blood tests, correlated with behavior. They also noted that participants often consumed less than the UK reference nutrient intakes despite availability (especially Se, Mg, K, I, and Zn), and that young offenders in the community are likely to consume even worse diets than regular meals offered during custody, related to social breakdown of the family and its function of providing food (Gesch 2002).

A three-year trial, scheduled to start in 2010, conducted by principle investigator Bernard Gesch hopes to replicate the above results showing that nutritional supplements can reduce violence among prisoners. The proposed study includes before and after blood chemistry analysis and a battery of computer-based behavioral and cognitive tests.

The nutritional supplement trial involving more than 1000 prisoners from three UK prisons is funded by UK's Wellcome Trust (Bohannon 2009). Called The Three Prisons Study, early results in publication Autumn, 2012 (See, http://ifbb.org.uk/three-prisons-study).

Reported incidences reduced, but not aggressiveness —level II evidence

A similar study reported on a study of 221 young adult prisoners who received nutritional supplements containing vitamins, minerals, essential fatty acids, or placebos, over a period of 1 to 3 months. Reported incidents concerned aggressive and rule-breaking behavior observed by the prison staff. In the supplemented group, reported incidents were significantly reduced compared with the placebo group. Other assessments, however, revealed no statistically significant reductions in aggressiveness or psychiatric symptoms. Since no significant improvements were found in a number of other self-reported outcome measures, the authors commented results should be interpreted with caution (Zaalberg 2010).

Juvenile delinquency and vitamin–mineral supplementation—level I evidence

In a stratified, randomized, double-blind, placebo-controlled, four-month trial, a daily vitamin-mineral supplement was provided to students 6 to 12 years of age. The supplement provided 50% of the U.S. Recommended Daily Allowance, designed to raise the intake of the participants to the levels recommended by the National Academy of Sciences for children 6 to 11 years of age. Pre- and postsupplement measures of antisocial behaviors were utilized and outcome measures of disciplinary actions for violent and nonviolent delinquency were reported.

Of the children who had previously been disciplined at least once and who received the supplement, the rate of disciplinary action was 47% lower than the children who had previously been disciplined at least once and who had not received the supplement. The study also confirmed a modest increase of 2 to 3 points of nonverbal intelligence in the group that received the supplement (Schoenthaler 1999, 2000). An earlier study involved three months of supplements given to confined juvenile delinquents and measured both blood levels of vitamins and violations of rules concerning violent and nonviolent behavior. Participants who experienced a correction of low blood levels had a decrease of violent acts from 131 to 11 incidents during the intervention. Those who maintained normal or low blood levels of vitamins showed no change in violent acts: 39 acts during baseline, 37 during the intervention (Schoenthaler 1997).

Rage and labile mood in two children—level IV evidence

An ABAB design studied two medication-free boys, ages 8 and 12, who were diagnosed with atypical obsessive-compulsive disorder and pervasive developmental delay, respectively. Symptoms of angry outbursts, moodiness, and obsessive-compulsive disorder remitted while taking a micronutrient supplement, returned when the supplement was discontinued, and remitted when the supplement was resumed. Both boys were stable on the supplements while followed for over two years (Kaplan 2002).

Other

Hostility, BMI, waist–hip ratio, calorie intake, and lipids—level II evidence

The associations between hostility and cardiovascular risk factors related to metabolic syndrome in 1081 older men indicated that a total hostility (Ho) score was positively associated with waist-hip ratio, BMI, total caloric intake, fasting insulin level, and serum triglycerides. The Ho score was inversely related to education and high-density lipoprotein (HDL) cholesterol concentration. Effects of hostility on insulin, triglycerides, and HDL cholesterol were mediated by its effects on BMI and waist-hip ratio, which, in turn, exerted their effects on lipids and blood pressure through insulin (Niaura 2000).

As illustrated by this and other studies, scientists find the complexity of the human body is mirrored by the complexity needed in research methods and statistical analyses.

Anger and metabolic syndrome—level II evidence

For 7.4 years, 425 women (42 to 50 years old at baseline) were monitored for psychological risk factors including depression, anxiety, tension, currently perceived stress, and anger. Biological components of metabolic syndrome were also monitored: plasma glucose level, plasma triglyceride level, waist circumference, and blood pressure were measured at baseline and 1 to 8 years after menopause. Women who exhibited high levels of depression, tension, and anger at baseline and who showed an increase in anger during follow-up had a significantly higher risk of developing metabolic syndrome. The reciprocal was also true: subjects with metabolic syndrome at baseline showed significant increases in anger and anxiety during follow-up (Raikkonen 2002). For an educational booklet from the World Federation for Mental Health, see http://www.diabetesanddepression.org/.

Cholesterol, violent behavior, and genes—level IV evidence

Three generations of male relatives whose histories included suicidal behavior and homicide were found to have a novel gene mutation of apolipoprotein B. The odds ratio for association of hypocholesterolemia and violent behavior in this family was 16.9 (Edgar 2007).

Conclusions

1. Impulsive behaviors such as criminal and aggressive acts, suicide attempts, along with an apparent genetic susceptibility for early onset alcoholism appear to be linked by reduced functions of serotonin. Serotonin levels may be linked to intake of essential fatty acids (DHA), carbohydrates, and the amino acid tryptophan.
2. Observational studies have shown a decrease of aggressive behavior, violence, and hostility with vitamins and mineral supplementation in adolescents and children.
3. Hostility has been linked to cardiovascular disease and metabolic syndrome, which are linked to BMI and dietary intake.
4. Genetic variation has been implicated in the links between aggressive behavior and possible nutritional influences (See also Chapter 7 "Mood Disorders").

References

Bohannon, J. The theory? Diet causes violence. The lab? Prison. *Science* 2009; 325(5948):1614–1616.

Buydens-Branchey, L, M Branchey, and JR Hibblen. Associations between increases in plasma N-3 polyunsaturated fatty acids following supplementation and decreases in anger and anxiety in substance abusers. *Prog Neuropsychopharmacol Biol Psychiatry*. 2008; 32(2):568–575.

Edgar, PF, AJ Hooper, NR Poa, and JR Burnett. Violent behavior associated with hypocholesterolemia due to a novel APOB gene mutation. *Mol Psychiatry* 2007; 12(3):258–263.

Fernstrom, JD and RJ Wurtman. Brain serotonin content: Physiological dependence on plasma tryptophan levels. *Science* 1971; 173:149–152.

Gesch, CB, SM Hammond, SE Hampson, et al. Influence of supplementary vitamins, minerals and essential fatty acids on the antisocial behavior of young adult prisoners. *Br J Psychiatry* 2002; 18:22–28.

Golomb, BA, T Kane, and JE Dimsdale. Severe irritability associated with statin cholesterol-lowering drugs. *QJM Int'l J Med* 2004; 97(4):229–235.

Golomb, BA, H Stattin, and S Mednick. Low cholesterol and violent crime. *J Psychiatr Res* 2000; 34(4–5):301–9.

Hatchcock, JN, and JI Rader. Food Additives, Contaminants and Natural Toxins. In *Modern Nutrition in Health and Disease*. 1593. Ed. ME Shils, JA Olson, and M Shike. Philadelphia: Lea & Febiger, 1994.

Hibbeln, JR, M Linnoila, JC Umhau, et al. Essential fatty acids predict metabolites of serotonin and dopamine in cerebrospinal fluid among healthy control subjects, and early- and late-onset alcoholics. *Biol Psychiatry* 1998; 44(4):235–242.

Hibbeln, JR, JC Umhau, M Linnoila, et al. A replication study of violent and nonviolent subjects: Cerebrospinal fluid metabolites of serotonin and dopamine are predicted by plasma essential fatty acids. *Biol Psychiatry* 1998; 44(4):243–249.

Iribarren, C, JH Markovitz, DR Jacobs Jr, et al. Dietary intake of n–3, n–6 and fish: relationship with hostility in young adults-the CARDIA study. *Euro J Clin Nutr* 2006; 60(7):882–888.

Kaplan, BJ, SG Crawford, B Gardner, and G Farrelly. Treatment of mood lability and explosive rage with minerals and vitamins: Two case studies in children. *Child Adolesc Psychopharmacol*. 2002 Fall; 12(3):205–219.

Lonsdale, D and RJ Shamberger. Red cell transketolase as an indicator of nutritional deficiency. *Am J Clin Nutr* 1980; 33(2):205–211.

Mascitelli, L, F Pezzetta, and MR Goldstein. Low cholesterol, delinquency, and suicidality. *Prim Care Companion J Clin Psychiatry* 2008; 10(5):413–414.

Niaura, R, SM Banks, KD Ward, et al. Hostility and the metabolic syndrome in older males: The Normative Aging Study. *Psychosomatic Med* 2000; 62:7–16.

Raikkonen, K, KA Matthews, and LH Kuller. The Relationship between psychological risk attributes and the metabolic syndrome in healthy women: Antecedent or consequence? *Metab: Clin Exper* 2002; 51:1572–1577.

Shoenthaler, S, S Amos, W Doraz, et al. The effect of randomized vitamin-mineral supplementation on violent and non-violent antisocial behavior among incarcerated juveniles. *J Nutr & Environ Med* 1997; 7(4):343–352.

Schoenthaler, SJ, ID Bier, K Young, et al. The effect of vitamin-mineral supplementation on juvenile delinquency among American Schoolchildren: A randomized, double-blind placebo-controlled trial. *J Altern Complement Med* 2000; 6(1):19–29.

Schoenthaler, SJ and ID Bier. Vitamin-mineral intake and intelligence: A macrolevel analysis of randomized controlled trials. *J Alterm Complement Med* 1999; 5(2):125–134.

Stasiak, EA. Nutritional approaches to altering criminal behavior. *Corrective Social Psychiatry and J Behav Techny Methods & Ther* 1982; 28(4):110–114. http://www.ncjrs.gov/App/Publications/abstract.aspx?ID=85790

Walsh, WJ., HR Isaacson, F Rehman, and A Hall. Elevated blood copper/zinc ratios in assaultive young males. *Physiology & Behavior* 1997; 62(2):237–329.

Zaalberg, A, H Nijman, E Bulten, et al. Effects of nutritional supplements on aggression, rule-breaking, and psychopathology among young adult prisoners. *Aggressive Behav* 2010; 36(2):117–126.

chapter four

Autism spectrum disorders (ASDs) and attention deficit hyperactivity disorder (ADHD)

Introduction

Attention deficit hyperactivity disorder (ADHD), autism spectrum disorders (ASDs), dyslexia, dyspraxia, and pervasive developmental disorder (PDD) are labels for particular patterns of behavioral and learning difficulty. Differences between individuals are substantial and symptoms may overlap. Scientists and clinicians are beginning to move beyond simply describing these neurobehavioral syndromes toward understanding the biologic, neurologic, and genetic components. There is evidence of familial/genetic association between these conditions (Eichler 2008). A direct link between chromosomal deletions and ADHD was reported in 2010 (Williams 2010).

Autism spectrum disorders

Autism was first described by Leo Kanner in 1943 (Kanner 1943). The Centers for Disease Control (CDC) reported a tenfold increase in autism in Atlanta from 1986 to 1996. In 2007, it was estimated an average of 1 in 150 children in the U.S. were affected by an ASD. Part of this apparent increase may be due to inconsistencies in data-reporting methods and in changes in diagnostic criteria in the past 15 years. Boys are four times more likely than girls are to be diagnosed with autism (CDC 2007).

Since there is no medical test for ASDs, a diagnosis is typically made after a thorough evaluation by a qualified professional. Such an evaluation frequently includes clinical observations, parent interviews, developmental histories, psychological testing, speech and language assessments, and possibly the use of one or more autism diagnostic tests.

ASDs typically begin before a child is 3 years of age and last throughout a person's life. The ASD syndrome is described as having impairments in three domains: social interaction, language, and range of interests. Individuals handle information in their brain differently than others do, and might have unusual ways of learning, paying attention, and reacting to different sensations. Some have seizures; some eat nonfood substances such as sand, paint, dirt, or paper. Chronic constipation or diarrhea is not uncommon. Hypersensitivity to sound, touch, taste, or smells is also observed. Variation in pain threshold may also occur (Saenger 2005).

Pica: the eating of nonfood materials.

Treatment may include improvement in the following: (1) skin or gastrointestinal symptoms, (2) allergic symptoms, (3) visual symptoms, (4) ability to pay attention (distractibility, poor concentration and memory), (5) emotional sensitivity (depression, excessive mood swings, or anxiety), or (6) sleep problems (Richardson 2003).

Although everyone with an ASD has significant challenges in certain areas of life, some might be gifted in other areas. Some individuals might have relatively good verbal skills, but have difficulty or little interest in interacting with other people. Children with ASDs often do not take part in pretend play, have a hard time starting social interactions, and engage in unusual or repetitive behaviors (e.g., flapping hands, making unusual noises, rocking from side to side, or toe walking).

Children with ASDs might have developmental disabilities such as intellectual disability, seizure disorder, fragile X syndrome, Down syndrome, or tuberous sclerosis. Some children might have mental health problems, such as depression or anxiety (CDC 2006; Kinney 2010).

Nutrients and autism—level V evidence

Attempts to link autism with nutrition or nutrients have covered many facets of nutrition. The possibility of malabsorption resulting in general or selective malnutrition has led to recommendations that growing children get not only adequate, but also mega-doses of nutrients to overcome malabsorption. Specific nutrients that have been investigated included omega-3 fatty acids, omega-6 fatty acids, gamma linoleic acid (GLA), magnesium, zinc, calcium, iron, and vitamins A, B_6, B_{12}, C, D, and E, as well as the amino acids tyrosine, lysine, glutamine, and methionine. Researchers have also questioned whether (1) the metabolism of sulfates, (2) high levels of copper or lead, or lactose, (4) abnormal thyroid levels, (5) T cells in the immune system, (6) neurotransmitter levels (such as serotonin), or (7) food allergies could play a role in the development of autism. Practitioners report on analysis of blood, serum, urine, feces, and hair for determining levels of various nutrients for deficient or toxic levels. Supplements of numerous products such as probiotics, evening primrose oil, digestive enzymes, and other products have been recommended by some to compensate for possible alterations in metabolism (McGinnis 1999; McCandles 2003).

A summary of treatments and results from a survey by the Autism Research Institute are summarized by James B. Adams. The survey included reported effects of dietary factors such as sugar, gluten, allergens, essential fatty acids, etc. In general, many treatments are said to help about half those who tried them and did not help the other half who tried them (www.autism.com 2010).

A review of nutritional intake and interventions was published in 2010 by Infant, Child, and Adolescent Nutrition (ICAN) in Canada. Problem eating behaviors such as (1) picky eating, (2) selective food refusal (related to texture, specific food group, unfamiliarity, etc.), (3) insistence on nonfunctional routines (such as one food not touching another), and (4) tantrums occur in 46 to 89% of children with autism, compromising nutrient intake. These behaviors are common in children who do not have autism, but are not as intense, are of shorter duration, and may occur singly rather than jointly. Side effects of medications commonly taken by those with autism can decrease appetite, or cause nausea, vomiting, and insomnia. Other medications may result in increased appetite, weight gain, and sedation.

Children with autism frequently avoid beef, pork, poultry, and fish, which leaves them potentially low in iron and zinc. Gluten-free foods used in diets that avoid gluten and casein are not fortified or enriched, and therefore do not provide the same vitamins and minerals provided by common breads and cereals. The authors list common micronutrient deficiencies of vitamins A, D, E, K, pantothenic acid, calcium, and zinc in children with autism. One survey found 83% of families used at least one supplement daily, 50% used at least two supplements. Non-IgE mediated food sensitivities and allergies are common, but are likely to be outgrown. In addition to nutritional assessments, assessments for iron deficiency, bone density, and dental caries are recommended (Geraghty 2010, parts 1 and 2). The online version of these comprehensive reports can be found at http://can.sagepub.com/content/2/1/62 and http://can.sagepub.com/content/2/2/120.

For overall health maintenance and normal growth, a child (with or without autism) needs to have an adequate, but not excessive, supply of calories, protein, carbohydrates, essential fatty acids, vitamins and minerals, fiber, and water. "Adequate" needs to take into consideration any individual conditions that decrease absorption of nutrients.

Consensus report and evaluation guidelines for pediatric ASD digestive problems—level III evidence

In 2008, 25 experts met in Boston to review medical research on ASD and write a consensus report. Gastrointestinal disorders such as abdominal pain, chronic constipation, and gastroesophageal reflux disease are commonly reported in individuals with ASD. The new report calls for more rigorous research into the prevalence of digestive problems and whether special diets might help some children. The expert opinion was that individuals with ASD deserve the same standard of care in the diagnosis and treatment of gastrointestinal concerns as should occur for patients without ASD and can benefit from adaptation of the general pediatric guidelines for diagnostic evaluation. These guidelines may be found at http://pediatrics.aappublications.org/cgi/reprint/125/Supplement_1/S19.

Although nearly one in five children with ASD is reported by parents to be on a special diet, the report did not support this practice. Most common are diets that eliminate gluten (a protein found in many grains) and casein (a protein found in milk). The 2008 report advises doctors to watch for nutritional deficiencies in patients with autism. It also recommends involvement of a nutritionist if an individual follows a diet with food omissions and limitations (Buie 2010).

Environmental and genetic factors in autism—level IV evidence

In a shift in scientific thinking, environmental factors are being investigated as possible triggers for autism in susceptible children with alterations/mutations in 5 to 15 genes. Environmental factors being studied include nutrients, drugs, vaccines, viruses, and toxic substances such as lead and mercury. Environmental influences may occur *in utero*. The nervous system and immune system are developed before birth and for two years after birth, increasing vulnerability to damage.

Methylation is the addition of a 1-carbon atom with its attached hydrogen atoms (a methyl group). Addition of this 1-carbon unit may activate or inhibit the protein, enzyme, hormone, or DNA gene to which the methyl group has been added. Removal of a methyl group has the opposite effect.

Folic acid is a vitamin commonly supplying 1-carbon methylation groups used during metabolism. A dietary deficiency of folic acid is one nutrient known to cause birth defects known as neural tube defects.

Supplementation of grain foods with folic acid in the United States was approved in 1994 to counteract the effects of potentially inadequate intakes of folic acid in the population.

A biochemical pathway has been described that is critical to normal gene expression and could cause neurodevelopmental disorders such as autism (Waly 2004). A child who has a mutation in genes that encode methylation-related enzymes is at higher risk of damage from toxins.

Methionine synthase, an enzyme critical to methylation, uses an active form of vitamin B_{12} to function. Thimerosal interferes with the conversion of dietary vitamin B_{12} to the active form, which links to folic acid function. This impedes DNA methylation and disrupts normal gene action. In human cells exposed to thimerosal, the process of methylation dropped significantly. Thimerosal, a mercury-based preservative in childhood vaccines, has been phased out of most vaccines, and no conclusive evidence linking thimerosal and autism has been confirmed. Neuronal cells exposed to lead, ethanol, and aluminum also disrupt methylation, but via different mechanisms.

> The widely controversial claims of Andrew Wakefield concerning vaccines and autism were declared fatally flawed both ethically and scientifically in 2011. (See Godlee, F. et al. Wakefield's article linking MMR vaccine and autism was fraudulent. *BMJ* 2011; 342:c7452.)
>
> Others report that Professor Walker-Smith and Dr. Amar Dhillon described seven similar cases of behavior changes following MMR vaccine. (See http://www.naturalnews.com/031116_Dr_Andrew_Wakefield_British_Medical_Journal.html, accessed January 7, 2012).

Autism and environmental factors—level IV evidence

Risk factors for ASD also include advanced parental age, oxidative stress, neuroinflammation, and mitochondrial dysfunction related to air pollution or organophosphates. Herbert suggests that some of the underlying biochemical disturbances can be reversed by targeted nutritional interventions; for example, abnormalities in glutathione (Herbert 2010).

Autism and nongenetic risk factors—level V evidence

Kinney and colleagues hypothesize that genetic risk factors for autism are de novo mutations. Risk factors for these mutations include exposure to mercury, cadmium, nickel, trichloroethylene, and vinyl chloride, which are established mutagens. Additional risks include vitamin D deficiency, which has been related to decreased sun exposure due to living in urbanized areas, at higher latitudes, or in areas with high levels of precipitation. Vitamin D is important for repairing DNA damage and protecting against oxidative stress (Kinney 2010).

> De novo mutations: a genetic mutation that is neither possessed nor transmitted by a parent.

Lipids

Essential fatty acids have been the subject of research in this population. Increased need for highly unsaturated fatty acids (HUFA) could result from differences in an individual's metabolism. Differences could include (1) altered conversion of simple essential fatty acids (EFA) into HUFA, (2) higher than usual metabolism, breakdown, and excretion of HUFA, and (3) alteration in transporting or incorporating HUFA into cell membranes.

Fatty acid: Mental retardation vs. autism—level II evidence

In 2001, a study in France examined plasma phospholipids in subjects with autism and compared the levels to control subjects who were mentally retarded. Results showed a reduction of omega-3 fatty acids in subjects with autism, but without a reduction in omega-6 levels. This resulted in a statistically significant decrease of 25% in the omega-6:omega-3 ratio. The authors felt this difference might open a new area of investigation for biological indices in autism (Vancassell 2001).

Fatty acids in erythrocytes and plasma lipids—level II evidence

Forty-five children with autism and 38 children who experienced developmental delay were matched with typically developing children and erythrocyte fatty acids were compared in a case control study. Participants with autism who had consumed fish oil supplements and those who had not were also compared. Some differences were observed, but were not statistically significant. Saturated fatty acids and monounsaturated fatty acids (components of axonal myelin sheaths) were decreased in those with developmental delays. Subjects who had consumed fish oil showed reduced erythrocyte arachidonic acid (AA) and omega-6 fatty acids, and increased EPA and AA:EPA ratios. The authors felt this indicated that autism subjects did not have a phospholipid disorder, but may have an imbalance of essential highly unsaturated fatty acids (Bell 2010).

Carbohydrates

Galactose is the simple sugar derived from lactose in milk. Omission of milk from the diet eliminates this major source of galactose. In healthy people, galactose is made in the liver to fulfill the important functions of galactose in cell membranes, glycolipids, glycoproteins, sulfation pathways, and phenol degradation. Low sulfate levels, an elevated cysteine-to-sulfate ratio, and high phenol levels have been linked by some to autism. Research is needed to clarify these associations (Waring 2007).

Gene expression of enzymes involved in carbohydrate metabolism—level II evidence

Impaired carbohydrate metabolism in children with autism appears related to altered genetic expression of five genes involving three enzymes that metabolize disaccharides, which in turn changed the type and ratios of common intestinal bacteria. These changes influenced symptoms of bloating and mucus in stools, although diarrhea, stool frequency, and consistency were not different between participants with autism and those without autism (Williams 2011). Research using mice has shown that composition and activity of bacteria in the gut influences brain development, behavior, and anxiety (Heijtz 2011).

Protein

Gluten- and casein-free diet intervention—level II evidence

A 24-month, two-stage, randomized, controlled trial of 72 children, using a gluten-free diet, casein-free diet, or no-diet approach and using assessment of core autism behaviors, found significant improvement of the diet group scores on several measures. Assessments were done at baseline, 8, and 12 months. Researchers noted conclusions might have been influenced by the absence of a placebo group in the research (Whitely 2010).

Casein- and gluten-free diets: A Cochrane review—level III evidence

Reviewers found two small randomized clinical trials (RCTs) for a total N = 35. Studies showed three outcome measures in favor of diet intervention: overall autistic traits, social isolation, and overall ability to communicate and interact. Three additional outcomes showed no difference between the treatment and control groups. No harm from the intervention was reported (Milward 2008).

Vitamins

Vitamin D and autism: Hypothesis—level V evidence

In *Medical Hypotheses* online, Cannell summarizes literature concerning vitamin D and autism. The apparent increase in the prevalence of autism over the last 20 years corresponds with increasing medical advice to avoid the sun, advice that has probably lowered vitamin D levels and would theoretically greatly lower activated vitamin D (calcitriol) levels in developing brains. Animal data has repeatedly shown that severe vitamin D deficiency during gestation deregulates dozens of proteins involved in brain development and causes increased brain size and enlarged ventricles. Children with vitamin D deficient rickets have several autistic markers that are reputed to disappear with high-dose vitamin D treatment. Consumption of fish providing vitamin D during pregnancy reduces autistic symptoms in animal offspring. Autism is more common in areas of impaired UVB penetration such as latitudes closer to the poles, urban areas, areas with high air pollution, and areas of high precipitation. Autism is more common in dark-skinned persons and severe maternal vitamin D deficiency is exceptionally common. Cannell concludes that normal distributions of the enzyme that activates neural calcitriol combined with widespread gestational or early childhood vitamin D deficiency may explain both the genetics and epidemiology of autism. If so, he concludes, much of the disease is iatrogenic, brought on by medical advice to avoid the sun (Cannell 2007).

UVA – wavelength 315–400 nm*
UVB – wavelength 280–315 nm
UVB affects the outer layer of the skin, causing sunburn, and stimulates vitamin D production.
UVC – wavelength 100–280 nm
*nm = nanometer = 1 millionth of a millimeter
Canadian Centre for Occupational Health and Safety, http://www.ccohs.ca/oshanswers/phys_agents/ultravioletradiation.html, accessed October 16, 2010.

Vitamin B_2, B_6, and magnesium supplements and dicarboxylic acids in children with autism—level II evidence

Children with autism receiving supplements of 20 mg vitamin B_2, 500 mg B_6, and 200 mg of magnesium had a decrease of succinic acid, adipic acid, and suberic acid in the urine. These acids may be connected to potential problems with energy production and intestinal dysbiosis of individual children with autism (Kaluzna-Czaplinska 2011).

Minerals

Fatal overdose of magnesium—level IV evidence

McGuire and colleagues reported a case of a child's death from hypermagnesemia after treatment with a high dose of magnesium oxide (McGuire 2000). Magnesium compounds are often found in vitamin-mineral supplements, antacids, and laxatives. The 2001 Dietary Reference Intake (DRI) for adults is 300 to 420 mg/day and for children is 80 to 130 mg/day, dependent on age.

Magnesium and vitamin B_6: A Cochrane review—level III evidence

Reviewers found three small studies (total N = 33) that addressed use of supplements of magnesium and vitamin B_6 (pyridoxine) with children with autism. One study had insufficient data; another yielded no significant difference in social interaction, communication, compulsivity, impulsivity, or hyperactivity between treatment and placebo groups. A third study (n = 8) addressed pervasive developmental disorders (PDD). Change in scores of IQ and "Social Quotient" were measured and a statistically significant benefit of the supplement for IQ scores was found. No recommendations regarding use of magnesium and B_6 as a treatment for autism were made (Nye 2005).

Supplements

Vitamin/mineral supplement—level I evidence

A multivitamin/mineral supplement provided to children who had medical, nutritional, biochemical, and behavioral assessment before and after supplementation reported significant changes compared to a placebo group in a 3-month study (Table 4.1). Dosage levels were significantly above the RDA, but not higher than the tolerable upper limits. The supplement did not contain vitamin K, copper, iron, or lipoic acid. It did contain lithium at amounts similar to a typical natural daily intake and < 1% of lithium intake when used as a medication. The supplement also included Coenzyme Q and N-acetyl-cysteine.

Table 4.1 Changes Following Vitamin/Mineral Supplementation of Children with Autism

Initially low	
Biotin	
SAM (S-adenosylmethionine),	
ATP (adenosine-5′-triphosphate)	
NADH (nicotinamide adenine dinucleotide)	improved from 25% below normal to near normal
NADPH (nicotinamide adenine dinucleotide phosphate)	
Sulfate	
Plasma Tryptophan	
Initially high	
Oxidative stress biomarkers (nitrotyrosine, ratio of oxidized: reduced glutathione)	improved to near normal
Uridine, (impaired methylation)	
Highest correlations with parental global impressions—revised scores	
Supplement group showed greater improvement on all subscales than control group	
Vitamin K (initially similar to control group; no change in either)	
Biotin (placebo and control group had similar changes)	
Lipoic acid (negative correlation with PGI-R scores)	

Source: Adams et al. 2011.

The treatment group had significant (p < 0.002) increases in vitamins B_1, B_6, B_{12}, E, whole blood manganese, calcium, and lithium. Sulfate status improved, but was still low.

Among other conclusions, investigators stated that (1) children who were initially low were more likely to have the most improvement with supplementation, as shown by variability in response, (2) bacteria in the gut, being significant sources of vitamin K and biotin and affected by antibiotics, may affect levels of these nutrients in this population, (3) biotin and vitamin K may be biomarkers for overall nutritional status, and (4) it may take longer than 3 months for improvements in metabolic status to manifest as improvements in behavioral symptoms (Adams 2011).

Lower baseline values normalized with supplements—level II evidence

A comparison of 33 controls and 20 children with autism found children with autism had significantly lower baseline values of plasma levels of methionine, S-adenosylmethionine (SAM), adenosine, and oxidized gluthione. This suggests an impaired capacity for methylation and increased oxidative stress. An intervention targeting the lower metabolites in a subset of 8 children with autism normalized the metabolic imbalance. Interventions included 800 μg folinic acid and 1000 μg betaine (anhydrous trimethylglycine) for 3 months. For the following month, a supplement of 75 μg/kg injectible methylcobalamin (B_{12}) was added (James 2004).

A systematic review: Folate metabolites, interventions, and genes—level III evidence

A review of 18 studies of folate metabolites, interventions, and genes concluded that conflicting findings could be attributed to heterogeneity between subjects, sampling issues, a wide range of analytic techniques, and the need for adequate statistical power before definitive conclusions could be made. They also questioned whether functional benefits occur when apparent deficits in folate-methionine metabolism are corrected (Main 2010).

Other

Research on aspartame—level I evidence

There has been much controversy over the possible effects of consuming aspartame on mental health. The scientific reports of research into these effects do not confirm commonly held suspicions. Studies that compared aspartame with sucrose and with saccharin, studies conducted under double-blind, randomized condition, and studies with parents, teachers, and scientists making structured observations for cognitive and behavioral changes do not show significant differences in children who consume aspartame (Wolraich 1994; Shaywitz 1994; Spiers 1998). The referenced studies by Wolraich et al. and Spiers et al. were especially well controlled and analyzed for numerous factors often thought to be sources of bias in experiments related to behavioral effects of aspartame and sugar. (See also Chapter 5 "Genetics, Nutrition, and Inherited Disorders of Metabolism.")

Common diets reported—level IV evidence

The diets commonly reported by parents of children with autism are reviewed in detail by psychologist Kenneth J. Aitken in *Dietary Interventions in Autism Spectrum Disorders: Why They Work When They Do, Why They Don't When They Don't* (including the Feingold diet, diets addressing specific carbohydrates, gluten and casein, oxalates, glutamate, aspartate, phenylalanine, phenols, and a rotation diet). He describes a systematic method for a trial of a dietary approach and the observations required for reaching conclusions. He also details specific nutritional deficiencies that may occur when using the diets reviewed and recommends involvement of a nutritionist as well as use of appropriate supplements. Nutrients potentially deficient include thiamin (B_1), pyridoxine (B_6), B_{12}, vitamin A, calcium, magnesium, iron, zinc, selenium, and essential fatty acids, as well as the simple carbohydrate galactose.

Aitken describes the theory behind various diets and the potential pitfalls. He notes that there is a difference between having evidence that dietary interventions do not work and not having evidence that dietary interventions do work. He states there is an overall lack of scientific evidence on the subject of the efficacy of interventions using nutrients and overall diet in treating ASD. He believes, like others, that a single etiology or single treatment is not likely to be found for the continuum of ASD. Subgroups need to be established, and what appears to affect symptoms and behavior of these subgroups needs to be investigated using scientific methods (Aitken 2009).

Digestive enzyme supplementation—level II evidence

Observations of 43 children by parents and therapists after using a digestive enzyme supplement for more than 6 months were analyzed and showed no clinically significant improvement in (1) behavior, (2) gastrointestinal symptoms, (3) sleep quality, (4) engagement with the therapist, or (5) improved scores on tests of vocabulary or sentence complexity. A small statistically significant improvement in food variety was found (Munasinghe 2010).

Substitutive and dietary approaches—level III evidence

Reviewers of treatment approaches report that parents of children with autism are greatly interested in dietetic and alternative treatment approaches and professionals need to be informed about such approaches. Reviewers include the following opinions:

1. Known genetically caused enzymatic and metabolic errors need to be under the supervision of specialized medical teams.
2. Contradictory results from various methods of study of gluten- and casein-free diets and the possibility of protein malnutrition suggest caution in the use of this diet.
3. Observed lack of some amino acids may have been caused by food selectivity on the part of the children.
4. True food allergy is a possibility for some children with autism, who would have a favorable outcome from omitting the allergen, although some of the published food restrictions are drastic and "ethically debatable."
5. Not enough studies have been done to draw firm conclusions concerning use of ketogenic diets for seizure control.
6. Vitamin deficiencies and toxicity from heavy metals are the most-often reported, placebo-controlled areas of research and more well-designed studies are recommended. Also recommended are studies of the eating habits of children with autism because avoidance and food selectivity are commonly reported (Hjiej 2008).

Management using micronutrients and medication—level II evidence
Using natural preferences of parents for choosing micronutrients without pharmaceuticals or medication as treatment choices, two groups of 44 each of patients 2 to 28 years old were studied by Mehl-Madrona et al. (2010). Both groups showed statistically significantly improvement in Aberrant Behavior Checklist scores; self-injurious behavior (SIB), aggression, and tantrums were lower, and improvement on the Clinical Global Impressions scale was observed. Advantages to treatment with micronutrients were reported to include lower activity level, less social withdrawal, less anger, more spontaneity with the examiner, less irritability, and less weight gain. Advantage of medication management included insurance coverage, fewer pills, and less frequent dosing (Mehl-Medrona 2010).

Parental perceptions and treatment choices—level V evidence
In testing the validity of an assessment tool, "Revised Illness-Perception Questionnaire," scores of 89 parents indicated that a higher sense of personal control was associated with reduced use of nutritional or pharmaceutical treatments. The researcher felt the assessment tool was a reliable means for exploring the concept of illness held by parents of children with ASD (Al Anbar 2010).

Attention deficit hyperactivity disorder (ADHD)

Attention deficit hyperactivity disorder (ADHD) is one of the most common psychiatric disorders, affecting approximately 8 to 9% of school-aged children and 4.4% of adults (Kollins 2008). Family, adoption, and twin studies all support the notion that genetics contribute to the development of ADHD to a degree that is as high as, if not higher than, nearly all other psychiatric disorders.

A promising area of recent work has been to examine how certain genes increase the likelihood of developing the disorder when combined with specific environmental factors. Included are genes related to neurotransmitters dopamine and serotonin, and other compounds as well as their transporters and receptors.

Brain imaging studies of ADHD, including positron emission tomography (PET) studies, structural magnetic resonance imaging (MRI) studies, and functional magnetic resonance imaging (fMRI) studies, have enhanced understanding not only of the neurobiologic basis of ADHD, but also of the mechanisms underlying pharmacologic treatment response. These studies have been successful at identifying between-group differences, but are not used at this time for diagnostic purposes.

Lipids

Effect of high-dose olive, flax, or fish oil on phospholipids—level I evidence

Young et al. (2005) investigated the effect of 60 g doses of olive oil, flax oil, or fish oil given for 12 weeks to a randomized group of 30 adults diagnosed with ADHD. Serum phospholipids were measured at baseline and 12 weeks. Flax oil supplementation produced an increase in alpha linoleic acid (α-LNA) and a slight decrease in the AA-to-EPA ratio. Fish oil supplementation increased EPA, DHA, and total omega-3 fatty acids, and decreased the AA-to-EPA ratio. It was felt that future study of the correction of fatty acids levels would result in therapeutic benefit for this population.

Omega-3 assessment and effect of supplement on behavior—level II evidence

In a two-phase study of 26 children with ADHD, baseline values indicated the group was not deficient in EFA. In phase one, 8 children received a supplement of 20 to 25 mg/kg EPA and 8.5 to 10.5 mg/kg DHA (which resulted in doses of 500, 750, or 1000 µg total depending on the child's weight). The supplement also included 25 to 75 mg phospholipids and 3.75 U alpha tocopherol. The placebo given to the second group in phase one was 500 mg sunflower oil. In phase two, the second group of 8 children was also given the supplement. Detailed plasma fatty acid profiles indicated changes in blood levels of EFA. There were no adverse effects of supplement use and families of children preferred to continue the use of omega-3 supplements. Researchers commented that some changes might be from the placebo effect, that the effect of omega-6 polyunsaturated fatty acids (PUFA) could not be excluded, and suggested ALA be included in supplements for studies of this type.

Despite efforts of investigators, questionnaires reporting behavior were frequently not returned by teachers and parents. Participants showed significant improvement in hyperactivity/impulsive, inattention, and global scores (Belanger 2009).

Carbohydrates

Sugar and hyperactivity—level I evidence

Comisarow reviews reasons for continued misperceptions that sugar intake increases hyperactivity despite evidence to the contrary. Parents, teachers, physicians, and others may be responding with "the illusory correlation" that is related to situational changes in behavior (birthday parties are high-activity events with or without sugar), or attribution to external factors (sugar) instead of personal factors (genetics or upbringing). He suggests that it may be inappropriate and burdensome to pursue a sugar-free diet to control a child's behavior, as withholding sweets may provoke as much disruptive behavior (Comisarow 1996).

An example:
Children aged 3 to 8 years are estimated to need ~1650 to 1750 kcal/day.
10% of 1700 kcal = 170 kcal = 43 g refined carbohydrate
25% of 1700 calories = 428 calories = 107 g refined carbohydrate
50 g sugar = 10.6 tsp.
4.7 g sugar = 1 tsp.
1 C Cheerios = 1 g sugar
12 Honey Teddy Grahams = 4 g sugar
1 Betty Crocker Blueberry Muffin = 14 g sugar
1 12-oz. can regular soft drink = 40 to 50 g sugar

An excessive percent of daily calories from low/no nutrient foods (such as refined sugar) can replace nutrient sources and lead to a nutritionally adequate diet, sometimes called "high-calorie malnutrition." The World Health Organization recommends no more than 10% of calories should come from free sugar (WHO/FAO 2009). The U.S. Institute of Medicine recommends no more than 25% of calories from added sugars (Murphy 2003). The 2005 U.S. Dietary Guidelines recommend limiting discretionary calories (which includes both added sugar and solid fat) to 13% of energy requirement (U.S. Department of Health & Human Services 2005).

Vitamins

Vitamin D and ADHD—level III evidence

A medical chart review of 117 patients found patients with ADHD had unexpectedly low intact parathyroid (iPTH) levels. In 56.4% of patients, 25-hydroxy vitamin D levels were below 50 nmol/L, which may be linked with unfavorable health outcomes. A diagnosis of ASD or schizophrenia predicted low 25-hydroxy vitamin D levels. Authors noted considerable psychiatric improvement in those who were treated with 4000 iu (100 µg) of vitamin D (Humble 2010).

Minerals

Zinc and ADHD—level I (Post-Hoc) evidence

Zinc is an important cofactor for metabolism of neurotransmitters, fatty acids, prostaglandins, and melatonin, and it indirectly influences dopamine metabolism, which is involved in ADHD. The authors of one study reanalyzed data from an 18-subject, double-blind, placebo-controlled crossover treatment comparison of d-amphetamine and Efamol (evening primrose oil, rich in gamma-linolenic acid—GLA). Subjects were categorized as zinc-adequate (n = 5), borderline zinc (n = 5), and zinc-deficient (n = 8) by hair, red cell, and urine zinc levels. Placebo-controlled d-amphetamine response appeared linear with zinc nutrition. Efamol response to zinc appeared U-shaped; Efamol benefit was evident only with borderline zinc levels. The study's authors commented that this post-hoc exploration suggests that zinc nutrition may be important for treatment of ADHD. They suggest that if Efamol benefits ADHD, it likely does so by improving or compensating for borderline zinc nutrition (Arnold 2000).

Supplements

A review of nutritional supplements by the Canadian Pediatric Society—level III evidence

In 2002, the Canadian Pediatric Society published a review of trials investigating the effect of nutritional interventions on behavior of children with ADHD. In general, their conclusions included (1) inconsistent changes reported using the Feingold diet, (2) inconsistent changes reported with use of allergen diets, and (3) inconsistent effects with use of multivitamin supplements and use of single-supplements of Mg, Fe, B_6, Zn, and EFA, and that methodological flaws in most studies compromised the ability to recommend these interventions (Bernard-Bonnin 2002).

Other

Western diet patterns, adolescents, and ADHD—level II evidence

A study of 1700 adolescents tracked individuals who from birth reported on their dietary patterns. Of these adolescents, 115 (91 boys, 24 girls) had been diagnosed by age 14 years with ADHD. A diet high in the "Western" pattern of food intake had more than double the risk of having an ADHD diagnosis than those following the "Healthy" diet pattern. The "Western" diet pattern was described as being high in takeaway foods, processed meats, red meat, high-fat dairy products, and confectionery. The "Healthy" diet pattern was described as being high in fresh fruits and vegetables, whole grains, and fish, tending to be higher in omega-3 fatty acids, folate, and fiber. The authors commented that this cross-sectional study is unable to conclude whether a poor diet leads to ADHD or whether ADHD and impulsive food choices may lead to poor dietary intake (Howard 2010). In a separate report, this group also reported that for every additional food group included in breakfast there was an improvement in scores on the Child Behavior Checklist (O'Sullivan 2009).

Self-medication with nicotine—level V evidence

Because nicotine is known to modulate dopamine activity in relevant brain regions, some have speculated that smoking provides a means of self-medication in patients with ADHD. Research has established that individuals with ADHD smoke more than their nondiagnosed peers smoke, start smoking earlier, and experience more withdrawal symptoms that are significant when they try to quit. Conners and colleagues demonstrated that acutely administered nicotine improves cognitive performance and ADHD symptoms in patients (Conners 1996). Development of agents that can have the same effects on cognition without the accompanying addictive potential of nicotine might be promising. Nicotine raises the need for vitamin C and may influence the nervous system through oxidative stress.

Conclusions

1. A high percentage of children with ASD have gastrointestinal problems (altered absorption, diarrhea, and constipation) as well as behaviors and communication difficulties that can contribute to eating and nutritional problems.
2. For overall health maintenance and normal growth, a child (with or without autism) needs an adequate, but not excessive, supply of calories, protein, carbohydrates, essential fatty acids, vitamins and minerals, fiber, and water. "Adequate" needs to be taken into consideration with any individual's conditions that decrease absorption of nutrients.

3. Potentially deficient nutrients in diets commonly tried for autism include thiamin (B_1), pyridoxine (B_6), B_{12}, vitamin A, calcium, magnesium, iron, zinc, selenium, and essential fatty acids, as well as the simple carbohydrate galactose. Common micronutrient deficiencies of vitamins A, D, E, K, pantothenic acid, calcium, and zinc are observed in children with autism.
4. Gluten-free foods used in diets that avoid gluten and casein are not fortified or enriched, and do not provide the same vitamins and minerals provided by common enriched breads and cereals.
5. One survey found 83% of families used at least one supplement daily, 50% used at least two supplements. Supplement use of individuals should be assessed because nutrients in excess may have undesirable effects.
6. Trials on many nutrients yield variable reports on effectiveness in changing behavior patterns or other parental concerns. This indicates the need for increased attention to group definition, methods for study, and levels of observation. Defined subgroups for studying symptoms and behaviors related to nutritional interventions need to be investigated using scientific methods.
7. Nutritional assessment and assistance for parents to assure nutritional adequacy for children with ASD and ADHD is a relevant part of treatment and health.

References

Adams, JB, T Audhya, S McDonough, et al. Nutritional and metabolic status of children with autism vs. neurotypical children, and the association with autism severity. *Nutrition & Metabolism* 2011; 8:34. Also J BioMed Central: June 8, 2011.

Aitken, KJ. *Dietary Interventions in Autism Spectrum Disorders: Why they work when they do, Why they don't when they don't.* Jessica Kingsley Publishers. Philadelphia. 2009.

Al Anbar, NN, RM Dardennes, A Prado-Netto, et al. Treatment choices in autism spectrum disorder: The role of parental illness perceptions. *Res Dev Disabil* 2010; 31(3):817–28. http://www.ncbi.nlm.nih.gov/pubmed/20299185

Arnold, LE, SM Pinkham, and N Votolato. Does zinc moderate essential fatty acid and amphetamine treatment of attention-deficit/hyperactivity disorder? *J Child Adol Psychopharmacology* 2000; 10(2):111–117.

Belanger, SA., M Vanesse, S Schohraya, et al. Omega-3 fatty acid treatment of children with attention-deficit hyperactivity disorder: A randomized, double-blind, placebo-controlled study. *Paediatr Child Health* 2009; 14(2):89–98.

Bell, JG, D Miller, DJ MacDonald, et al. The fatty acid compositions of erythrocyte and plasma polar lipids in children with autism, developmental delay or typically developing controls and the effect of fish oil intake. *Br J Nutr.* 2010; 103(8):1160–7. http://www.ncbi.nlm.nih.gov/pubmed/19995470

Bernard-Bonnin, A-C, Psychosocial Paediatrics Committee, Canadian Paediatric Society. The use of alternative therapies in treating children with attention deficit hyperactivity disorder. *Paediatrics and Child Health* 2002; 7(10):710–718. http://www.cps.ca/english/statements/PP/pp02-03.htm accessed 11/1/10.

Buie, T, DB Campbell, GJ Fuchs III, et al. Evaluation, diagnosis, and treatment of gastrointestinal disorders in individuals with ASDs: A consensus report. *Pediatrics*. 2010; 125 Suppl 1:S1–18. http://www.ncbi.nlm.nih.gov/pubmed/20048083 and ... 8084

Cannell, J. J. Autism and Vitamin D. *Med Hypotheses*. 2008; 70(4):750–759. Epub Oct 24, 2007. PubMed 17920208.

Centers for Disease Control and Prevention. Prevalence of Autism Spectrum Disorders—Autism and Developmental Disabilities Monitoring Network, United States, 2006. MMWR Surveill Summ 2009; 58:6.

Centers for Disease Control and Prevention. Prevalence of Autism Spectrum Disorders—Autism and Developmental Disabilities Monitoring Network, United States MMWR Surveillance Summ 2009; 58(SS–10).

Comisrow, J. Can Sweets Drive Kids Crazy? Sugar and Hyperactivity in Children. *Nutrition Bytes*: 1996; 2(1) Article 2. http://repositories.cdlib.org/uclabiolchem/nutritionbytes/vol2/iss1/art2

Conners, CK, ED Levin, E Sparrow, et al. Nicotine and attention in adult attention deficit hyperactivity disorder ADHD). *Psychopharmacol Bull* 1996; 32:67–73.

Eichler, EE. and AW Zimmerman. A hot spot of genetic instability in autism. *NEJM* 2008; 358:737.

Geraghty, ME, GM Depasquale, and A Lane. Nutritional interventions and therapies in autism: A spectrum of what we know: Part 1. *ICAN: Infant, Child & Adolescent Nutrition* 2010; 2:62.

Geraghty, ME, J Bates-Wall, K Ratliff-Schaub, and AE Lane. Nutritional interventions and therapies in autism: A spectrum of what we know: Part 2. *ICAN: Infant, Child & Adolescent Nutrition* 2010; 2:120.

Hjiej, H, C Doyen, C Couprie, et al. Substitutive and dietetic approaches in childhood autistic disorder: Interests and limits. *Encephale* 2008; 34(5):496–503. Epub 2008 Mar 4 http://www.ncbi.nlm.nih.gov/pubmed/19068339 translated from French.

Heijtz, RD, S Wang, F Annar, et al. Normal gut microbiota modulates brain development and behavior. *Proc Natl Acad Sci USA* 2011; 108:3047–3052.

Herbert, MR. Contributions of the environment and environmentally vulnerable physiology to autism spectrum disorders. *Curr Opin Neurol*. 2010; 23(2):103–10. http://www.ncbi.nlm.nih.gov/pubmed/20087183

Howard, AL, M Robinson, GJ Smith, et al. ADHD is associated with a "Western" dietary pattern in adolescents. *J Attn Dis* 2010; 15(5) :403–411. http://www.autism.com/pdf/providers/adams_biomed_summary.pdf (accessed 10/16/10).

Humble, MB, S Gustafsson, and S Bejerot. Low serum levels of 25-hydroxyvitamin D (25-OHD) among psychiatric out-patients in Sweden: Relations with season, age, ethnic origin and psychiatric diagnosis. *J Steroid Biochem Mol Biol*. 2010; 121(1–2):467–70. http://www.ncbi.nlm.nih.gov/pubmed/20214992

James, SJ, P Cutler, S Melnyk, et al. Metabolic biomarkers of increased oxidative stress and impaired methylation capacity in children with autism. *Am J Clin Nutr* 2004; 80(6):1611–1617.

Kaluzna-Czaplinska, J, E.Socha and J Rynkowski. B vitamin supplementation reduces excretion of urinary dicarboxylic acids in autistic children. *Nutr Res* 2011; 31(7):497–502.

Kanner, Leo. Autistic disturbances of affective contact. *Nervous Child*. 1943; 2:217–50.

Kinney, DK, Barch DH, Chayka B, Napoleon S, Munir KM. Environmental risk factors for autism: do they help cause de novo genetic mutations that contribute to the disorder? *Med Hypotheses*. 2010 Jan; 74(1):102–6. http://www.ncbi.nlm.nih.gov/pubmed/19699591

Kollins, SH. Genetics, neurobiology, and neuropharmacology. *Medscape Psychiatry & Mental Health*. 2008. http://cme.medscape.com/viewarticle/580423 accessed 10/16/2010.

Main, PAE, MT Angley, P Thomas, et al. Folate and methionine metabolism in autism: A systematic review. *Am J Clin Nutr* 2010; 91(6):1598–1620.

McCandles, Jacqueline. *Children with Starving Brains: A Medical Treatment Guide for Autism Spectrum Disorder*. Putney, Vermont: Bramble Books, 2003.

McGinnis, WR. Nutritional perspectives and the behavioral child. 1999. http://www.autism.org/mcginnis.html

McGuire, JK., MS Kulkarni, and HP Baden. Fatal hypermagnesemia in a child treated with megavitamin/megamineral therapy. *Pediatrics* 2000; 105(2):e18.

Mehl-Madrona, L, B Leung, C Kennedy, et al. Micronutrients versus standard medication management in autism: A naturalistic case-control study. *J Child Adolesc Psychopharmacol* 2010; 20(2):95–103. http://www.ncbi.nlm.nih.gov/pubmed/20415604

Milward, C, M Ferriter, S Calver, and G Connell-Jones. Gluten-and casein-free diets for autistic spectrum disorder. *Cochrane Database Syst Rev*. 2008; 16(2):CD003498 http://www.ncbi.nlm.nih.gov/pubmed/18425890

Munasinghe, SA, C Oliff, J Finn, and JA Wray. Digestive enzyme supplementation for autism spectrum disorders: A double-blind randomized controlled trial. *J Autism Dev Disord*. 2010 Mar 5. http://www.ncbi.nlm.nih.gov/pubmed/20204691

Murphy, SJ and RK Johnson. The scientific basis of recent US guidance on sugars intake. *Am J Clin Nutr* 2003; 78(suppl):827S–833S.

Nye, C, and A Brice. Combined vitamin B6-magnesium treatment in autism spectrum disorder. *Cochrane Database Syst Rev*. 2005; 19;(4):CD003497. http://www.ncbi.nlm.nih.gov/pubmed/16235322

O'Sullivan, TA, M Robinson, GE Kendall, et al. A good-quality breakfast is associated with better mental health in adolescence. *Pub Hlth Nutr*. 2009; 12(2):249–58.

Richardson, A. Fatty acids in dyslexia, dyspraxia, ADHD and the autism dpectrum. 2003. *Food and Behavior Research*. www.fabrearch.org

Saenger, E. Autism: An expert interview with Eric Hollander, MD. *Medscape Psychiatry & Mental Health*. Feb. 2005. http://www.medscape.org/viewarticle/497979 accessed 3/21/11.

Shaywitz, BA, CM Sullivan, GM Anderson, et al. Aspartame, behavior, and cognitive function in children with attention deficit disorder. *Pediatrics*. 1994; 93(1):70–5.

Spiers PA, L Sabounjian, A Reiner, et al. Aspartame: Neuropsychologic and neurophysiologic evaluation of acute and chronic effects. *Am J Clin Nutr* 1998; 68(3):531–7.

U.S. Department of Health and Human Services, U.S.D.A. Dietary Guidelines for Americans 2005. 2005. see http://www.health.gov/dietaryguidelines/dga2005/document/pdf/dga2005.pdf

Vancassel, S, G Durand, C Barthelemy, et al. Plasma fatty acid levels in autistic children. *Prostaglan, Leukot Essen Fatty Acids* 2001; 65(1):1–7.

Waly, M., H Olteanu, R Banerjee, et al. Activation of methionine synthase by insulin-like growth factor-1 and dopamine: A target for neurodevelopmental toxins and thimerosal. *Mol Psychiatry* 2004; 9:358–370.

Waring, RH, LV Klovrza and RM Harris. Diet and individuality in detoxification. *J Nutr & Envir Med* 2007; 16(2):95–105.

Whiteley, P, D Haracopos, AM Knivsberg, et al. The ScanBrit randomised, controlled, single-blind study of a gluten- and casein-free dietary intervention for children with autism spectrum disorders. *Nutr Neurosci.* 2010; 13(2):87–100.

WHO/FAO Release Independent Expert Report on Diet and Chronic Disease http://www.wpro.who.int/media_centre/press_releases/pr_20030304.htm accessed 2/1/09.

Williams, BL, M Hornig, T Buie, et al. Impaired carbohydrate digestion and transport and mucosal dysbiosis in the Intestines of children with autism and gastrointestinal disturbances. *PLoS.* 2011; 6(9):1–20. e24585.

Williams, NM, I Zaharievaa, A Martina, et al. Rare chromosomal deletions and duplications in attention-deficit hyperactivity disorder: A genome-wide analysis. *Lancet* 2010; 376(9750):1401–1408.

Wolraich, ML, SD Lindgren, PJ Stumbo, et al. Effects of diets high in sucrose or aspartame on the behavior and cognitive performance of children. *N Engl J Med* 1994; 330:301–307.

Young, GS, JA Conquer, and T Rene'. Effect of randomized supplementation with high dose olive, flax or fish oil on serum phospholipid fatty acid levels in adults with attention deficit hyperactivity disorder. *Reprod Nutr Dev.* 2005; 45(5):549–58.

chapter five

Genetics, nutrition, and inherited disorders of metabolism

Nutrient–gene interaction

The Human Genome Project found that a human has about 25,000 genes. Not all genes are expressed in all cells at all times. Conditional expression of genes implies existence of controls that determine when codes are transcribed and translated into gene products. Metabolic control mechanisms involve hormones, metabolites, ions, and second messengers that modify the individual's (phenotypic) expression of these genes. These control mechanisms may be metabolic products of biochemical reactions, nutrients, or both.

Aberrations in genes may come from the environment (natural or manufactured radiation, X-ray, etc.) or aberrations may be inherited. Genetic change may not manifest in noticeable metabolic defects because such changes can be heterozygous or recessive.

Three concepts for thinking about nutrient–gene interactions are described by Zeisel (2007): (1) direct interactions in which a nutrient binds to a receptor, then binds to DNA and acutely induces gene expression, (2) epigenetic interactions in which a nutrient can alter the structure of DNA and gene expression is chronically altered, and (3) common genetic variations in single-nucleotide polymorphisms (SNPs), which can alter the expression or function of genes.

Even if an individual has a gene for a metabolic disorder, the manifestation and consequences may be influenced by environment. The environment includes (among many other factors) diet, individual nutrients, and weight status. Nonnutrient components in the diet, such as fiber and individual phytochemicals, may also influence whether a gene manifests a disease or health condition. Contaminants from agriculture, manufacturing, and other parts of the environment may also influence gene expression.

Vitamins and genetic stability—*level IV evidence*

The following explanations by Ames and colleagues, at the University of California at Berkeley, are helpful in considering how genetic mutations can interact with nutrition. Although mutation is a sudden event, it can produce almost any degree of effect from those barely detectable to those too extreme for the cell to survive. Altered enzyme activity, which may arise from dietary nutritional insufficiency, will ultimately affect cell function. When one factor is inadequate in the metabolic network, repercussions are felt in a large number of systems and can lead to degenerative disease. This may, for example, result in (1) an increase in DNA damage (and possibly cancer), (2) neuron decay (and possibly cognitive dysfunction), or (3) mitochondrial decay (and possibly accelerated aging and degenerative diseases).

Deficiency of vitamins B_{12}, folic acid, B_6, niacin, C, and E, or the minerals iron and zinc appears to mimic radiation by damaging DNA, by causing single-strand and double-strand breaks, oxidative lesions, or both. Inadequate intake of folic acid, vitamin B_6, and B_{12} causes millions of uracils to be incorporated into the DNA of each cell, causing associated chromosome breaks, essentially mimicking radiation damage. Changes such as the substitution of cytosine for thymine alters derivatives of niacin such as nicotinamide adenine dinucleotide phosphate (NADP), and cofactor flavin adenine dinucleotide (FAD), and are linked to migraines and raging behavior as well as cardiovascular disease. Ames also discusses the mutation substituting lysine (an essential amino acid) in a crucial location in a protein, which influences alcohol tolerance and Alzheimer's disease. Oxidation during aging, which may result in changes in enzyme conformation, may respond to higher levels of vitamin coenzymes and increase vitamin binding, thus maintaining enzyme function during the aging process.

The four nucleotide bases that bond and form the double-helix structure of DNA:
Adenine pairs with thymine
Guanine pairs with cytosine
In RNA, uracil is the base that pairs with adenine during transcription.

Genetic polymorphism and vitamin-dependent enzymes—level IV evidence

Polymorphism refers to a specific genetic change in >1% of the population. A genetic change may alter which amino acid is incorporated into a protein during gene expression. This may change the shape/conformation of an enzyme and decrease its affinity and binding with metabolic products with which it reacts. Decreased affinity often lowers the activity or function of the enzyme. For vitamin-dependent enzymes, this decrease in affinity may be overcome by providing high-dose supplements of the vitamin. One-third or more of such mutations respond to high concentrations of a vitamin.

Approximately 50 human genetic diseases involve defective enzymes that can be remedied by high concentrations of the vitamin component of the coenzyme. There have been 3827 enzymes cataloged; 22% use a cofactor. Ames discusses vitamins that are known to be cofactors: thiamin, riboflavin, niacin, pyridoxine, folic acid, pantothenic acid, and B_{12}. Minerals may also function as cofactors (magnesium, calcium, zinc, iron, and potassium). Mineral concentrations are more closely regulated by the body than vitamins are, and toxicity is possible at lower concentrations. For instance, the human zinc requirement is approximately 10 mg/day; toxicity can occur at intakes of 100 mg/day.

Ames discusses pyridoxine in treating developmental delay and autism. Some studies report improvement; some do not. Use of pyridoxine to reduce symptoms of tardive dyskinesia in patients with schizophrenia has been shown to be useful and is being explored (Ames 2002).

Biomarker for genome stability influenced by vitamin and mineral intake—level II evidence

Using the micronucleus (MN) assay to measure genome damage in lymphocytes, Fenech and colleagues studied the association of dietary intake and genome damage. The MN index serves as a sensitive biomarker; its formation is associated with chromosomal instability and has good correlation with estimated dietary intake of nutrients. Dietary intake was determined using a food frequency questionnaire. Subjects were 190 healthy males averaging 47.8 years of age.

They also investigated whether a 6-month randomized trial of supplementation with beta-carotene, vitamins C and E, and zinc improves genome stability. The highest tertile intake of vitamin E, retinol, folic acid, nicotinic acid, and calcium is associated with significant *reductions* of MN, signifying genomic stability. The highest tertile of riboflavin, pantothenic acid, and biotin were associated with significant *increases* of MN, indicating less genomic stability. Supplementation with combined A, C, E, and zinc significantly reduced the MN index, showing a protective effect of these nutrients. The destabilizing effect of riboflavin status is influenced by low folate intake (Fenech 2005).

Alcoholism and genetics: Polymorphism—level IV evidence

Acetaldehyde is a product of the metabolism of alcohol. One gene on chromosome 12 has been linked to the aldehyde dehydrogenase (ALDH) enzyme. Two genes have been linked to the enzyme alcohol dehydrogenase (ADH) on chromosome 4 (ADH1B and ADH1C). Polymorphisms of these genes have been shown to alter enzyme function in the metabolism of alcohol. ADH1B and ALDH2 are hypothesized to cause an accumulation of acetaldehyde due to more rapid production and slower rate of removal. This accumulation causes more intense reactions to alcohol and often leads to a lower alcohol intake. Data support the hypothesis that lower intake is thought to reduce risk for developing alcoholism in individuals with this genetic combination (Wall 2005). These and other polymorphisms of genes involved in alcohol metabolism vary in cultural groups such as Asians, Polynesians, Native Americans, and others.

Folate and genetic testing

One example of gene–nutrient interaction involves folic acid. A genetic change from cytosine to thymine in DNA can cause moderate hyperhomocysteinemia—a risk factor for cardiovascular disease. Interestingly, 10% of people of Northern European descent and 15% of those with Southern Europeans ancestors are estimated to have this genetic substitution.

Fifteen genes and five conditions linked to folic acid are described on the National Institutes of Health (NIH) Genetics Home Reference (GHR) site (http://ghr.nlm.nih.gov/search?query=Folic+Acid&Search). Genetic testing for DNA sequences that are linked to risk of disease or influenced by a nutrient may be ordered by a health care professional. Some testing is offered direct to consumer (DTC) by testing laboratories. Consumers should be aware that (1) laboratories may or may not be certified by the Clinical Laboratory Improvement Amendments (CLIA), (2) tests may indicate risk for, but not certainty of, developing a condition, (3) the scientific evidence regarding consequences of a gene change may not be conclusive, and (4) there may not be a proven treatment strategy available even if a genetic change is discovered and it is clearly linked to a health issue. Due to these complexities, genetic counseling is available for interpretation of gene testing.

Inherited disorders of metabolism

One of the earliest known links between a nutrient and mental health were the links between the altered metabolism of an amino acid or a carbohydrate and the effects of these alterations. Previously called "inborn errors of metabolism," inherited disorders of metabolism refer to these conditions and the numerous new alterations in metabolism that are being discovered.

Three hundred genetic disorders have been reported that result in accumulation, deficiency, or overproduction of normal physiological substances (Elsas 2006). These alterations are present from conception and may mean a change in the metabolism of just one dietary component, which has the potential for disrupting the development of the child after birth when it no longer relies on the maternal metabolic systems. Some changes in metabolism cause severe mental retardation or death, if an intervention cannot compensate for the genetic metabolic change.

Phenylketonuria (PKU)

Phenylketonuria (PKU) is one example of an inherited disorder of metabolism. There are 400 known gene mutations relevant to PKU. This results in over 80,000 possible genetic combinations that may affect phenylalanine metabolism (Guttler 2006). Individuals with PKU are lacking or have inadequate function of the enzyme phenylalanine hydroxylase (PAH), which is necessary to metabolize the essential amino acid phenylalanine (phe). If the enzyme is absent, insufficient, or not functioning, this causes elevation of phe in the blood. Increasing levels of phe in the blood and tissues can prevent normal brain development and mental retardation in most, but not all, of individuals with the gene mutations.

Levels of phe in the brain may be a more critical value than blood levels. The correlation between blood and brain phe is not very high (Baumeister 2006). Elevated levels of phe and the often-decreased levels of tyrosine are followed by an imbalance of distribution of many amino acids across cellular membranes and the blood-brain barrier. This impairs synthesis of neurotransmitters (Scriver 2006).

Protein and a minimum amount of phenylalanine are essential for normal growth and repair. Adequate intake of protein, essential amino acids, and calories must be provided, especially during fast-growth periods of infancy, childhood, adolescence, and during pregnancy. If inadequate amounts of dietary protein or calories are consumed, a breakdown of body tissues for fuel can cause elevation of phenylalanine in the blood and body tissues. Too little or too much phenylalanine is not conducive to growth and health.

Phenylalanine requirement as indicated by amino acid oxidation

Using an indicator of amino acid oxidation with lysine as the indicator, and breath $^{13}CO_2$ as the endpoint, Courtney-Martin and colleagues concluded that the maximal phenylalanine intake for children with phenylketonuria should be 14 to 20 mg/kg/day, but no higher than 20 mg/kg/day (Courtney-Martin 2002).

Screening, diagnosis, signs, and symptoms

Forty-eight states have laws requiring screening of newborns' blood for phenylalanine levels during the first week of life using the Guthrie screening test. Other states may screen for PKU, but it is not mandated by law. Most states define classical PKU as blood phenylalanine >20 mg/dl or >1200 μmol/L. Others use criteria of 4 to 15 mg/dl. Normal blood level

Table 5.1 Classifications of Altered Phenylalanine Metabolism

Normal blood phenylalanine	50–110 µmol/L
Mild phenylalaninaemia	120–600 µmol/L
Mild phenylketonuria	600–1200 µmol/L
Moderate phenylketonuria	900–1200 µmol/L
Classical phenylketonuria	>1200 µmol/L

Source: Blau et al. 2010.

for phe is ~0.8 to 1 mg/dl or 50 to 110 µmol/L. The maximum normal level has also been defined as 0.125 mM/L (Table 5.1) (Scriver 2006).

The Guthrie screening test is a serum test that detects abnormal levels through the growth rate of *bacillus subtilis*—an organism that needs phenylalanine to thrive. It must be performed after three to four full days of milk or formula feeding or may result in a false negative result.

Within a few weeks or months, a baby with PKU may have a musty odor of skin, hair, and urine, weight loss from vomiting and diarrhea, "fussy" behavior, and may be sensitive to light. Older children may exhibit behavior such as screaming episodes, repetitive rocking, head banging, and arm biting. They may have developmental delays and loss of skills and abilities. Since PKU affects synthesis of melanin, they often have blond hair, fair skin, and blue eyes (http://www.webmd.com/parenting/baby/tc/phenylketonuria-pku-symptoms).

Based on phe tolerance, patients can be assigned to one of four categories: classical PKU, moderate PKU, mild PKU, and mild hyperphenylalanemia (MHP). (See Table 5.1)

Classical PKU is a complete block in the breakdown of phenylalanine. This is treated with a low-phenylalanine diet. Normally phenylalanine metabolism produces tyrosine, also an essential amino acid. With phenylalanine restriction, another source of tyrosine must be assured.

With some gene combinations, a woman who may not need a special diet most of the time may need to restrict phe during pregnancy. In pregnancy, a woman will have higher levels of phenylalanine in her body, and this can cause serious problems for the unborn baby. These women's babies may be mentally retarded, quite small, or have birth defects such as heart problems. It is recommended that women with any type of PKU start on a low-phenylalanine diet several months before becoming pregnant.

Treatment

A specially formulated drink is available for children that supplies most of the essential protein, vitamins, and minerals with little or no phenylalanine. Children with PKU acquire a taste for the drink, although it might be objectionable to those not accustomed to it. All other foods need to be carefully measured and monitored. It is debated whether those with plasma phenylalanine levels below 600 μmol/L (10 mg/dL) require dietary treatment. Many believe there is no safe age when dietary restrictions can be discontinued (Sullivan 1999, 2001). Guidelines now recommend more aggressive treatment: earlier in onset, more stringent in restoring euphenylalaninemia, and longer in duration—perhaps for life. Premature termination of therapy is frequently associated with neurophysiologic and psychological dysfunction and perhaps a decline in cognitive function (Medical Research Council Working Party on Phenylketonuria 1993).

Assessment through adulthood is recommended for phe tolerance in relation to increases in body mass. Increased intake may be necessary to support protein synthesis and prevent catabolism as individuals grow or change weight (McLeod 2009). Treatment should reduce blood phenylalanine to not more than 2 mg/dl (120 mmol/L) to allow normal mental development. Alternately, in some clinical settings, treatment goals may be 10 mg/dl.

Those with classical or moderate PKU have been found to have IQs in the mid 80s. Mild PKU may result in IQs in the mid 90s. Those treated for more than 6 years had IQs of 10 points above the typical IQ in their category. A study of 108 Danish patients who had been on dietary therapy of 10 to 14 years had a median IQ in the normal range (Guttler 2006).

Instituting dietary control of phe intake has been shown to decrease problem behaviors seen in patients with PKU. It takes at least 2 months before blood levels reach 3 to 10 mg/dL, the level needed for positive behavioral changes (Baumeister 2006).

A residual phenylalanine/tyrosine imbalance may be the cause of subtle cognitive deficits in infants and children treated with a diet low in phenylalanine (Wainwright 2005). Sharman et al. report that the phe-to-tyr ratio is more strongly associated with executive function than phe measures alone. Children with phe:tyr <6 had normal executive function. Children with a lifetime phe:tyr of >6, on average, had clinically impaired executive function (Sharman 2010).

Another treatment approach involves tetrahydrobiopterin therapy (5 or 10 mg/kg body weight). Tetrahydrobiopterin (BH4) (sapropterin) is used to convert phenylalanine to dopamine, as this is an alternate metabolic pathway for phenylalanine. Ames et al. (2002) reported this therapy as effective in four patients who had mild hyperphenylalaninemia.

Sapropterin is a synthetic analog of tetrahydrobiopterin.

Assessment of response to BH4 therapy—level II evidence
During a BH4 challenge test for 557 newborns and children with PKU with blood phe levels of 301 to 4753 μmol/L, blood levels were checked at 8 hours and 24 hours after treatment. A 30% decrease in blood phe

was defined as a response to treatment. Of patients with MHP, 79 to 83% responded to treatment, 49 to 60% with mild PKU responded to treatment, and 7 to 10% with classical PKU responded to treatment (Fiege 2007).

BH4 and patients with psychiatric illness—level II evidence
Koch and associates (2002) reported a case of a 25-year-old woman with MHP, depression, and panic attacks. A dosage of 100 mg/day of tetrahydrobiopterin resulted in significant improvement and allowed discontinuation of psychotropic medications. Another group of researchers led by Richardson (2005) reported finding a significant decrease of fasting plasma total biopterins (34%) in 23 patients with schizophrenia compared with 21 controls.

Lipids

Supplementation with DHA—level II evidence
Supplementation with fish oil for 3 months, equivalent to 15 mg/kg of DHA per day, increased DHA and EPA and decreased AA in 24 patients with PKU over 4 years of age. Motor skills were improved as measured on the Rostock-Oseretzky Scale (Beblo 2007).

Plasma DHA and EPA associated with bone mineral density—level II evidence
Noting that dietary treatment of PKU results in an extremely low intake of omega-3 fatty acids, Lage et al. (2010) found plasma DHA and EPA and total omega-3 fatty acids were significantly below healthy controls and had positive association with bone mineral density. There was no association with phe and calcium levels or intake, 25-hydroxyvitamin D, and bone mineral density.

Minerals

Zinc, selenium, and copper not correlated with dietary formula—level II evidence
Due to dietary restrictions of protein, Barretto et al. (2008) analyzed diet and blood samples in 32 children and adolescents with PKU. Erythrocyte zinc (37% of subjects) and serum selenium (90% of subjects) were below normal. Plasma copper levels were normal. There was no correlation between the blood levels and amounts supplied in dietary formulas.

Other

The artificial sweetener aspartame is metabolized to phenylalanine and aspartic acid. Food labeling laws now require aspartame be listed on any food or beverage that contains aspartame.

Some research is investigating administration of the lacking enzymes to individuals so that phenylalanine metabolism can approach normal. Oral administration of an enzyme is not effective because the enzyme is digested like other proteins. Other means of administering drugs include inhalers, skin patches, sublingual (beneath the tongue) absorption, injections directly into the bloodstream, or possibly a stem cell or genetic alteration method.

Galactosemia

Galactosemia is an inherited disorder of carbohydrate metabolism. Galactose is a monosaccharide that makes up half the molecule of the sugar lactose. The sugar/carbohydrate in milk is 50% galactose and 50% glucose (Sunehad 2003). Dietary sources of galactose are largely converted to glucose. A metabolic path for de novo production of galactose supplies galactose in the absence of a dietary source. (See also Chapter 4.)

Individuals with galactosemia are missing an enzyme that converts galactose to glucose. Usually this is the enzyme galactose-1-phosphate uridyl transferase. Other enzymes potentially involved are galactokinase or UDP-galactose-4-epimerase.

Many states routinely check newborns' blood during the first week of life for galactosemia. The incidence of galactosemia is 1 in 50,000 in Caucasian parents. If left untreated, galactose accumulates in the blood, resulting in vomiting, diarrhea, and failure to gain weight. Continued accumulation of galactose results in jaundice, enlarged liver, cataracts, mental retardation, and possible death. Mental retardation is evident in 6 to 12 months after birth (Isselbacher 1994).

A normal level is 18.5 to 28.5 units of galactose per gram of hemoglobin. Less than 5 U/g of galactose in hemoglobin indicates galactosemia; 5 to 18 U/g may indicate the individual is a carrier of the aberrant gene. If both parents are carriers, it is important for their baby to be screened at birth.

Children with galactosemia have the same nutritional needs as other children do except their diet must omit milk or foods with lactose. Treatment includes a galactose-free formula, the rigid restriction of galactose for one year, and then 1 cup of milk per day (or an equal amount of lactose in another form). After this age, children are usually given a soy-milk replacement (Elsas 1994).

The enzyme UDP-galactose-4-epimerase may be deficient in red blood cells only and individuals may not need dietary intervention. If the enzyme is deficient in all tissues, a restricted diet is necessary.

If an individual lacks galactokinase, dietary intervention is needed to prevent mental retardation, cataracts, liver failure, and death. Treatment calls for the omission of dietary lactose to obtain a blood level of 0 mg/dl of galactose (Springhouse 1991, 1994).

Maple syrup urine disease (MSUD)

Maple syrup urine disease (MSUD) is lack of the enzyme necessary for metabolizing the branched chain amino acids (BCAAs): leucine, isoleucine, and valine. These amino acids are found in all food proteins and without the enzyme, BCAAs build up in blood, spinal fluid, and urine. This produces a distinctive sweet smell.

If untreated, high levels of BCAA interfere with normal brain development, resulting in mental retardation and various life-threatening complications. Many states screen newborns during the first week after birth. A specially formulated beverage is available for children with MSUD who need adequate protein, vitamins, and minerals for growth.

Ames and associates (2002) described a thiamine-dependent enzyme complex related to a mutation in the E2 area that is present in MSUD and responds to high-dose thiamin supplements. Related to this and other genetic factors, Ames suggested therapy with thiamin, lipoic acid, riboflavin, nicotinamide, and adequate potassium for stabilizing the enzyme system (Ames 2002).

Homocystinuria

Homocystinuria is the lack of an enzyme that breaks down a group of amino acids, one of which is methionine. Accumulation of methionine can lead to abnormal accumulation in blood and urine of other amino acids, including homocystine. Accumulation interferes with normal brain development, causing mental retardation and other serious side effects. Many states screen newborns for homocystinuria. Like many other similar conditions, a special diet is needed that includes a formula to supply adequate levels of nutrients for normal growth.

Additional information on inherited disorders of metabolism

Current concepts of inherited metabolic diseases are based largely on investigations of amino acid disorders and inherited changes in the metabolism of lipids. Normally up to 200 mg of amino acids may be excreted in urine per 24 hours. Abnormal metabolism causes excess amounts to appear in plasma. As the renal threshold is exceeded, amino acids appear in the urine. More than 70 disorders of amino acid metabolism are now known. Approximately 60 of these are catabolic defects; approximately 10 are transport defects. The incidence is 1 in 10,000 for cytinuria or PKU and 1 in 200,000 for homocystinuria or alkaptonuria. Collectively, an inherited metabolic disorder occurs in 1/500 to 1/1000 live births. Biochemical and genetic heterogeneity in genetic errors are common: there are five forms of hyperphenylalaninemia, seven forms of homocystinuria, and seven types of methylmalonic acidemia.

> Trimethylaminuria is a disorder in which the body is unable to break down trimethylamine. It causes affected people to give off a fish-like odor in their sweat, urine, and breath. Although gene mutations account for most cases of trimethylaminuria, the condition can also be caused by excess of certain proteins in the diet, an increase in certain gut bacteria, or liver or kidney disease.
> Avoiding foods such as eggs, legumes, certain meats, fish, and other foods that contain choline, carnitine, nitrogen, sulfur, and lecithin may help in decreasing the odor. (From http://ghr.nlm.nih.gov/condition=trimethylaminuria. accessed January 7, 2012.)

Central nervous system dysfunction in the form of developmental retardation, seizures, alterations of the senses, or behavioral disturbances occurs in more than half the disorders.

Tests include chromatography analysis of urine or plasma (Springhouse 1991, 1994).

Hartnup disease is a mutation involving neutral amino acids. It is produced from the mutation of a shared transport protein. A genetic alteration produces a constant neutral aminoaciduria, resulting in intermittent symptoms of pellagra.

> Amino acids known to exhibit genetic alteration:
> Arginine*, argininosuccinic acid
> Cysteine
> Citrulline
> Histidine
> Homocysteine
> Isoleucine, leucine
> Lysine*
> Ornithine*
> Phenylalanine
> Tryptophan
> Tyrosine
> Valine
> *Defective transport

The malabsorption of tryptophan (one of the essential amino acids) can be affected by the mutation of transport protein producing indoluria and hyercalcemia.

Methionine malabsorption is caused by the mutation of the transport protein resulting in white hair, mental retardation, convulsions, and edema. Histidinuria results in the mutation of a transport protein and results in mental retardation.

> Two interesting symptoms resulting in altered metabolism of the amino acid leucine are (1) sweaty feet odor (the mutation of two different enzymes) and (2) the production of the odor of cat's urine. Either may be very disturbing to the individuals affected.

DNA-based diagnostics can identify substitutions, deletions, and insertions to DNA and help diagnose and describe these disorders. Some can be diagnosed prenatally with amniotic cells or fluid. Clinical manifestations in many of these conditions can be prevented or successfully treated if diagnosis is achieved early and appropriate treatment instituted promptly (Rosenberg 1994).

Conclusions

1. Hundreds of genetic alterations have been tied to changes in nutrient metabolism. Alterations include (1) indirect interactions through receptors, (2) acute or chronic alteration of DNA structure or function, or (3) expression of genes through RNA. Genetic counseling is available to interpret gene analysis.
2. State laws mandate testing for some genetic alterations soon after birth.
3. The consequences of genetic alterations that involve nutrition may be as serious as mental retardation or death if not identified and treated. Other alterations may interact with diet and other environmental factors, increasing an individual's need for a nutrient or the risk of developing a chronic disease.
4. Treatments for inherited disorders of metabolism may require omission or limitation of an amino acid, a simple carbohydrate, or other nutrient while assuring ample supplies of other nutrients for growth. Reassessment during growth, pregnancy, or weight gain may be necessary. Future treatments may include supplementation with other large neutral amino acids competing for transport into the brain, enzymes that influence phenylalanine metabolism, and gene therapy.

References

Ames, BN., I Elson-Schwan, and EA Silver. High-dose vitamin therapy stimulates variant enzymes with decreased coenzyme binding affinity (increased Km): Relevance to genetic disease and polymorphisms. *Am J Clin Nutr* 2002; 75(4):616–658.

Baretto, JR, LR Silva, ME Leite, et al. Poor zinc and selenium status in phenylketonuric children and adolescents in Brazil. *Nutr Res* 2008; 28(3):208–211.

Baumeister, A. Dietary Treatment of Phenylketonuria With Behavior Disorders. NIH Committee on Consensus Development Conference on Phenylketonuria (PKU): Screening and management last update 8/28/2006. http://www.nichd.nih.gov/publications/pubs/pku/sub16.cfm

Beblo, S, H Reinhardt, H Demmelmair, et al. Effect of fish oil supplementation on fatty acid status, coordination and fine motor skills in children with phenylketonuria. *J Pediatr* 2007; 150(5):479–484.

Blau, Nenad, FJ van Spransen and HL Levy. Phenylketonuria. *Lancet* 2010; 376:1417–1427.

Courtney-Martin, G., R Bross, M Raffi, et al. Phenylalanine requirements in children with classical PKU determined by indicator amino acid oxidation. *Am J Physio Endocrinol Metab* 2002; 283(6):E 1249–1256.

Elsas II, LJ, and PB Acosta. Inherited Metabolic Disease: Amino Acids, Organic Acids and Galactose. *Modern Nutrition in Health and Disease.* ed. ME Shils, M Shike, AC Ross, B Caballero, and RJ Cousins. 909–959. Lippincott Williams & Wilkins, Philadelphia, 2006.

Elsas II, LJ, and PB Acosta. Nutrition Support of Inherited Metabolic Disease. *Modern Nutrition in Health & Disease.* Edited by ME Shils, J A Olson, and M Shike. 1,198–1,202. Lea & Febiger, Malvern PA 1994.

Fenech, M, P Baghurst, W Luderer, et al. Low intake of calcium, folate, nicotinic acid, vitamin E, retinol, β-carotene and high intake of pantothenic acid, biotin and riboflavin are significantly associated with increased genome instability-results for a dietary intake and micronucleus index survey in South Australia. *Carcinogenesis* 2005; 26(5):991–999.

Fiege, B and N Blau. Assessment of tetrahydrobiopterin (BH4) responsiveness in phenylketonuria. *J Pediatr* 2007; 150(6):627–630.

Guttler, F. The use of mutation analysis to anticipate dietary requirements in phenylketonuria. NIH Committee on Consensus Development Conference on Phenylketonuria (PKU): Screening and management. Last update 8/28/2006 http://www.nichd.nih.gov/publications/pubs/pages/index.aspx accessed 5/7/13.

Isselbacher, K J. Galactosemia, Galactokinase Deficiency, and Other Rare Disorders of Carbohydrate Metabolism. *Harrison's Principles of Internal Medicine.* ed. KJ Isselbacher, E Braunwald, JD. Wilson, JB Martin, AS Fauci, and DL Kasper. 2131–2132. McGraw-Hill, Inc., San Francisco 1994.

Koch, R, F Guttler, and N Blau. Mental illness in mild PKU responds to biopterin. *Molecular Genetics and Metabolism* 75(3) (2002):284–286.

Lage, S., M. Bueno, F. Andrade, et al. Fatty acid profile in patients with phenylketonuria and its relationship withbone mineral density. *J Inherit Metab Dis* epub Sept 10, 2010.

MacLeod, IL, ST Gleason, SC van Calcar, and DM Ney. Reassessment of phenylalanine tolerance in adults with phenylketonuria is needed as body mass changes. *Mol Genet Metab* 2009; 98(4):331–337.

Report of the Medical Research Council Working Party on Phenylketonuria. Recommendations on the dietary management of phenylketonuria. *Arch Dis Child* 1993; 68:426–7.

Richardson, MA, LL Read, CL Taylor, et al. Evidence for a tetrahydrobiopterin deficit in schizophrenia. *Neuropsychobiology* 2005; 52(4):190–201.

Rosenberg, LE. Inherited Disorders of Amino Acid Metabolism and Storage, *Harrison's Principles of Internal Medicine*. ed. KJ Isselbacher, E Braunwald, JD Wilson, JB Martin, AS Fauci, and DL Kasper. 2,117–2,125. McGraw-Hill, Inc., San Francisco 1994.

Scriver, CR. Phenylketonuria: Paradigm for a treatable genetic disease...? NIH Planning Committee on Consensus Development Conference on Phenylketonuria (PKU): Screening and Management last update 8/28/2006. http://www.nichd.nih.gov/publications/pubs/pku/sub7.cfm accessed 10/29/2010.

Sharman, R, K Sullivan, R Young, and J McGill. A preliminary investigation of the role of the phenylalanine:tyrosine ratio of children with early and continuously treated phenylketonuria: toward identification of "safe" levels. *Dev Neuropsychol* 2010; 35(1):57–65.

Springhouse Corporation. *Clinical Laboratory Tests*. Springhouse Pennsylvania:1991.

Springhouse Corporation. Other Metabolic Amino Acid Disorders. *Illustrated Guide to Diagnostic Tests*, 123–125 and 426–442. Springhouse Pennsylvania:1994.

Sullivan, JE. Emotional outcome of adolescents and young adults with early and continuously treated phenylketonuria. *J Ped Psychology* 2001; 26(8):477–484.

Sullivan, JE.and P Chang. Review: Emotional and behavioral functioning in phenylketonuria. J. *Ped Psychology* 1999; 24(3):281–299.

Sunehad, A, S Tigas and M Haymond. Contribution of Plasma galactose and glucose to milk lactose synthesis during galactose ingestion. *J Clin Endocrin & Metab*. 2003; 88(1):225–229.

Wainwright, PE and D Martin. Role of dietary polyunsaturated acids in brain and cognitive function: Perspective of a developmental psychobiologist. *Nutritional Neuroscience*. Ed. HR. Lieberman, RB. Kanarek, and C Prasad. 167. New York: CRC Press, Taylor & Francis Group LLC., 2005.

Wall, TL. Genetic associations of alcohol and aldehyde dehydrogenase with alcohol dependence and their mechanisms of action. *Ther Drug Monitor* 2005; 27(6):700–703.

Web M.D. http://www.webmd.com/parenting/baby/tc/phenylketonuria-pku-systoms.

Zeisel, SH. Nutrigenomics and metabolomics will change clinical nutrition and public health practice: insights from studies on dietary requirements for choline. *Am J Clin Nutr* 2007; 86:542–548.

chapter six

Intellect, cognition, and dementia

Introduction

A poll of 1000 Americans found that more than twice as many Americans (62%) fear losing their mental capacity as they age than those who fear a diminished physical capacity (29%). A reflection of the finding is captured in the quip, "Of all the things I've lost, I miss my mind the most" (*Parade* 2006). Mental capacity refers to specific aspects of mental functioning.

Intellect refers to the ability to learn, reason, think abstractly, and the capacity for knowledge and understanding. Intelligence refers to the ability to acquire and apply knowledge, to comprehend. Fluid intelligence refers to reasoning, the capacity to solve complex problems, the ability to learn, and abstract thinking ability.

Cognition refers to operation of the mind by which we become aware of objects of thought or perception; it includes all aspects of perceiving, thinking, reasoning, and remembering.

Dementia refers to an organic mental syndrome characterized by a general loss of intellectual ability involving impairment of memory, judgment, and abstract thinking as well as changes of personality. It does not include loss of intellectual functioning caused by clouding of consciousness (as in delirium) nor that caused by depression or other functional mental disorder.

Intellect

Lipids

Infants and essential fatty acids—level IV evidence

Alteration in diet intake during early development may produce changes in brain function that may not be fully expressed until around puberty. Especially in the third trimester of pregnancy and during early breastfeeding, essential fatty acids (EFA) and preformed DHA, EPA, AA, and GLA are absolutely necessary to brain and retinal development. An infant is supplied with optimum nutrition when a well-nourished mother breastfeeds her baby. Jones and Sidwell (2001) report that an infant will rob a mother of these compounds if conditions are not adequate for both. They comment that with an inadequate intake of EFA, a breastfeeding mother can lose up to 5% of her brain weight. The authors comment further that insufficient EFA can provoke postpartum depression. Essential long-chain fatty acids such as DHA, usually derived from fish oil in the diet, make up 30% of the fatty acids of the neurons and appear to be the factors in breast milk that account for the increased intelligence of children fed breast milk compared to formula. In 2002, the U.S. Food and Drug Administration (FDA) approved supplementation of infant formula with both omega-3 and omega-6 fatty acids (Jones and Sidwell 2001). A review by Oh (2005) reports on an increase of intelligence in children at 4 years of age when their mothers were supplemented with fish oil from 18 weeks of pregnancy to 3 months postpartum.

> Biomagnification refers to the transport of long-chain fatty acids from mother to fetus at the expense of maternal nutritional status (Kuipers, 2011).

Supplementation during pregnancy with essential fatty acids has shown mixed results. With 12 to 20% of women experiencing perinatal depression, studies are needed that are designed with a primary outcome defined as the influence of nutrients on depression, with adequate group size, durations including ante- and postpartum depression, and which account for environmental and genetic influences (Jans 2010; Doornbos 2009; Leung 2009).

Vitamins

Vitamin B_{12} and infants—level IV evidence

Symptoms of a vitamin B_{12} deficiency in infants are likely to occur within 3 to 6 months after birth. Infants born to and breastfed by a mother with low serum levels of vitamin B_{12} are at the greatest risk for a B_{12} deficiency. Babies of mothers with pernicious anemia may develop symptoms of delay in neurological development at about 6 to 10 weeks of age. Since vitamin B_{12} is involved in (1) formation and maintenance of the myelin sheath surrounding neurons, (2) the metabolism of lipids, (3) the metabolism of methionine, and (4) the development of pernicious anemia, the consequences of a B_{12} deficiency in a newborn are potentially diverse and subtle. Possible consequences include abnormal neurological development and possibly a delay in mental development even after vitamin therapy (Centers for Disease Control 2003).

Fluid intelligence: Vegan diets, B_{12}, and adolescents—level II evidence

Data suggest that vitamin B_{12} deficiency, even in the absence of changes in blood levels, may lead to impaired cognitive performance in adolescents. Adolescents who consumed macrobiotic diets (a vegan type of diet) up to about age 6, who then changed to lacto-vegetarian or omnivorous diets, were compared to adolescents who had consumed omnivorous diets from birth. Of the 48 students who had consumed a macrobiotic diet, 31 were deficient in methylmalonic acid (MMA), a measure of vitamin B_{12} status. Seventeen of previously macrobiotic students and all controls had normal B_{12} status as assessed by cobalamin status. Psychological tests showed a significant correlation between vitamin B_{12} deficiency and fluid intelligence. Additional tests indicated better performance by control subjects, but results were not statistically significant (Louwman 2000).

> Fluid intelligence involves reasoning, the capacity to solve complex problems, abstract thinking ability, and the ability to learn.

Minerals

Lead toxicity in adults—level II evidence

In a prospective study of lead neurotoxicity, exposed foundry workers were evaluated using neurobehavioral and other tests. Results showed increased rates of depression, confusion, anger, fatigue, and tension among workers with blood lead levels over 40 µg/dl. Other aspects of neurobehavioral function, including verbal concept formation, memory, and visual/motor performance, were also impaired (Baker 1983).

Lead toxicity in children—level II evidence

Exposure to even low levels of lead has been shown to produce effects that are subtle and permanent in children. Subtle symptoms may include irritability and fatigue as well as diarrhea. Children may lose cognitive ability, verbal and perceptual ability, and develop learning disabilities, sleep disturbances, seizures, and behavioral problems. More severe, longer-term exposure can lead to nerve damage, paralysis, mental retardation, and even death. Low intakes of calcium, zinc, and vitamins C or D, as well as iron-deficiency anemia, increase potential lead toxicity. Lead-induced anemia in children is reported to occur at blood levels of 1.93 µmol/L (40 µg/dl).

To determine whether the effects of low-level lead exposure persist, an 11-year follow-up was conducted of children studied as primary school children from 1975 to 1986. Of the original 270 children studied, 132 were reexamined. Continued impairment in neurobehavioral function was compared to records of lead content of teeth shed at the ages of 6 and 7. Those with initially higher lead content in lost teeth had a markedly higher risk of (1) dropping out of high school, (2) having a reading disability, (3) a lower class standing in high school, (4) increased absenteeism, (5) lower vocabulary and grammatical-reasoning scores, (6) poorer hand-eye coordination, and (7) longer reaction times and slower finger tapping. The authors concluded that exposure to lead in childhood is associated with deficits in central nervous system functioning that persist into young adulthood (Needleman 1991).

Lead levels in children and effect on children's achievement in school—level II evidence

An examination of the relationship of relatively low blood concentrations of lead (<10 µg/dl; mean of 1.9 µg/dl) in cognitive functioning in children and adolescents found an inverse relationship between blood lead concentration and arithmetic and reading scores. For children with blood lead concentrations lower than 5.0 µg/dl, for every increase of 1 µg/dl in blood lead concentrations there was (1) a 0.7-point decrease in average arithmetic scores, (2) an approximately 1-point decrease in mean reading scores, (3) a 0.1-point decrease in average scores on a measure of nonverbal reasoning, and (4) a 0.5-point decrease in average scores on a measure of short-term memory. Out of 4853 children ages 6 to 16 years, 172 (2.1%) had blood lead concentrations ≥10 µg/dl. The authors concluded that deficits in cognitive and academic skills associated with lead exposure occur at blood lead concentrations lower than 5 µg/dl rather than at the higher level of 10 µg/dl previously used as a cutoff (Lanphear 2000).

Lower threshold for lead exposure—level II evidence

The "behavioral signatures" of lead exposure (intelligence, reaction time, visual-motor integration, fine motor skills, attention, executive function, off-task behaviors, and teacher-reported withdrawal behaviors) all demonstrate a relationship to blood lead concentration with levels as low as 3 µg/dl. There is no apparent lower threshold for effects of childhood lead exposure (Chiodo 2004). The authors comment it is advisable to consider whether the criteria for acceptable blood level for lead should be reduced to below the 10 mcg/dl used in the past (Chiodo 2007).

Surkan et al. (2007) found that children with 5 to 10 mcg/dl had IQ scores 5.0 points lower than children with lead levels of 1 to 2 mcg/dl. Lead levels of 5 to 10 mcg/dl were associated with decreased attention, memory, spatial attention, and executive function.

Lead in spices from India—level II evidence

Although many are acquainted with the risk of lead poisoning related to old paint, significant lead poisoning continues to be associated with imported nonpaint products. Levels of <3 mcg/dl have been shown to influence children's development. Four cases of pediatric lead poisoning in Boston from Indian spices or cultural powders are described by Lin et al. (2010). Eighty-six spices and 71 cultural powders imported from India were analyzed for lead using x-ray fluorescence spectroscopy. Of 86 spices and foodstuff products, 22 contained 2.6 to 7.6 mcg/dl. Of 71 cultural products, 46 contained 5.2 to 41.4 mcg/dl. Three sindoor products contained >47% lead.

Lead in Mexican pottery in Oklahoma—level II evidence
Ceramic ware collected from a Hispanic community in Oklahoma City was assessed for lead content, and the amount of lead that leached into foods cooked in those vessels was quantified. Lab results were combined with consumer intake levels for foods and compared with the provisional tolerable total intake level (PTTIL) for lead.

Of the vessels tested, 52% exceeded the FDA action level for ceramic ware. Consumption of a low-pH food (tomatoes) cooked in 23 of 25 vessels would result in a dose of lead exceeding the PTTIL. For higher-pH foods, 3 of 25 vessels used to cook hominy and 5 of 25 vessels used to cook beans exceeded the PTTIL. Lead glazed ceramics (LGC) continue to represent a significant public health concern (Lynch 2008).

Lead, arsenic, and mercury in Ayurvedic medications—level II evidence
Ayurvedic remedies may by design contain herbs, minerals, metals (lead, arsenic, mercury), or animal products. Remedies in the form of pills, powders, and syrups are used for numerous health problems. One analysis reported a medication contained 73,900 ppm lead in a pill recommended to be taken four times a day. Common symptoms of lead toxicity include abdominal pain, nausea, vomiting, and others, and result in anemia. Cases of lead toxicity from Ayurvedic remedies were reported from 35 states by the CDC Blood Level Epidemiology and Surveillance in 2004, numbering 10,568 cases (Araujo 2004; Kales 2007).

A survey of Ayurvedic medicine products in the Boston area found that 20% of the products tested contained lead, mercury, or arsenic. If consumed in the amounts recommended, it would result in intakes above safety standards (Saper 2004).

Iron deficiency—level IV evidence
Iron is an essential nutrient, but it may also be a potent toxin. Childhood death due to iron poisoning from adult supplements is not infrequent. Functions of iron specific to neurological activity include synthesis of dopamine, serotonin, catecholamines, and possibly gamma aminobutyric acid (GABA) and myelin formation. Iron uptake is maximal during rapid brain growth and coincides with development of the myelin sheath.

White matter throughout the brain, largely in the myelin sheath, is a major site of iron concentration. The highest levels of iron in the brain occur in the basal ganglia.

Iron deficiency in children 9 to 26 months of age have lower psychomotor scores, which normalized with a short period of iron treatment. In older children, iron deficiency results in decreased attentiveness, narrow attention spans, perceptual restriction, and poor performance on the Bayley Mental Index (Beard 1996). Normal values for hemoglobin for children 6 months to 2 years of age are 12.5 mg/dl (Alpers 1995).

Iodine and selenium

Iodine deficiency leads to dramatic impairment in brain development when it occurs during pregnancy, resulting in permanent neurological damage of the fetus. At 10 to 18 weeks gestational age, iodine deficiency may result in mental deficiency, deafness, mutism, motor disorders, and basal ganglia calcification. Cretinism refers to this congenital lack of thyroid. This condition, widespread in some parts of the world, is often related to iodine levels in the soil where food is produced. It has been successfully addressed in many areas by adding iodine to salt.

A combination of iodine and selenium deficiency may play a role in development of myxedema (the name given to adult hypothyroidism). Symptoms include dry, waxy swelling of the skin and non-pitting type edema, with swollen lips and thickened nose. When selenium is deficient in the diet, there is a hierarchy of consequences. The activity of the enzyme 5-deiodinase activity is maintained at the expense of selenium-dependent glutathione peroxidase (Se-GPX) activity. Despite a functional state of hypothyroidism elsewhere in the body, the brain may be protected from selenium deficiency. Iodine deficiency can result in fatigue severe enough to alter the lifestyle of individuals affected. (See also Chapter 11.)

Zinc

Zinc is one of the minerals functioning in many body systems, participating in the activity of over 200 enzymes. Zinc deficiency leads to both primary and secondary alterations in brain development and growth. It is utilized for incorporation of thymine into the structure of DNA. Inadequate zinc produces alteration in gene transcription, an example being primary neural tube defect.

Zinc is more abundant in gray matter of the brain than in white matter. It is present in three distinct pools: free zinc, vesicular zinc, and protein-bound zinc. The highest levels of zinc are observed in the hippocampus. Normal distribution of zinc may be altered by magnesium deficiency. Concentration of zinc in the hippocampus and amygdala are assumed to be necessary for experiential learning. Deprivation effects are seen most strongly in poor growth of cerebellum and Purkinje cells.

Indirect effects of zinc deficiency may be observed in the lipid content of the developing brain. Alterations of brain function due to changes in membrane characteristics may be related to zinc. Changes in brain zinc concentration are also observed in Alzheimer's disease (AD), Down syndrome, epilepsy, multiple sclerosis, retinal dystrophy, and schizophrenia (Beard 1996).

Supplements

Vitamin–mineral supplementation and intelligence in school children—level I evidence

In a randomized, double-blind, placebo-controlled study, a low-dose supplement (50% of the American DRI) was provided to schoolchildren for 3 months. An increase in nonverbal intelligence was statistically significant in the supplemented schoolchildren compared to the placebo group. The mean difference was 2.5 IQ points, although in some children the increase was 15 to 16 IQ points for the 3-month period. The researchers noted that nonverbal intelligence is closely associated with academic performance. They recommended that parents of children who aren't performing well ensure that the child is consuming a healthy diet and have the child's nutritional status assessed by a physician trained in nutrition (Schoenthaler 2000).

Supplements and academic performance in schoolchildren—level I evidence

A 9-month study of 684 children ages 8 to 12 years old assessed the effect of vitamin-mineral supplementation on change in scores on a standardized achievement test. Also observed were days absent, tardiness, and grade point average. In this double-blind, placebo-controlled, clinical trial, supplements were provided to children at multiple schools Monday through Friday. The supplement provided 100% of the recommended daily values for ages 4 to 21 *except* for (1) calcium (12.5% of the recommended daily value), (2) magnesium (18% of the recommended daily value), (3) copper (50% of the recommended daily value), and (4) iron (50% of the recommended daily value). A baseline dietary intake report of a subset of participants indicated that ≥25% did not meet the RDA for iron. Seventy-six percent of children ages 8 years or younger and 93% of children ages 9 and older did not meet the Adequate Intake (AI) for calcium. Blood values were not obtained.

No significant improvements were observed for total or sectional scores on the achievement test. No substantial improvements were observed in days absent, tardiness, and grade point average. Authors concluded that the results did not support the routine administration of a standard vitamin-mineral supplement to children of this age. It was suggested that future research should include assessment of micronutrients from the diet and blood serum level of participants. Populations with documented deficiencies should be studied (Perlman 2010).

Cognition

Lipids

Brain membranes contain many different fatty acids in the form of phospholipids. DHA is abundant and important in the development and maintenance of cognitive function. Breast-fed infants and infants of a formula supplemented with DHA and AA for the first 17 weeks of life had higher scores on the Bayley Mental Development Index at 1 year old or later. In participants with an average age of 76 years, tracked in the Framingham Heart Study, high intakes of fish or DHA were shown to be protective against cognitive decline and dementia.

Cognition and essential fatty acids: DHA, EPA, and AA—level IV evidence

The beneficial effects of DHA include (1) maintaining the fluidity of neuron membranes, which affects receptors, ion channels, and imbedded proteins, (2) transformation to active metabolites, (3) differentiation of neuron stem cells, and (4) decreasing the AA content of brain phospholipids. DHA and eicosapentaenoic acid (EPA), but not by AA have also been shown to promote synthesis of synaptic membranes (Wurtman 2008; Canseva 2008).

Aging, cognition, and fish oil supplements: Assessment at 11 and 64 years of age—level II evidence

A group in Scotland, born in 1936, were tested for mental ability using the Mini-Mental State Examination (MMSE) in 1947 (at 11 years old) and retested in 2000–2001 (at 64 years old). Assessments included cognitive function, dietary habits, vascular risk factors, physical examination, and blood samples. Specific cognitive advantages at the age of 64 were found in users of supplements. Supplement users were categorized as (1) users of fish-oil supplements with or without vitamins A, D, or E, (2) vitamins without fish oil, and (3) other. The authors note that these categories were correlated with the digit symbol subtest, which is sensitive to cognitive aging and AD. This test and the block design test together predict functional capacity in dementia. The authors commented that users of fish-oil supplements also consumed more vegetables than fruits, which resulted in a higher vitamin C intake, confounding the conclusions that can be drawn from the study. They also noted that measures of erythrocyte omega-3 fatty acids are more informative than reports of fish-oil consumption (Whalley 2004).

DHA and cognition in midlife adults free of neuropsychiatric disorders—level II evidence

A study of 280 adults, ages of 35 to 54 years, with no neuropsychiatric disorders, investigated the association of omega-3 fatty acids (ALA, EPA, and DHA) in serum phospholipids with measures of cognition. Higher DHA was related to better performance on (1) nonverbal reasoning, (2) mental flexibility, (3) working memory, and (4) vocabulary evaluation. Associations were generally linear. The relationship of DHA levels with nonverbal reasoning and working memory persisted when adjustments for education and vocabulary were made. Neither ALA nor EPA showed the same relationships (Muldoon 2010).

Carbohydrate

Observation of hypoglycemia and hyperglycemia using magnetic resonance imaging—level II evidence

It is known that different brain functions have different susceptibilities to acute hypoglycemia. For example, autonomic activation occurs in response to a quite modest decrease in circulating glucose levels. Functional MRI (fMRI) detects regional changes in brain oxygenation state during activation by a task. Significant cortical dysfunction requires more profound glucose deprivation. Brain activation in healthy volunteers (N = 8) using fMRI examined the effect of acute hypoglycemia while participants performed cognitive tasks commonly used in hypoglycemia research.

Fasting plasma or serum glucose of 70 to 115 mg/dl = 3.8 to 6.4 mmol/l.

In one part of the study (N = 6), plasma glucose was maintained at euglycemic levels of 5 mmol/L (90 mg/dl) throughout assigned tasks. Plasma glucose was reduced to 2.5 mmol/l (45 mg/dl) for the second set of tasks. Performance of tasks resulted in maps of specific brain activation areas. Hypoglycemia is associated with task-specific localized reductions in brain activation and impairs simple brain functions. For a task with greater cognitive load, signals in planning areas of the brain appear to recruit other brain regions in an attempt to limit dysfunction. Researchers commented that this knowledge might be helpful to individuals on insulin because these levels would not occur naturally in healthy individuals. In health, autonomic activation occurs at plasma glucose of 3.0 to 3.6 mmol/L (54 to 64 mg/dl). Reaction times involving choice and performance of Stroop tests become slower at ~3 mmol/L; short-term memory deteriorates at 2.5 mmol/L (Rosenthal 2001).

Vitamins

Homocysteine, B vitamins, brain atrophy, and cognitive impairment—level I evidence

Reports indicate that 16% of those over 70 years old have mild cognitive impairment and half of these develop AD. In a study of 271 70-year-old subjects with mild cognitive impairment, half were given folic acid (0.08 mg/d), vitamin B_6 (20 mg/d), and vitamin B_{12} (0.5 mg/d) for 24 months. Researchers compared the rate of cognitive decline with 83 participants who received a placebo. One hundred sixty-eight participants (half subjects, half control group) completed an MRI. The mean rate of cognitive decline in the treatment group was 0.76% per year. In the placebo group, the rate was 1.08% (significant at $p = 0.001$). The rate of atrophy in participants with homocysteine >13 mmol/L was 53% lower in the active treatment group ($P = 0.001$).

Cognitive impairment and vitamin B_{12}—level II evidence

As reported by British researchers, 125 patients seen at a memory disorders clinic were found to have low serum B_{12}. Based on neuropsychological testing, 66 of these had dementia and 22 had cognitive impairment. The patients with low B_{12} were treated with B_{12} supplementation. Their test scores were compared before and after treatment, and with age-matched control patients from the clinic. The majority of patients with low serum B_{12} had normal blood levels after supplementation. Cognitive impairment improved significantly in test scores for verbal fluency ($p < 0.01$) after treatment, compared to the age-matched control patients. There was no significant improvement in dementia patients after treatment (Eastley 2000).

Folic acid, B_{12} supplements, and cognitive decline—level I evidence

Supplements of 400 μg folic acid and 100 μg B_{12} per day for 24 months resulted in improved short and delayed recall, but not orientation, attention, semantic memory, processing speed, or informant reports in community-dwelling adults aged 60 to 74 years. Homocysteine (which is thought to affect DNA repair and induce oxidative stress) increased less than the control group (an increase of 8% vs. 22%). Assessment and follow-up with participants was conducted via telephone over the length of the study (Walker 2012).

Vitamin D and cognition—level II evidence

Assessment of 60 subjects (30 African Americans and 30 European Americans) over 55 years old for cognition, depression, serum vitamin D, and additional physical attributes found 76% of the African Americans and 27% of the European Americans were vitamin D deficient (using criteria of 20 ng/ml). Subjects with vitamin D deficiency scored worse on the Short Blessed Test (SBT) of cognition. There was no difference in scores on the MMSE for mood disorder (Wilkins 2009).

Evidence report by Agency for Healthcare Research and Quality (AHRQ)—level III evidence

An evidence report in 2006 prepared by the Agency for Healthcare Research and Quality (AHRQ) concluded that folic acid alone or with vitamin B_{12} or vitamin B_6 had no significant effects on cognitive function (Johns Hopkins University 2006). (See also Chapter 2.)

Minerals

Boron, brain function, and cognitive performance—level II evidence

A research protocol compared brain activity with a low boron diet (0.23 mg/2000 calories), supplemented with (1) a low boron supplement (0.25 mg/day/2000 calories) or (2) a high boron supplement (3.25 mg/day/2000 calories) and included cognitive testing. Among other findings, a low boron intake was related to poorer performance in tests of attention, perception, and long- and short-term memory. Significant increases in low-frequency brain activity and decrease in high-frequency brain activity were similar to changes observed in general malnutrition and metal toxicity. The basic menu rotation used for a consistent low dietary intake of boron included chicken, beef, pork, potatoes, rice, bread, and milk, with low intake of vegetables and fruits (Penland 1994).

A presentation on this research included the following comment:

> The findings also underscore the need to examine a variety of cognitive processes and response characteristics when assessing the impact of nutritional intervention or suboptimal nutrition on cognitive function. ... We don't know the biochemical basis ... we can't always anticipate mechanisms of action, so we may be limiting ourselves if we only look first to the biochemical function and then to match a behavioral outcome. (Penland 1997)

Free plasma copper vs. bound copper—level II evidence

A study of 64 women whose mental status had been assessed for cognitive function found that free copper levels in plasma were significantly negatively correlated with scores on the MMSE and attention-related tests. Bound copper levels did not correlate with scores in the cognitive domain.

Researchers suggested free copper might be a risk factor in the development of impaired cognition (Saluski 2010).

Risks of copper toxicity

Use of copper tubing in water systems and copper in nutritional supplements may be major contributors to copper intake. Copper in food (organic) is absorbed and sequestered in the liver. Inorganic copper from copper tubing and some supplements enter the free copper pool in the blood and can penetrate the blood–brain barrier and effect mental status (Brewer 2009, 2010). (See also "Dementia" section later in this chapter and Chapter 11.)

High copper, high saturated fat, and trans fats—level II evidence

A correlation between diets with a total copper intake of 1.5 mg/day, along with high saturated fat and high trans fat was found for persons over 65 years old, but not younger persons. Individuals with this intake had a higher rate of cognitive decline (143%), as if 19 years had been added to their age. There was greater association between cognitive change and copper from supplements than from copper intake from food. This data from The Chicago Health and Aging Project, included over 3700 adults 65 years or older, followed for ~5.5 years. Sixteen percent, or 604 individuals, fell into the highest quintile of copper and had the highest intake of saturated and trans fats (Morris 2006).

Effect of vitamin C, E, beta-carotene, zinc, and copper on cognition—level II evidence

Over 2100 participants in the Age-Related Eye Disease Study were randomly assigned to a treatment group receiving 500 mg vitamin C, 400 IU vitamin E, 15 mg beta-carotene, 80 mg zinc, and 2 mg copper and followed for ~7 years. The supplemented group did not differ on any of six tests of cognition when compared to a group not receiving the supplements (Yaffe 2005).

Other

Resveratrol and cognition—level I evidence

Resveratrol, the ingredient in grapes and red wine said to improve health, was investigated to determine the effects on blood flow and cognition. The randomized, double-blind, placebo-controlled, crossover study involved 22 healthy adults who were assigned to receive placebo and either 250 or 500 mg of trans-resveratrol. Forty-five minutes after the dose, the blood flow and cognitive performance of the participants was measured over a 36-minute period. Resveratrol produced a dose-dependent increase in cerebral blood flow; there was no increase in the placebo group. The researchers also noted an increase in levels of deoxyhemoglobin after both doses of resveratrol, which indicates increased oxygen extraction and utilization. No effect on cognitive function was noted. The dose used in this study may produce problems over the long term because trans-resveratrol is a chelator of copper and could potentially induce a copper deficiency. Resveratrol is not suitable for children or pregnant women (Kennedy 2010; Daniels 2010).

Cognition and diabetes

Metabolic syndrome, also called Syndrome X, is characterized by an individual having at least three of the following health problems: insulin resistance, elevated cholesterol, elevated low-density lipoproteins (LDL), low high-density lipoproteins (HDL), obesity, and elevated blood pressure. Type 2 diabetes commonly includes these symptoms. Cognitive impairment and changes in quality of life are often associated with these conditions.

Cognitive impairment in diabetes—level II evidence

A study of 33 insulin-dependent diabetics (IDDM), 135 non-insulin-dependent diabetics (NIDDM), and over 2000 control subjects, ages 50 to 90 years, compared scores on two tests of general intellectual ability and three tests of verbal memory. The combined IDDM and NIDDM groups had significantly lower average scores on all cognitive tasks. The most cognitively impaired was the NIDDM group managed by hypoglycemic drugs. The effects were not related to depression or other factors. Early detection and effective treatment of diabetes may prevent or delay cognitive decline and improve quality of life for this population (Bent 2001).

Dementia

Preventing Alzheimer's disease and cognitive decline: 2010 NIH conference statement

The 2010 National Institute of Health (NIH) conference statement on Preventing Alzheimer's Disease and Cognitive Decline (http://consensus.nih.gov/2010/alz.htm) stated several dietary and lifestyle factors as well as medications have been linked to a decreased risk of Alzheimer's disease (AD). These include adequate folic acid intake, low saturated fat consumption, high fruit and vegetable consumption, and light to moderate alcohol consumption. However, the quality of evidence for the association of these factors with AD was reported to be low, the primary limitation being most evidence indicates association rather than causality. The most consistent evidence is available for longer chain omega-3 fatty acids (often measured as fish consumption), with several longitudinal studies showing an association with reduced risk for cognitive decline.

No consistent associations were found for beta-carotene, flavonoids, multivitamins, vitamins B_{12}, C, and E, plasma homocysteine level, obesity and body mass index, metabolic syndrome, blood pressure, antihypertensive medications, nonsteroidal anti-inflammatory drugs, gonadal steroids, or exposures to solvents, electromagnetic fields, lead, aluminum, or gingko biloba.

Depression and depressive symptoms have been consistently found to be associated with mild cognitive impairment and cognitive decline. Much of the available evidence comes from studies that were originally designed and conducted to investigate other conditions, such as cardiovascular disease and cancer. Evidence from studies conducted to date is limited by methodological issues in the assessment of the outcome (cognitive decline) or exposures (risk factors).

Several randomized clinical trials (RCTs) reviewed did not find a role of vitamin supplementation in preventing cognitive decline. However, these trials used varying doses of the nutrients, did not uniformly measure and monitor patients' cognitive function and baseline nutritional status, had short and variable follow-up, and mostly measured cognitive decline as a secondary or tertiary outcome. Thus, these trials may have been underpowered.

The conference statement concluded that light to moderate alcohol intake is associated with reduced risk of AD, but findings for cognitive decline are inconsistent (NIH 2010; see also Chapter 5.)

Lipids

Oral doses of DHA and changes in synaptic characteristics in animals—level IV evidence

Canseva and colleagues (2008) comment there seems to be little doubt that the numbers of brain synapses in patients with AD diminish and that this is a major factor causing patients to develop cognitive disturbances. In general, nutrients or drugs that modify brain function or behavior do so by affecting synaptic transmission, usually by changing the quantities of particular neurotransmitters present within synapses or by acting directly on neurotransmitter receptors or signal-transduction molecules.

These authors questioned if it were possible to cause the surviving neurons in damaged brain regions to make more or larger synapses, would this restore neurotransmission and would it alter the behavioral symptoms of the disease? Canseva writes it has never been possible to test this proposition because no method has been known that reliably increases synaptic number or size. Now a treatment has been identified in animal studies that increases the quantities of synaptic membrane in and the numbers of dendritic spines on hippocampal cells of normal animals. Treatment involves DHA and EPA, but not AA. Phosphatide precursors can also increase neurotransmitter release (acetylcholine, dopamine) and affect behavior in animals. The authors hypothesize if similar increases occur in human brains, then this treatment might have use in patients with the synaptic loss that characterizes AD, other neurodegenerative diseases, or loss that occurs after stroke or brain injury (Canseva 2008).

Vitamins

Alzheimer's disease and niacin—level II evidence

To assess the relationship of niacin to the development of AD, over 3700 residents of three south Chicago neighborhoods were followed for 5½ years. Residents in the lowest quintile of niacin intake consumed 12.6 mg niacin/day; the highest quintile of intake was 22.4 mg/day from foods. Those in the highest quintile had an 80% reduction in risk of incident AD than those in the lowest quintile. Higher intake of niacin *from food* was linearly associated with lower rate of cognitive decline. *Total* niacin (food, supplements, and niacin equivalents) intake had no association with cognitive change. This suggested that the protective benefit was related to niacin from food rather than tryptophan and endogenous production (Morris 2004). The DRI is 14 mg/day for women and 16 mg/day for men. As a comparison, intake of 8.8 mg of niacin equivalents per 2000 calories per day intake is associated with symptoms of confusion, psychosis, and encephalopathy in pellagra and severe alcoholism.

Alzheimer's disease and vitamin E—level IV evidence

Animal and laboratory studies show that AD involves oxidative and inflammatory processes. Whether these are a cause or effect is not known. The brain is the site of high metabolic activity and the generation of free radicals. Although free radicals are a normal part of human metabolism, excess free radicals can disrupt neuron cell functioning and signaling, which in turn can lead to neuronal cell death. Protection against this disruption is the role of antioxidants.

> Free radicals are very reactive unpaired electrons that can be toxic to cell tissue.

Vitamin E is a potent antioxidant in cell membranes. It also has anti-inflammatory functions. Animals fed antioxidants have superior learning ability and memory retention. Vitamin C, present in the circulating plasma, restores vitamin E when the body needs antioxidant capacity recharged. Consuming extra vitamin E from food has not been shown to be harmful and may be beneficial in preventing a number of conditions related to oxidative stress. However, supplements of vitamin E and vitamin C have not been associated with less risk of AD. The form of vitamin E may cause variation in study results. Gamma tocopherol is highest in the American diet; alpha tocopherol is more biologically active and often used in research studies (Morris 2004; Jiang 2001). The richest sources of vitamin E from food are vegetable oils, margarine, nuts (especially almonds), and seeds (especially sunflower seeds). The 2010 Dietary Guidelines report that moderate sources of vitamin E include whole grains, egg yolk, collard greens, avocados, apples, and melons.

Brain atrophy and folate—level II evidence
Using serum levels and MRI or CT images, Yang et al. (2007) found that 69% of a group of elderly poststroke patients (N = 89) were below serum levels of 6 ng/mL folate. Inadequate level was defined as <6 ng/mL. Low folate was defined as <3.0 ng/mL. The odds ratio of brain atrophy was 9.6 in the group with low folate.

Minerals

Dementia and minerals/metals—level V evidence
One theory regarding AD is that amyloid protein formed in brain capillaries alters permeability of the blood–brain barrier, increasing permeability to aluminum, iron, and mercury. These metals displace zinc from enzymes, leading to abnormal DNA synthesis, abnormal protein production, paired helical filaments, and neurofibrillary tangles. Production of amyloid protein is altered in zinc deficiency (Beard 1996).

Amyloid aggregation and toxicity—level II evidence
Different minerals have differing aggregation properties and potential for toxicity. Zinc(II) and copper(II) ions completely prevented the formation of soluble fibrillary aggregates; Al (aluminum) can effectively interact with amyloid β, forming aggregates with peculiar properties that are associated with a high neurotoxicity (Bolognin 2011).

Dementia and copper—level IV evidence
Copper (Cu) is a required mineral. Wilson's disease is an autosomal recessive genetic disorder of copper storage. Copper accumulates in the liver and neuronal tissue when the gene ATP7B is defective. The Wilson protein (also called ATP7B or ATPase) is required for copper transport and removal from tissue. Copper accumulation in the liver, brain, and cornea of the eye can cause neurologic damage and cirrhosis. One third of patients present with psychiatric symptoms such as depression, labile mood, impulsiveness, disinhibition, self-injurious behavior, or psychosis. Dementia presenting in those under age 40 is treatable with chelating

agents (Brown 1994). Use of a chelating agent along with avoidance of foods high in copper reduces copper stores. Zinc therapy can, over time, block absorption of copper and prevent it from entering the bloodstream (See also Chapter 11.)

> Foods high in copper include: liver, broccoli, legumes, chocolate, nuts, mushrooms, and shellfish, particularly oysters, crab, and lobster.

Copper and cognition in Alzheimer's disease—level I evidence
In a 12-month prospective, randomized, double-blind, placebo-controlled clinical trial involving patients with mild AD participants were given 51.62 mg of verum (copper-II orotate dehydrate), which provided 8 mg copper/day and was well tolerated. In the treated group, Cu levels were maintained; mean plasma Cu levels decreased in the placebo group over the year. Cognitive abilities worsened progressively in both groups. Plasma Cu levels are generally found to be lower in those with AD (Kessler 2008).

> Copper DRI for adults = 0.9 mg/day
> TUIL for copper = 10.0 mg/day

Plasma levels of Cu, Fe, and Zn and cognitive function differ in men and women—level II evidence
Concentrations of plasma Cu, Fe, and Zn (Table 6.1) and scores of 12 tests of cognitive function differed significantly in 602 men and 849 women ~75 years old living in their community. In men, there were significant associations at both high and low iron levels and performance on some tests (an inverted U-shaped curve). In women, iron and copper had an inverse linear association with some test scores; zinc had a positive association with some scores (Lam 2008).

Iron and dementia—level IV evidence
In AD, brain dysfunction may relate to iron content. Iron is a significant component of senile plaques and iron encrustation of the brain's blood

Table 6.1 Low, Intermediate, and High Mineral Levels in Plasma

	Low Levels	Intermediate Levels	High Levels
Iron	<40 µg/dl	40 µg/dl–215 µg/dl	>215 µg/dl
Copper	<90 µg/dl	90 µg/dl–215 µg/dl	>215 µg/dl
Zinc	<55 µg/dl	55 µg/dl–100 µg/dl	>100 µg/dl

Source: Lam et al. 2008.

vessels, and is found in the blood vessels of senile patients. Elevated levels of iron are also observed in other parts of the brain: the hippocampus, the amygdala, the basal nucleus of Maynert, and the cerebral cortex. Iron reactivity with hydrogen peroxide and oxygen can initiate lipid peroxidation, membrane damage, and ultimately cell death. Increase of free radical production has been noted in AD (Morris 2004).

Dementia and mercury toxicity—level IV evidence

Diagnosis of mercury toxicity requires consideration of factors such as (1) chemical form, (2) route of exposure, (3) dose, and (4) patient factors such as age, genetics, environmental aspects, and nutritional status. Patient factors are responsible for different individual responses to similar doses. When blood and urine are collected to evaluate exposure, the results are influenced by (1) specimen collection, (2) analysis, and (3) the time elapsed from exposure. Research in this area needs to consider and report on all these factors (Nutall 2004). Dementia has been linked to exposure to heavy metals such as lead, mercury, and aluminum as well as other causes (Schofield 2005; Penland 1997).

Mercury in older adults—level II evidence

Since current recommendations encourage increased fish consumption, and questioning the possibility that this may increase mercury levels, Weil and colleagues (2005) conducted a study to determine the effect of mercury levels on neurobehavior of adults aged 50 to 70 years. They completed a cross-sectional analysis of data from 474 randomly selected participants in the Baltimore Memory Study in Baltimore. Neurobehavioral domains assessed included nonverbal reasoning and intelligence, verbal memory, visual memory, visuoconstruction and visuoperception, motor dexterity, and manual dexterity. The median blood mercury level was 2.1 μg/L with a range of 0 to 16 μg/L. Overall, the data did not provide strong evidence that blood mercury levels are associated with decline of neurobehavioral performance in this population of older urban adults (Weil et al. 2005).

Conclusions

1. Inadequacies, excess, and imbalance of vitamins (B_{12}, B_6, choline, folate, and niacin), minerals (aluminum, boron, copper, iodine, lead, mercury, selenium, and zinc), and essential fatty acids (AA, DHA, EPA, and GLA) have been observationally linked to intellect, cognition, and dementia.
2. Influences may occur during pregnancy through old age.
3. Dietary choices such as vegan diets, choice of fish species, imported spices, Ayurvedic medications, and paints on pottery used to prepare or serve food can contribute to alterations in mental status.

4. It may take weeks or months for some effects to develop, but nutritional influences should be included in the search for causes of change in mental status.
5. Investigators need to carefully define and report methods used. Mental status, behavior, and nutritional status should be included as primary outcomes.

References

Alpers, DH, WF Stenson, and DM Bier. *Manual of Nutritional Therapeutics* 3rd ed. Table 7–26, p. 236. Little, Brown and Company, Boston, 1995.

Araujo, J, AP Beelen, LD Lewis, et al. Lead poisoning associated with Ayurvedic medications-five states 2000–2003. *MMWR.* 2004; 53(26):582–584.

Baker, EL, RG Feldman, RF White, and JP Harley. The role of occupational lead exposure in the genesis of psychiatric and behavioral disturbances. *ACTA Psychiatrica Scan. Suppl* 1983; 303:38–48.

Beard, J. Nutrient Status and Central Nervous System Function. In *Present Knowledge in Nutrition,* 7th ed., by EE Ziegler and LJ Filer, Jr., 612–622(613). Washington, DC: International Life Sciences Institute (ILSI) Press, 1996.

Bent, N, P Rabbitt, and D Metcalfe. Diabetes mellitus and the rate of cognitive aging. *Brit J Clin Psychiatry* 2001; 39(part 4):349–362.

Bolognin, S, L Messori, D Drago, et al. Aluminum, copper, iron and zinc differentially alter amyloid-Ab(1–42) aggregation and toxicity. *Int J Biochem Biol* 2011; 43(6):877–885.

Brewer, GJ. The risks of copper toxicity contributing to cognitive decline in the aging population and to Alzheimer's disease. *J Am Coll Nutr* 2009; 28(3):238–242.

Brewer, GJ. Risks of copper and iron toxicity during aging in humans. *Chem Res Toxicol* 2010; 12(2):319–326.

Brown, MM. and VC Hachinski. Acute confusional states, amnesia and dementia. *Harrison's Principles of Internal Medicine.* 137–146. Ed. KJ Isselbacher, E Braunwald, JD Wilson, JB Martin, AS Fauci, and DL Kasper. McGraw Hill, New York. 1994.

Canseva, M, RJ Wurtman, T Sakamoto, and IH Ulusa. Oral administration of circulating precursors for membrane phosphatides can promote the synthesis of new brain synapses. *Alzheimers Dement.* 2008; 4(Supplement 1):153–168.

Centers for Disease Control & Prevention. Neurologic impairment in children associated with maternal dietary deficiency of cobalamin. *MMWR* 2003; 52(4):61–64.

Chiodo, LM, C Covington, RJ Sokol, et al. Blood levels and specific attention effects in young children. *Neurotoxicol Teratol* 2007; 29(5):538–546.

Chiodo, LM, SW Jacobson, and JL Jacobson. Neurodevelopmental effects of postnatal lead exposure at very low levels. *Neurotoxicol Teratol* 2004; 26(3):359–371.

Daniels, S. Resveratrol may boost blood flow in the brain. http://www.nutraingredients-usa.com/Research/Resveratrol-may-boost-blood-flow-in-the-brain-Study. Accessed Sept 2010.

Doornbos, B, SA van Goor, DA Dijck-Brouwer, et al. Supplementation of a low dose of DHA or DAH+AA does not prevent peripartum depressive symptoms in a small population based sample. *Prog Neuropsychopharmacol Biol Psychiatry* 2009; 33(1):49–52.

Eastley, R, GK Wilcock, and RS Bucks. Vitamin B-12 deficiency in dementia and cognitive impairment: The effects of treatment on neuropsychological function. *Intl J Geriatric Psychiatry* 2000; 15(3):226–33.

Jans, LA, EJ Geltay, and AJ Van der Does. The efficacy of n-3 fatty acids DHA and EPA (fish oil) for perinatal depression. *Br J Nutr* 2010; 104(11):1577–1585.

Jiang, Q, S Christen, MK Shigenaga, and BN Ames. Gamma-tocopherol, the major form of vitamin E in the US diet, deserves more attention. *Am J Clin Nutr* 2001; 74(6):714–722.

The John Hopkins University Evidence-based Practice Center, Multivitamin/Mineral supplements and prevention of chronic disease. Evidence Report/Technology Assessment Number 139. AHQR Publication No.06–E012. 2006: v. http://www.ahqr.gov/downloads/pub/evidence/pdf/multivit/multivit.pdf

Jones, JW and M Sidwell. Essential fatty acids and treatment of psychiatric diseases. *Orig Intern.* 2001; 8:5.

Kales, SN, CA Christophi, and RD Saper. Hematopoietic toxicity from lead-containing Ayurvedic medications. *Med Sci Monit* 2007; 13(7):CR 295–8.

Kennedy, DO, EL Wightman, JL Reay, et al. Effects of resveratrol on cerebral blood flow variables and cognitive performance in humans: A double-blind, placebo-controlled, crossover investigation. *Amer J Clin Nutr* 2010; 1(6):1590–1597.

Kessler, H, TB Bayer, D Bach, et al. Intake of copper has no effect on cognition in patients with mild Alzheimer's Disease: A pilot phase 2 clinical trial. *J Neural Transm* 2008; 115:1181–1187. http://www.ncbi.nlm.nih.gov/pmc/articles/PMC2516533/?tool=pubmed

Kuipers, D et al. Maternal DHA equilibrium during pregnancy and lactation is reached at an erythrocyte DHA content of 8 g/100 g fatty acids. *J Nutr* 2011; 141(3):418–27.

Lam, PK, D Kritz-Silverstein, E Barrett-Conner, et al. Plasma trace elements and cognitive function in older men and women: The Rancho Bernardo study. *J Nutr Health Aging* 2008; 12(1):22–27.

Lanphear, BP, K Dietrich, P Auinger, and C Cox. Cognitive deficits associated with blood lead concentrations <10 microg/dL in U.S. children and adolescents. *Pub Hlth Rep* 2000; 115(6):521–529.

Leung, BMY and BJ Kaplan. Perinatal Depression: Prevalence, risks, and the nutrition link-A review of the literature. *J Amer Diet Assoc* 2009; 109(9):1566–1575.

Lin, CG, LA Schaider, DJ Brabander, and AD Woolf. Pediatric lead exposure from imported Indian spices and cultural powders. *Pediatrics.* 2010; 125(4):e828–835. http://www.ncbi.nlm.nih.gov/pubmed/20231190

Louwman, MWJ, M van Dusseldorp, JR Fons, et al Signs of impaired function in adolescents with marginal cobalamin status. *Amer J Clin Nutr 2000;* 72 (3):762–769.

Lynch, R, B Elledge, and C Peters. An assessment of lead leachability from lead-glazed ceramic cooking vessels. *J Environ Health.* 2008; 70(9):36–40, 53. http://www.ncbi.nlm.nih.gov/pubmed/18517152

Morris, MC, DA Evans, CC Tangney, et al. Dietary copper and high saturated and trans fat intakes associated with cognitive decline. *Arch Neurol* 2006; 63(8):1085–1088. http://www.medscape.com/view article/542990

Morris, MC, DA Evans, JL Bienias, et al. Dietary niacin and risk of incident Alzheimer's Disease and of cognitive decline. *J Neurology, Neurosurgery and Psychiatry* 2004; 75:1093–1099.

Morris, MC. Diet and Alzheimer's Disease: What the evidence shows. *Medscape General Medicine* 2004;6(1):e8. http://www.medscape.com/viewarticle/466037

Muldoon, MF, CM Ryan, L Sheu, et al. Serum phospholipid docosahexanonic acid is associated with cognitive functioning during middle adulthood. *J Nutr* 2010; 140(4):848–853.

National Institutes of Health. *State-of-the-Science Conference Statement: Preventing Alzheimer's Disease and Cognitive Decline.* 2010; 27(4). http://consensus.nih.gov/2010/docs/alz/ALZ_Final_Statement.pdf (accessed 1/8/12)

Needleman, HL, A Schell, D Bellinger, et al. The long-term effects of exposure to low doses of lead in childhood: An 11-year follow-up report. *New Engl J Med* 1991; 324(6):415–418.

Nuttall, KL. Interpreting mercury in blood and urine of individual patients. *Ann Clin Lab Sci* 2004; 34:235–250.

Oh, R. Practical applications of fish oil (Omega-3 fatty acids) in primary care. *J Amer Board Fam Pract* 2005; 18(1):28–36.

Parade Publications. What Americans Think about Aging and Health. *Parade Magazine.* New York, February 5, 2006:11.

Penland, James G. Dietary boron, brain function, and cognitive performance. *Environ Health Perspect* 1994; 102(Suppl 7):65–72.

Penland, JG. 1997. Trace elements, brain function and behavior: Effects of zinc and boron. In *Trace Elements in Man and Animals – 9: Proceedings of the Ninth International Symposium on Trace Elements in Man and Animals.* 213–216. Edited by PF Fischer, MR L'Abee, KA Cockell R, S.Gibson. NRC Research Press, Ottawa, Canada.

Perlman, AI, J Worobey, J O'Sullivan Maillet, et al. Multivitamin/Mineral supplementation does not affect standardized assessment of academic performance in elementary school children. *J Amer Diet Assoc* 2010; 110(7):1089–1093.

Rosenthal, JM, SA Amiel, L Yágüez, et al. The effect of acute hypoglycemia on brain function and activation: A functional Magnetic Resonance Imaging study. *Diabetes* 2001; 50(7):1618–1626.

Saluski, C, G Barbati, R Ghidoni, et al. Is cognitive function linked to serum free copper levels? A cohort study in a normal population. *Clin Neurophysiol* 2010; 121(4):502–507.

Saper, RB., SN Kales, J Paquin, et al. Heavy metal content of Ayurvedic herbal medicine products. *J Amer Med Assoc.* 2004; 292(23):2868–2873.

Schofield, P. Dementia associated with toxic causes and autoimmune disease. *International Psychogeriatrics* 2005; 17 (Suppl 1):S129–S147.

Schoenthaler, S.J., I D Bier, K Young, et al. The effect of vitamin-mineral supplementation on the intelligence of American schoolchildren: A randomized, double-blind placebo-controlled trial. *J Altern Compl Med* 2000; 6(1):31–35.

Smith, DA, SM Smith, CA deJager, et al. Homocysteine-lowering by B-vitamins slows the rate of accelerated brain atrophy in mild cognitive impairment: A randomized controlled trial. *PLoS One* 2010; 5(9):e12244. www.plosone.org

Surkan, PJ, A Zhang, F Trachtenberg, et al. Neuropsychological function in children with blood levels <10 microg/dL. *Neurotoxicol* 2007; 28(6):1170–1177.

Walker, JG, PJ Batterham, AJ Mackinnon, et al. Oral folic acid and vitamin B-12 supplementation to prevent cognitive decline in community-dwelling older adults with depressive symptoms-the Beyond Ageing Project: A randomized controlled trial. *Am J Clin Nutr* 2012; 95:194–203.

Whalley, LJ., HC Fox, KW Wahle, et al. Cognitive aging, childhood intelligence, and the use of food supplements: Possible involvement of n-3 fatty acids. *Amer J Clin Nutr* 2004; 80:1650–1657.

Weil, M, J Bressler, P Parsons, et al. Blood mercury levels and neurobehavioral function. *J Amer Med Assoc* 2005; 293(15):1875–1882.

Wilkins, CH, SJ Birge, YI Sheline, and JC Morris. Vitamin D deficiency is associated with worse cognitive performance and lower bone density in older African Americans. *J Natl Med Assoc* 2009; 101(4):349–354.

Wurtman, RJ. Synapse formation and cognitive brain development: Effect of docosahexaenoic (DHA) and other dietary constituents. *Metabolism.* 2008; 57(Suppl 2):S6–10.

Yaffe, K, TE Clemons, WL McBee, and AS Lindblad. Impact of antioxidants, zinc and copper on cognition in the elderly: A randomized, controlled trial. *Neurology* 2005; 63(9):1705–1707.

Yang, LK, KC Wong, MY Wu, et al. Correlations between folate, B12, homocysteine levels and radiological markers of neuropathology in elderly post-stroke patients. *J Am Coll Nutr* 2007; 26(3):272–278.

chapter seven

Mood disorders: Depression, bipolar disorder, and suicide

Depression

Introduction

Approximately 20.9 million American adults, or about 9.5% of the U.S. population age 18 and older, have a mood disorder in a given year (NIH 2011). The median age of onset for mood disorders is 30 years (Kessler 2005).

Mood disorders include major depressive disorder (MDD), dysthymic disorder, bipolar disorder, mood disorder due to a general medical condition, substance-induced mood disorder, and atypical depression plus additional subcategories (DSM-IV 1994). Mood disorders are a prolonged experience of feelings that color the entire emotional life. They may be a short-term episode or may persist for weeks or years. Mood disorders may be accompanied by symptoms such as depression, anxiety, agitation, irritability, phobias, and sleep and appetite disturbances, or physical symptoms such as fatigue, headaches, or abdominal pain. They may be associated with a positive family history, substance abuse, eating disorders, medications, suicide, and suicide attempts (Jarvis 1992).

Atypical depression is a common form of major depression affecting up to 40 million Americans. It is characterized by a distinct combination of symptoms including mood swings, carbohydrate cravings, rejection sensitivity, lethargy, and weight gain. Weight gain, a common side effect of psychotropic medications, causes some individuals to discontinue taking prescribed medications.

Common symptoms occurring in major depression (Table 7.1) may directly affect nutritional intake as well as indirectly affect food procurement and preparation.

Lipids

Nerve dendrites and synapses are 80% lipid by weight. All cell membranes have high phospholipid content. Phospholipids link the essential fatty acids from the environment (diet) with enzymes, which are linked to DNA, providing a biochemical explanation for interaction between

Table 7.1 Prevalence of Associated Symptoms in Major Depression

Associated Symptoms	Prevalence in Major Depression (%)
Fatigue[a]/asthenia	73
Headache	33
Gastrointestinal symptoms[a]	34–57
Psychomotor slowing	59–65
Insomnia	63
Irritability	50
Arthralgia	31
Musculoskeletal pain	62–80
Abdominal pain[a]	21
Anorexia[a]	40
Anxiety	57
Poor concentration	51

Source: Su 2009.

[a] Of special interest regarding nutritional status.

genetic and environmental factors in psychiatric disorders. At the epidemiological level, studies indicate that populations with higher intakes of omega-3 fatty acids have lower rates of depression (Peet 2004).

> Phospholipid refers to a fatty acid with a phosphorous–oxygen duo, plus possible other molecular combinations linked to it.

EPA vs. DHA—level III evidence

A review that included 28 double-blind, placebo-controlled, randomized controlled trials concluded that (1) depression scores were reduced with omega-3 supplementation, (2) bipolar disorder and major depression were more influenced by supplementation than mild to moderate depression, (3) using omega-3 supplements was more effective as adjunctive therapy rather than given alone, and (4) a higher ratio of EPA to DHA was more effective. Factors that influenced study results reviewed included quality of methods used, sample size, and duration of study (Martins 2009).

Omega-3 fatty acids and depression—level II evidence

Early assessment of fatty acids in red blood cell membranes of 10 patients with depression and 14 controls showed a significant depletion of omega-3 fatty acids in patients with depression, which was not due to a reduced calorie intake. Severity of depression was correlated with fatty acid levels and with dietary intake data, although diet assessment methodology was thought to be a weak predictor (Edwards 1998).

Table 7.2 Fasting Serum Phospholipids as Percentage of Total Lipids

AA	8.84 (±1.66)
DHA	1.65 (±0.67)
EPA	0.51 (±0.43)
AA:DHA	6.03 (±2.23)
AA:EPA	23.11 (±11.81)

Source: Conklin et al. 2007.

Essential fatty acids, depressive symptoms, and neuroticism—level II evidence

Conklin and colleagues (2007) reported on the normative variation of essential fatty acid levels and their association with scores on the Beck Depression Inventory (BDI) and the NEO-Personality Inventory-Revised domain of neuroticism. Volunteers (N = 116) without hypercholesteremia, without psychopathology, and taking no omega-3 supplements or psychotropic, diabetic, or cardiovascular medications participated.

The neuroticism score was inversely associated with EPA and positively associated with AA levels, and ratios of AA to EPA and AA to DHA (Table 7.2). AA levels predicted BDI scores of ≥10. Investigators concluded results provided evidence that the relationship of omega-3 and omega-6 to mood extends to nonclinical variation in affect and not only to overt psychopathology. They noted due to the cross-sectional design, the association could not be interpreted as causal.

No effect of EPA and DHA on depression—level I evidence

A double-blind, randomized, controlled study of 190 individuals with mild or moderate depression provided a supplement of 1.5 g/day EPA + DHA or a placebo for 12 weeks. Participants were assessed using the depression subscale of the Depression, Anxiety and Stress Scales, BDI, and others. There were no significant beneficial or harmful effects resulting from use of this supplement (Rogers 2008).

Fatty acids in plasma and erythrocytes, and enzyme activity in depression—level II evidence

Using plasma and erythrocyte biochemistries, structured interviews, and physical assessment of BMI and waist-to-hip ratio (WHR), previously depressed patients in relapse or remission were compared to a matched recruited control group. The sum of plasma omega-3 fatty acids, EPA, and DHA was not different in subjects and controls.

In patients with depression in remission, in red cell membranes, the sums of total fatty acids and PUFAs were significantly lower while the sums of saturated fatty acids (SFA) and monounsaturated fatty acids (MUFAs) were similar. The sums of omega-6 fatty acids and linoleic acid were significantly higher in patients compared to a control group. Activity of elongase and saturase enzymes was deduced. While EPA levels were not different, the products of EPA and DHA elongase were lower in patients, implying lower enzyme activity. The products of the omega-6 desaturase were significantly higher in patients. Researchers concluded since no influence of current depressive status on fatty acid concentrations was observed, fatty acid alterations could represent a biological "trait" marker for recurrent depression. Genetically determined mitochondrial dysfunction and oxidative stress may also be an influence in such alterations (Assies 2010).

EPA supplements along with standard drugs—level I evidence
A 12-week randomized, double-blind, placebo-controlled study of 70 patients, who remained depressed while on doses of a standard antidepressant predicted to be adequate, were supplemented with doses of 1, 2, or 4 g/day of ethyl-eicosapentaenoate (EPA). Of the patients on 1 g doses, 53% achieved a 50% reduction on the Hamilton Depression Rating Scale Score. All of the individual items on three rating scales improved with the 1 g/day dose compared to the placebo group. Benefits included the effects on depression, anxiety, sleep, lassitude, libido, and suicidality. Individuals on the 4 g/day dose showed nonsignificant trends toward even greater improvement (Peet 2002).

EPA treatment for depression—level II evidence
In a study of 60 outpatients with major depressive disorder, a daily supplement of 1000 mg EPA was as effective as a selective serotonin reuptake inhibitor (SSRI) at alleviating symptoms of depression as measured on the Hamilton Depression Rating Scale. EPA, 20 mg/day fluoxetine, or a combination of the two, was randomly allocated over an 8-week period. A combination of EPA plus fluoxetine produced even greater results after week 4 (Jazayeri 2008).

Lipids, zinc, albumin, T cells, and depression—level II evidence
Maes and colleagues (1997) examined whether (1) major depression is accompanied by altered serum lipids or vitamin E, (2) suicidal attempts are related to lower serum HDL-C, and (3) whether there are significant associations between serum HDL-C and immune/inflammatory markers. A total of 36 subjects with major depression (28 of whom showed treatment resistance) as well as 28 normal control subjects had blood samples assayed for lipids, serum zinc, albumin, and ratio of T cell helper to T suppressor (CD4+:CD8+).

Serum HDL-C, total cholesterol, and ratio of HDL-C to total cholesterol were significantly lower in subjects with major depression than in normal controls. Serum HDL-C levels were significantly lower in men with depression who had at some time made serious suicidal attempts than in those without such suicidal behavior. Treatment with antidepressants for 5 weeks did not significantly alter either serum HDL-C or other lipid variables. Serum HDL-C levels were significantly and negatively correlated with the T cell ratio (CD4+:CD8+) and positively correlated with serum albumin and zinc. Maes commented that these results suggest that (1) lower serum HDL-C levels are a marker for major depression and suicidal behavior in men with depression, (2) lower serum HDL-C levels are probably induced by the immune/inflammatory response in depression, and (3) in depression there is impairment of reverse cholesterol transport from body tissues to the liver (Maes et al. 1997).

Essential fatty acids and depression: The Rotterdam study—level II evidence

In Rotterdam, over 3800 community-dwelling elderly ≥60 years of age were screened for depression. Those who screened positive for depression (N = 264) were interviewed and had blood tests for plasma phospholipids and ratio of omega-6 to omega-3 fatty acids. Also tested was the level of C-reactive protein (CRP), an indicator of inflammatory response. The investigators questioned whether atherosclerosis or an inflammatory response could be responsible for the relationship seen between fatty acids and depression.

Subjects with depressive disorders had a higher ratio of omega-6 to omega-3. Subjects with depression who had normal inflammatory ratings had lower *percentages* of omega-3 fatty acids and a higher *ratio* of omega-6 to omega-3 than control subjects. Tiemeier et al. (2003) concluded that the relationship between essential fatty acids and depression was not due to inflammation, atherosclerosis, or other factors. Researchers suggest a direct effect of fatty acid composition on mood.

Community-living adults and essential fatty acids—level II evidence

On the assumption that depression has been shown to compromise older adults' health status, diminish quality of life, increase the risk of disabilities, and even increase the risk of death, Ma and Taylor (2004) completed an in-depth assessment of 33 individuals. Of the 33 subjects, 10 were rated as depressed; 2 of these 10 had been diagnosed previously with clinical depression. Individuals who were rated as depressed using Ma and Taylor's criteria had significantly lower *levels* of total omega-6 and omega-3 fatty acids in red blood cells but no significant difference in the *ratios* of omega-6 to omega-3 fatty acids. Depressed individuals had lower levels of vitamin B_{12}. Other factors noted included that those rated as depressed had lower education, lower income, higher number of health conditions (coronary heart disease), and some were taking medications having the side effect of depression, illustrating the complexity of studying the relationship between nutrition and mental status.

Assessments Used by Ma and Taylor (2004)
Geriatric Depression Scale (10/30 = depression)
Mini-Mental State Examination (<20 = cognitive impairment)
Katz Activities of Daily Living Scale
Instrumental Activities of Daily Living screening questionnaire
Self-administered food frequency questionnaire

Elderly depressed women, omega-3 fatty acid supplements, and quality of life—level II evidence

Rondaneli and colleagues (2010) report on providing omega-3 supplementation to 22 women with depression compared to 24 women with depression in a control group. The supplement consisted of 1.67 g/day of EPA and 0.83 g/day of DHA (a ratio of 2:1) for eight weeks. Erythrocyte membrane phospholipids were monitored for compliance. In the treatment group, depression was significantly decreased and quality of life was significantly improved.

Cholesterol and the brain

The central nervous system contains 23% of total body cholesterol. The half-life of cholesterol in the brain is 6 months to 5 years, compared to a half-life of hours in plasma. Cholesterol in the brain is found in two pools: 70% is metabolically stable and is found densely packed in bilayers in the myelin sheath at a density of 40 mg/g of tissue. Thirty percent is found in the plasma and subcellular membranes of neurons and glial cells at a concentration of 8 mg/g of tissue. It is conjectured that loss of cholesterol from membranes could reduce serotonin receptors and be related to suicidal, aggressive, or criminal behavior (Uranga 2010).

Depression and low cholesterol—level II evidence

Steegmans et al. (2000) investigated whether middle-aged men with chronically low cholesterol levels had a higher risk of having depressive symptoms (indicated by scores on the BDI) when compared with a similar reference group. A similar comparison was also made for measures of anger, hostility, and impulsivity. Men with chronically low cholesterol levels showed a consistently higher risk of having depressive symptoms (a BDI score of 15–17) than the reference group, even after adjusting for age, energy intake, alcohol use, and presence of chronic diseases.

Cholesterol Values Defined (Steegmans et al. 2000)
Low cholesterol group: 4.5 mmol/L (≤174 mg/dl)
Reference group: 6–7 mmol/L (232–270 mg/dl)

In a similar study, over 1000 white males in Rancho Bernardo, California, were followed for 10 years. Investigators found that in men aged 70 or older, depression was three times more common in the group with low plasma cholesterol. Plasma cholesterol levels of 160 mg/dl (4.14 mmol/L) were associated with depression. Reduced cholesterol levels significantly reduced the incidence of coronary heart disease, but the investigators note that this is often offset by increases in mortality from other causes such as suicide, homicide, and accidents. Other investigators have reported similar associations (Morgan 1993).

Proteins—amino acids

Amino acids, immune-inflammatory response, and depression—level II evidence

Another study examined total serum tryptophan and five competing amino acids (CAA)—valine, leucine, tyrosine, phenylalanine, and isoleucine—as well as other nutrients. There were significant correlations between serum tryptophan and serum zinc, total serum protein, albumin, transferrin, iron, and HDL-C (all positive) and between the number of leukocytes and the CD4+:CD8+ T cell ratio (all negative). The tryptophan:CAA ratio was significantly and negatively related to the number of leukocytes and the CD4+:CD8+ T cell ratio. The results suggest that (1) treatment-resistant depression is characterized by lower availability of serum tryptophan, (2) decreased availability of tryptophan may remain despite clinical recovery, and (3) the lower availability of tryptophan is probably a marker of the immune-inflammatory response during major depression (Maes 1999).

Tryptophan and chronic insomnia—level I evidence

A double-blind, placebo-controlled study was conducted to compare effectiveness of intact protein as a source of tryptophan vs. pharmaceutical-grade tryptophan in influencing insomnia. Each source was combined with carbohydrate for two groups and a third group had carbohydrate alone. Forty-nine subjects completed the 3-week study. Protein source tryptophan (de-oiled gourd seed, a rich source of tryptophan) with carbohydrate significantly reduced time awake during the night. Pharmaceutical-grade tryptophan with carbohydrate also improved measures of insomnia, although improvement was not statistically significant. Carbohydrate alone did not improve measures of insomnia (Hudson 2005).

Tryptophan metabolism and dieting—level II evidence

Smith and others (2000) reported a study of 42 women, half of whom had a history of depression and half of whom did not. They found that on a 1000-calorie diet, tryptophan was one nutrient that was depleted. Regulation of tryptophan metabolism was altered in women previously depressed. When tryptophan was less available due to the low-calorie diet, their biochemical response did not adjust in the same way as women who had not been previously depressed.

Carbohydrates

Mixed meals and effect on mood—level II evidence

After researching the effect of meal content on the relative availability of tryptophan to the brain, Rogers and Lloyd (1994) concluded that in meals providing starch or sugar with protein, the magnitude of the effect is probably too small to produce functionally significant changes in brain serotoninergic activity. Changes observed reached their maximum effect in 2 to 3 hours after the meals, with almost no change in the first hour. They further commented that a small amount of protein (4% of calories) in a high-carbohydrate meal is sufficient to block any meal-induced increases in the ratio of plasma tryptophan to amino acids.

4% of an 800-calorie meal is 32 calories, approximately the calories provided by the protein in 1 oz. of meat, fish, or cheese. 1 oz. meat has ~7 grams protein × 4 kcal/gram = 28 kcal from the protein in the meat. The remainder of the calories in meat comes from the fat.

Vitamins

Ascorbic acid deficiency

In clinical scurvy, emotional changes occur along with the physical signs of (1) purpura (bleeding beneath the skin) of lower extremities, coalescing to become ecchymoses (large areas appearing similar to large bruises), (2) splinter hemorrhages (beneath the fingernails), (3) perifollicular papules (in which hairs become fragmented and buried), and (4) loosening of the teeth. With doses as small as 6.5 mg/day, the body pool eventually returns to normal. Symptoms do not improve until the body pool is replenished. Larger therapeutic doses produce more rapid repletion (Wilson 1994). The adult DRI for ascorbic acid is 75 to 90 mg/day.

Vitamin C is a required cofactor in conversion of dopamine to norepinephrine. A deficiency could result in norepinephrine depletion, thus providing a neurochemical rationale for the severe depression often seen in scurvy. Depression occurs before onset of psychomotor symptoms. Vitamin C may change receptor sensitivity, interfering with stimulation of dopamine-sensitive adenylate-cyclase.

Table 7.3 Changes Observed with Vitamin C Deficiency

Change in	Blood/100 ml	Body Pool
Mental; personality	1.21–1.17 mg	761–561 mg
Physical; scurvy	0.67–0.14 mg	190–63 mg

Source: Kinsman and Hood 1971.

Ascorbic acid, depression, and personality changes—level I evidence
In a study of five healthy volunteer prisoners, Kinsman and Hood (1971) tested physical, mental, biochemical, and personality changes at different levels of depletion of vitamin C over a 7-month period (Table 7.3). Four scales of the Minnesota Multiphasic Personality Inventory (MMPI) showed statistically significant changes that paralleled the decrease in ascorbic acid nutritional status. These four were the scales measuring hypochondriasis, hysteria, depression, and social introversion. Hypochondriasis, hysteria, and depression have been known as the "neurotic triad." Participants' subjective reports included fatigue, lassitude, and depression.

With repletion, scores returned to levels achieved before depletion. Personality changes occurred earlier and at less depletion than physical (psychomotor) changes.

Current terms for hypochondriasis and hysteria include somatization disorder, conversion disorder, and dissociative disorder or histrionic personality.

Ascorbic acid: Depression, self-induced starvation, and scurvy—a case study—level IV evidence
A 40-year-old male had been eating on a budget of $1/day for a year. His diet consisted of canned pork and beans, peanut butter, bread, fruit, cereals, and milk, which was deficient in vitamin C. In response to a depressed mood, he decided to starve himself to death. Within a month of his decision to starve himself, he had symptoms of scurvy: bruising, muscle soreness, and bleeding skin. After he remained in bed for an estimated one to two months, he was brought to the hospital. A dexamethasone suppression test was normal. Trials of halperidol, lithium, and desipramine were ineffective in bringing him out of asocial and amotivational states.

He was diagnosed with low protein, low iron, and folate deficiency, and he had an undetectable blood ascorbate (vitamin C) level. He was anxious, fearful, alert, oriented, and unresponsive to most questions, laying face down in the fetal position, refusing food and medicine.

Over 4 weeks, he was given vitamin and mineral supplements. Suicidal preoccupation and anxiety diminished, and his sleep and appetite returned to normal. He displayed more initiative and psychomotor activation, and his interest in his surroundings increased. He met the criteria for schizotypal personality. He did not meet the diagnostic criteria for affective disorder.

> Psychopathology and nutritional status are interdependent. A patient's mental state may contribute to unusual eating habits, leading to nutritional deficiencies, which may in turn lead to biochemical changes, which could exacerbate psychopathology already present. Nutritional status, eating habits and dietary intake is an important consideration in evaluating psychiatric patients. (Roy-Byrne 1983)

Vitamin C and mood—level II evidence

Hoffer estimates that 20% of acute-care hospitalized patients have vitamin C levels compatible with scurvy. Using a supplement of 500 mg of vitamin C twice a day, along with scores on the Profile of Mood States, it was determined that there was a 34% decrease in mood disturbance in hospitalized patients. A vitamin D supplement had no effect on mood (Gan 2007; Zhang 2011).

In acutely hospitalized patients, vitamin C supplements for one week improved leukocyte concentrations of vitamin C from a low level of 16.3 µmol/L to a normal level of 71.0 µmol/L. At the same time, measures of mood disturbance improved 33% (Evans-Olders 2010).

Biotin: Sequence of deficiency symptoms—level IV evidence

Biotin, a member of the family of B vitamins, is necessary to four biotin-dependent enzymes, and deficiency is not considered common. A biotin deficiency may be the result of altered intake, interference with metabolism, or increased demand. A deficiency may involve a genetic replacement in DNA and may be linked to altered fatty acid synthesis. It may be associated with treatments such as renal hemodialysis. Causes include (1) eating raw egg whites (which bind biotin and prevent absorption), (2) total parenteral feeding (TPF) of longer than one week with a formula not including biotin, and (3) treatment with anticonvulsants such as phenytoin, primidine, and carbamazepine. Drugs may inhibit transport across the intestinal mucosa. Anticonvulsants may accelerate biotin breakdown. Prolonged use of oral antibiotic therapy may alter intestinal flora and may cause biotin deficiency.

Regardless of the cause, the development of deficiency symptoms follows an identified sequence. In the first three to five weeks, a deficient intake results in dry skin; fungal infections; rashes; and fine, brittle hair or hair loss. Approximately one to two weeks later, neurologic symptoms develop. These include changes in mental status; mild depression, which may progress to profound lassitude; and somnolence. Nausea, vomiting, and anorexia may also be present (Scheinfeld 2011). Symptoms in patients with uremic poisoning are reported to include features of clinical neuropathy such as dizziness, memory failure, disorientation, confusion, and psychotic episodes (Yatzidis 1984).

Jones and Nidus (1991) report symptoms in chronic dialysis patients of "malignant hiccups" (long-term, intractable hiccups as often as one per second), which included depression, assaultive behavior, and suicidal ideation. Adults have responded with symptom relief with doses of 10 mg of biotin/day for three months (Jones 1991).

B_{12} and depression—level III evidence

In Rotterdam, of 3884 elderly people screened for depressive symptoms, 278 had results indicating depression. Those with positive results were tested for folate, vitamin B_{12}, and homocysteine (Hcy) blood levels. Hyperhomocysteinemia and vitamin B_{12} deficiency and, to a lesser extent, folate deficiency were related to depressive disorders. The depression from folate deficiency and hyperhomocysteinemia were partly attributed to the presence of cardiovascular disease and functional disability, but vitamin B_{12} appeared to independently relate to depression (Tiemeier 2002).

Folate and depression—level III evidence

Clinically reliable assays for folate became widely available in the 1960s, and evidence began to accumulate for an association between folate-deficiency states and depression.

> There is now substantial evidence of a common decrease in serum/red blood cell folate, serum Vitamin B_{12} and an increase in plasma homocysteine in depression. At the methyltetrahydrofolate reductase (MTHFR) C677T site of genetic alteration, polymorphism that impairs homocysteine metabolism is shown to be over-represented among depressive patients, which strengthens the association. On the basis of current data, it was suggested that oral doses of both folic acid (800 µg/day) and vitamin B_{12} (1 mg daily) should be tried to improve treatment outcome in depression. (Coppen 2005)

> *Folate* is the form of the vitamin found in natural foods.
> *Folic acid* is the form of the vitamin manufactured and found in vitamin supplements.
>
> There are differences in how the body responds to these two forms and findings have been published regarding possible undesirable effects of doses >800 mcg/day of folic acid (Smith 2008).
>
> *L-Methylate* is the only form of folate that can cross the blood–brain barrier and regulate neurotransmitters (Ginsberg 2010).

Folic acid: A Cochrane review—level III evidence
A 2003 Cochrane review discussed the use of folate in addition to other treatments of depression. Two of three studies reviewed showed that folate improved the Hamilton Depression Rating Scale scores 2.65 points beyond the scores with medication alone. One study used folate instead of trazadone and did not find a significant benefit from use of folate. There was no evidence of problems with its safety or acceptability. The review stated that it was unclear whether there would be a difference in effect of medication on individuals with or without a folate deficiency (Taylor 2003).

Folic acid: A population study—level II evidence
A study comparing a diverse population of 15- to 39-year-olds in the United States found individuals with a lifetime diagnosis of major depression had lower serum and red blood cell folate than individuals who had never been depressed. Individuals diagnosed with dysthymia or who had recently recovered from dysthymia were also folate deficient. The results controlled for B_{12} levels, alcohol consumption, overweight status, use of vitamin/mineral supplements, and use of cigarettes and illegal drugs (Morris 2003).

Folic acid, homocysteine, depression, and MRI scans—level II evidence
A random subsample of participants in a study of 412 persons aged 60 to 64 years underwent psychiatric assessment, physical assessments, and brain MRI scans. Results were controlled for gender, physical health, smoking, and creatinine. White matter intensities, but not other measures of the brain, had significant correlations with homocysteine and depressive symptoms. The effects of folic acid and homocysteine were overlapping but distinct. The author concluded that low folic acid and high homocysteine, but not low vitamin B_{12} levels, correlate with depressive symptoms in this population (Sachdev 2005).

Niacin deficiency—level IV evidence

Absorption capacity for niacin of humans is 3 to 4 g/day. A diet deficient of niacin and tryptophan for 1 to 2 months, with a urinary excretion of metabolites <1.5 mg/day will produce clinical deficiency symptoms. Mental symptoms include insomnia, apathy, and then encephalopathy with confusion, disorientation, hallucinations, loss of memory, and organic psychosis. Dysphagia and amenorrhea, along with glossitis, stomatitis, and diarrhea are among other physical symptoms (Wilson 1994).

Riboflavin deficiency—level IV evidence

Jurg Haller, nutrition scientist, reported on studies that associate severe riboflavin deficiency with changes of scores on the hypochondriasis, depression, hysteria, psychopathic-deviate, and hypomania scales on the MMPI assessment instrument. He concluded that "partial and acute vitamin deprivation suggest that the earliest impairments occur in measures of mood rather than mental performance" (Haller 2005).

Vitamin D and depression—level II evidence

Data from the third National Health and Nutrition Examination Survey (HANES) was used to assess the relationship between vitamin D and depression in 7970 U.S. residents aged 15 to 39 years. For those having a vitamin D level ≤50 nmol/L, there is significantly higher likelihood of having a current episode of depression than those whose serum vitamin D level was ≥75 nmol/L. Higher prevalence of vitamin D deficiency occurred in women, non-Hispanic blacks, persons living below the poverty line, those who do not consume supplements, those living in the south, west, and urban areas, persons with a higher BMI, and those experiencing a current depressive episode. Seventy percent of participants lived in the south and west and 30% lived in the northeast, which may have been a factor in results. Twenty-eight percent used supplements. Nearly 12% reported episodes of depression lasting longer than 2 years; 4% reported current depression (Ganji 2010).

Depression and vitamin D status in the elderly—level II evidence

A survey of 1281 community residents aged 65 to 95 years found that 38.8% of the men and 56.9% of the women had insufficient vitamin D status. The average blood level of vitamin D was 21 ng/ml. Of this community group, 26 had major depression and 169 had minor depression. Depression severity was significantly associated with decreased serum 25(OH)D levels ($P = .03$) and increased serum PTH (parathyroid) levels ($P = .008$) (Hoogendijk 2008).

Tocopherol and depression—level II evidence
In an investigation of vitamin E status and dietary intake in Australian patients with depression, patients had significantly lower plasma alpha-tocopherol (vitamin E) than levels reported for healthy Australians. Vitamin E levels were inversely related to depression level. A subset of patients had a diet history analysis, and researchers found that this subset met or exceeded the recommended intake. Low plasma levels did not appear to result from an inadequate intake of vitamin E in this group of patients with depression (Owen 2005).

Minerals
Chromium and depression—level I evidence
Chromium is involved in the metabolism of carbohydrate, and chromium supplements have been tentatively shown to reduce carbohydrate craving in individuals with atypical depression.

A double-blind, placebo-controlled human clinical trial of 600 mcg/day of chromium investigated the effect of picolinate, elemental chromium, or a placebo on patients with atypical depression. The supplement reduced some symptoms of atypical depression including appetite changes, eating habits, and carbohydrate craving. The placebo and supplemented groups both improved in scores on the Hamilton Depression Rating Scale (Docherty 2003). This study replicated work by a group of researchers led by Davidson (2003).

Chromium supplementation for depression:
Case studies—level IV evidence
McLeod and Golden (2000) reported on three cases in which supplementation with 400 to 600 μg/day of chromium was followed by improvement of depressive symptoms.

Patient 1. Diagnosed with bipolar disorder, this patient had a previously stabilized mood with lithium, but reported symptoms of breakthrough depressions, a 30-lb weight gain, and irritability. With a chromium supplement, the patient felt more relaxed, and his wife reported a change in his attitude. He eventually felt less hungry and more stable and energetic, gradually losing 23 lb. However, weight changes are often associated with use of lithium.

Patient 2. Diagnosed with major depression, after taking a chromium supplement this patient reported vivid dreams, a decrease in carbohydrate craving, an increase in energy level, improved concentration and cognition, as well as normalization of sleep.

Patient 3. Diagnosed with dysthymic disorder, panic attacks, and rage outbursts, this patient reported vivid dreams, dramatic improvement of mood and behavior, enjoyment of interactions with colleagues, and being less withdrawn after taking a chromium supplement. To test himself for a placebo effect, this patient substituted a vitamin B_{12} supplement for the chromium, which resulted in the return of his symptoms of sadness, anxiety, fatigue, hunger, and sleep disturbance. A repeat of the chromium supplement again improved symptoms.

The improvements reported by these patients lasted over months of follow-up. The form of chromium (chromium picolinate or chromium polynicotinate) affected different individuals differently. Individuals required different doses for effectiveness. The researchers hypothesized that chromium increased insulin sensitivity, allowing enhanced tryptophan availability and transport across the blood–brain barrier, and therefore availability for increased serotonin synthesis. They remarked that sustained remission is not typical of a placebo response.

Chromium supplements: Toxic or not?—level IV evidence

Levina and Lay (2008) warn about potential toxicity of chromium supplements, which may outweigh their possible benefits as insulin enhancers, especially Chromium(III) picolinate [Cr(pic)3]. At some concentrations, partial hydrolysis of Chromium III may be likened to the effects of Chromium IV, V, and VI, known to be cytotoxic. These products of Cr III metabolism may cause alteration of cell signaling in the cell membrane or in the cytoplasm as well as lesions of DNA in the cell nucleus.

However, based on a review of literature, Eastmond, Macgregor, and Slesinski (2008) concluded that even though there is conflicting information, *in vivo* data show that restricted access of Cr III in cells limits or prevents genotoxicity in humans or animals. Nutritional benefits appear to outweigh theoretical risk at normal nutritional or modestly elevated physiological intake levels. The DRI for chromium is 20 to 35 μg/day.

Electrolytes and mood—level II evidence

Thirty-eight women and 56 men, generally healthy and free of diagnosed mental health issues, participated in a study of diets for 2 weeks each following (1) a moderate sodium, high potassium, high calcium diet (DASH-type) (OD) and either a (2) low sodium, high potassium diet (LNAHK) or (3) high calcium diet (HC). Dietary electrolytes, blood pressure, cortisol levels in saliva (which relates to stress), urinary electrolytes, and the Profile of Mood States (POMS) were used in assessment. There was greater improvement in depression, tension, vigor, and the POMS global score for the LNAHK diet compared to OD ($P>0.05$) (Table 7.4) (Torres 2008).

Magnesium: Deficiency or excess—level IV evidence

Magnesium is an essential mineral that is involved in over 300 metabolic reactions. It is difficult to evaluate magnesium status, as serum magnesium may not reflect body stores. Symptoms of magnesium depletion may

Table 7.4 Associations of Modified Diet with Mood Changes

Diet	Mood Changes
Low sodium, high potassium diet	Small, but significant improvement in depression, tension, and vigor
Lower urinary sodium excretion	Lower anger and vigor
Lower sodium to potassium ratio	Lower anger and depression
Higher urinary potassium excretion	Lower fatigue
Higher urinary magnesium excretion	Higher vigor
Higher urinary calcium excretion	Lower fatigue; lower POMS score (less depression)

Source: Torres et al. 2008.

include depression and psychosis, as well as seizures and numerous physical signs. Excess magnesium may exhibit with confusion and lethargy in addition to symptoms of toxicity such as nausea, vomiting, EKG abnormalities, hypotension, respiratory depression, and cardiac arrest. Depression of the central nervous system may occur in the range of serum levels of 5 to 10 mEq/L. Normal serum magnesium range is 1.5 to 2.5 mEq/L (Rude 2005).

Selenium—level IV evidence

To date, mood is the clearest example of an aspect of psychological functioning that is modified by selenium intake. Five studies have reported that a low selenium intake was associated with poorer mood. The underlying mechanism is unclear, although a response to supplementation was found with doses greater than those needed to produce maximal activity of the selenoprotein glutathione peroxidase. Alterations expected could include changes in attention, arousal, and memory. Suggestions that oxidative injury plays a role in normal aging, schizophrenia, Parkinson's disease, and Alzheimer's disease imply a possible role for selenium in these conditions. Poor selenium status may alter neurotransmitter metabolism in an undetermined fashion (Benton 2002).

Selenium, mood, and quality of life—level I evidence

In a study of 448 participants, doses of 100, 200, or 300 μg Se per day as high-selenium yeast were compared to a control group who were given placebo yeast. Assessments included the POMS Bipolar Form, "Quality of Life" Short Form 36, and plasma Se levels. After 6 months, supplements raised plasma levels of Se from those at baseline; the increase correlated to the level of supplementation. Mood scores and quality of life scores were not changed and did not differ between any of the groups (Rayman 2006).

Selenium, depression, and nursing home residents—level I evidence

Supplements of selenium, vitamin C, and folate were provided to residents of a nursing home after a baseline assessment of these nutrients and two ratings of mood. Baseline values indicated 29% were depressed, and 24% had anxiety. Sixty-seven percent were low in vitamin C; all were in the reference range for selenium. After 8 weeks in a double-blind, randomized, controlled trial of a multivitamin/mineral supplement, depression was significantly related to selenium, but not folate or vitamin C. After supplementation, selenium levels increased and symptoms of depression improved. The supplement provided 60 µg of Se. This compares with the U.S. DRI of 55 µg for selenium (Gosney 2008).

Depression and zinc—level III evidence

In a brief critical review, Levenson (2006) reported observations that low serum zinc has been reported in patients with treatment-resistant depression. The severity of depression was correlated with the decrease in zinc levels. Studies have not determined whether zinc levels are a cause or an effect of depression or other possible factors. A human study indicated that 6 weeks of zinc supplementation improved response to antidepressant therapy by over 50% and was sustained over 12 weeks of the study. Investigators hypothesize that zinc may influence 5-HT uptake in the cell body, block reuptake at the synapse, or increase synaptic release. These functions might provide an explanation for the mechanism of how zinc affects depression.

Supplements

There has been increased attention to the potential antidepressant efficacy of agents marketed as dietary supplements or "nutraceuticals," such as S-adenosyl-methionine (SAM), hypericum (St. John's Wort), and 5-HTP (5-hydroxytryptophan). In the body, tryptophan is converted to 5-HTP, which is the compound used to synthesize serotonin. Serotonin may play a role in depression, insomnia, weight loss, migraine headaches, fibromyalgia, and anxiety. The amino acid tryptophan is readily available in animal protein foods.

> Hydroxytryptophan is a molecule of the amino acid tryptophan with an –OH fragment attached at the site designated by "5" used to form serotonin.

Depression and 5-HTP (5-hyroxytryptophan) supplement: A Cochrane review—level III evidence

A comprehensive Cochrane review investigated 5-HTP as a natural alternative to traditional antidepressants for use in treating unipolar depression and dysthymia (a form of chronic depression, although less severe than major depression). Although evidence available at that time suggested that 5-HTP is better than placebo at alleviating depression, the quality of the research was not reliable enough to warrant recommendation for clinical usefulness. The reviewers recommended further studies to evaluate the safety and efficacy of 5-HTP (Shaw 2003, 2008).

Caution regarding tryptophan supplements and drug interactions—level IV evidence

U.S. regulatory agencies have reversed a 20-year ban on the over-the-counter sale of the essential amino acid L-tryptophan, and now allows it to be sold as a dietary supplement. Because L-tryptophan increases serotonin levels, coadministration of L-tryptophan and antidepressants that increase serotonergic activity may increase both the efficacy and the toxicity of some drugs. Examples include selective serotonin-reuptake inhibitors (SSRIs), amitriptyline, or monoamine oxidase inhibitors (MAOIs).

If a patient is taking one of these medications, physician-nutritionist Alan Gaby warns, L-tryptophan should either be avoided completely or used with caution and in low doses (perhaps 500 to 1000 mg/day), while monitoring for signs of serotonin excess, also known as serotonin syndrome. Gaby details one exception to this caveat: when initiating SSRI treatment, administration of up to 2 g per day of L-tryptophan for up to 4 weeks may accelerate the onset of the drug's antidepressant effect and protect against the development of insomnia, a frequent side effect of SSRIs in the early stages of treatment. However, Gaby comments, patients should be monitored closely during this time because serotonin excess can be life threatening. Coadministration of L-tryptophan and a triptan (e.g., sumatriptan, zolmitriptan, eletriptan) may also lead to serotonin excess, and should be avoided. In addition, L-tryptophan and 5-hydroxytryptophan (5-HTP) each increase serotonin levels, and should not be taken together.

Taking L-tryptophan with a high-protein meal decreases its effectiveness, whereas taking it on an empty stomach increases its efficacy. L-tryptophan is most effective when taken on an empty stomach along with some carbohydrate (such as a small glass of orange juice or a piece of fruit).

L-tryptophan is a building block for protein synthesis and is metabolized to important compounds such as niacin and picolinic acid. Consequently, if a person is deficient in tryptophan, then supplementing with L-tryptophan would provide a broader spectrum of benefits than would treatment with 5-HTP (Gaby 2010).

Serotonin syndrome: Too much of a good thing?—level IV evidence

Serotonin syndrome might result from combining 5-HTP with drugs that raise serotonin levels, such as Prozac, Paxil, Zoloft, MAOIs (such as Nardil, Parnate, and Selegiline), and others. In her report on serotonin syndrome, Prator (2006) identified medications that may act as precipitating agents for the syndrome. These include MAOIs, SSRIs, weight-loss medications, and over-the-counter drugs, such as St John's Wort. She also references a diagnostic tool known as the Hunter Serotonin Toxicity Criteria (Prator 2006; Dunkley 2003).

Symptoms of serotonin syndrome are described as confusion, agitation, anxiety, hypomania, or, in severe cases, coma. Autonomic symptoms may include profuse sweating and hyperthermia. Muscle rigidity is another symptom of the syndrome (Young 2005; Litzinger 2008).

Safety of giving 5-HTP to children and pregnant/nursing women or individuals with liver and kidney disease has not been established. Individuals with Parkinson's disease should note that 5-HTP taken with the medication carbidopa might cause skin changes similar to those that develop in scleroderma. Kay reports the eosinophilia-myalgia syndrome was linked in the past to the contamination of a tryptophan supplement during the fermentation process. In another process, 5-HTP is extracted from the seed of the African *griffonia simplicifolia* plant, so it is unlikely the contaminant is present from these sources (Kay 2004).

Other

Vegetarian diets and mood states—level II evidence

Vegetarians often omit fish along with other animal proteins, and therefore lower their intake of omega-3 essential fatty acids EPA and DHA, while generally having a high intake of the plant-derived omega-6 fatty acid linoleic acid (LA). An average, omega-6:omega-3 ratio of 18.6 for vegans compared to a ratio of 9.9 for omnivores was reported by Kornsteiner et al. (2008).

A study of diet and self-reported moods compared those who follow a vegetarian diet with omnivores. Sixty vegetarians and 78 omnivores reported (1) dietary intake by Food Frequency Questionnaire (FFQ), (2) self-administered Profile of Mood States (POMS), and (3) Depression Anxiety Stress Scale (DASS) (Table 7.5).

Investigators concluded that participants with low intakes of EPA, DHA, and AA and high intakes of ALA and LA had more favorable mood states despite negligible sources of long-chain omega-3 fatty acids. They comment that high LA and ALA inhibit desaturase enzymes; as LA rises, AA does not. They note Sanders' (2009) conclusion that plasma ratios of EPA and DHA may be adequate as long as a high intake of ALA lowers the LA:ALA ratio, which in turn regulates conversion of LA to AA (Beezhold 2010).

Dieting and depression: 5-HTP supplements—level III evidence

Depression is often linked to overeating and lack of success in weight loss. Kay reported the typical dose of 5-HTP for combating depression was 300 to 900 mg/day, usually taken in two to three doses throughout the day. A double-blind crossover study found that 8 mg 5-HTP/kg of body weight

Table 7.5 Comparison of Vegetarians and Omnivores

	Vegetarians	Omnivores	
	Mean Calculated Fatty Acids Intake (g)	Mean Calculated Fatty Acids Intake (g)	
Intake EFA	0.005	0.093	
Intake DHA	0.015	0.162	
Intake AA	0.011	0.086	
ALA	2.86	1.48	
LA	14.83	9.16	
	Mean Mood Scores	Mean Mood Scores	Normative Scores
POMS (score represents number of negative mood states)	0.10	15.33	14.8–20.3
DASS (score represents level of stress/anxiety)	8.32	17.52	18.38

Source: Kornsteiner et al. 2008.

for 5 weeks led to a loss of more than 3 lb, with no conscious effort to eat less. The subjects reported a greater sense of satiety after eating. Other studies reported use of 750 and 900 mg 5-HTP/day. If nausea occurs, it is advisable to start at a low dose and gradually increase to desired intakes (Kay 2004).

For an individual weighing 150 lb, (8 mg × 68 kg), this would equal 545 mg of 5-HTP/day.

Depression and folic acid fortification—level II evidence

Ramos and colleagues (2004) investigated whether, in this era of folic acid fortification, low folate status was a determinant of depressive symptoms in a cohort of Latino participants over 60 years of age. Folate deficiency was defined as plasma folate ≤6.8 nmol/L. (Reference range: 7.0 to 39.7 nmol/L or 3.1 to 17.5 ng/ml.) Conclusions from the researchers' data indicate that, despite folic acid fortification, low folate status is associated with depressive symptoms in elderly Latina women (but not in elderly Latino men). Women in the lowest third of folate levels were twice as likely to score as depressed than were women in the highest third of folate levels.

Depression and anxiety in adolescents with diabetes—level II evidence

An 8-year longitudinal cohort study of adolescents who had diabetes found behavioral problems were significantly related to higher mean levels of glycosated hemoglobin (Hb_{A1c}). As a direct consequence of diabetes, one patient died and two patients became cognitively impaired. Especially in female patients, longitudinal Hb_{A1c} data did improve with age. Interestingly, there was a trend for emotional problems, such as anxiety and depression, to be associated with lower glycemic levels, which suggested that anxious children monitor more diligently and may take more effective action in response to signs of poor blood glucose control. Low self-esteem was associated with poor glycemic control, but this did not seem to mediate the relation between psychological problems and poor outcome. Recurrent hospital admissions for diabetic ketoacidosis were the only predictor of the Global Severity Index of Mental State score at follow-up. This suggests that when diabetes is significantly out of control, it raises the risk of psychological morbidity. None of the baseline variables at year one were related to recurrent hospital admissions for diabetic ketoacidosis (Bryden 2001).

The Hb_{A1c} test is a diagnostic test for monitoring diabetic therapy. The test measures a form of hemoglobin that has incorporated a glucose molecule. It indicates the average blood glucose level over the 4 to 6 weeks preceding the test.

Hb_{A1c} of 5 = blood glucose levels of 100 mg/dl or ~5.5 mmol/L; N = 4–6%.

Depression and osteoporosis—level III evidence

In a review of NIMH-funded published research, Cizza and associates (2001) found a strong association between depression and osteoporosis. Symptoms of depression included loss of interest or pleasure in activities that were once enjoyed, including sex; fatigue and decreased energy; difficulty concentrating, remembering, and making decisions; insomnia, early-morning awakening or oversleeping; appetite and weight loss or overeating and weight gain; thoughts of death or suicide; suicide attempts; restlessness and irritability; and persistent symptoms that do not respond to treatment, such as headaches, digestive disorders, and chronic pain.

Bone density at the lumbar spine was 15% lower in 80 men and women over 40 years of age with major depression compared to 57 men and women who were not depressed. Factors such as smoking, a history of excessive or inadequate exercise, or estrogen treatment did not affect the results, implying that depression, per se, had an effect on bone mass. Researchers concluded that a clinical evaluation of subjects with unexplained bone loss, especially in premenopausal women and young or middle-aged men, should include an assessment for depression.

Depression and eating patterns (Healthy Eating Index scores)—level II evidence

In a study of 1118 adults, scores on the Center for Epidemiologic Studies Depression (CES-D) scale were related to scores of diet quality as measured by the Healthy Eating Index (HEI)-2005. HEI-2005 is a revised version of the original Healthy Eating Index that was created by the Center for Nutrition Policy and Promotion in 1995.

The mean depression score was 11.64 out of a possible 40. Women had higher scores for depression than men, regardless of race. In this sample, the percentage at risk for depression was 19% for white men and 35% for white women. The mean score for diet quality was 51.17 out of a possible 100. The HEI-2005 score of typical diets in a U.S. population is approximately 58.

Diet quality was inversely and significantly associated with reported symptoms of depression. Investigators commented it is unclear whether the association is a reflection of the effects of diet on depression or the effects of depression on diet (Kuczmarski 2010).

Traditional diets vs. Western diets associated with depression and anxiety—level II evidence

Ten years of psychiatric and diet data showed that a traditional dietary pattern in Australia was associated with lower odds for major depression, dysthymia, and anxiety disorders. There was a positive association between eating a Western dietary pattern and odds for major depression or dysthymia. Unexpectedly, the modern dietary pattern was associated with a higher rate of depression. The researchers noted that the absolute amount of unhealthy food, rather than the proportion of overall diet, may most relevant to mental health (Jacka 2010).

Dietary Pattern	Foods in Each Pattern
Traditional	Vegetables, fruit, beef, lamb, fish, whole-grain foods
Western	Meats, pies, processed meats, pizza, chips, hamburgers, white bread, sugar, flavored milk drinks, beer
Modern	Fruits, salads, fish, tofu, beans, nuts, yogurt, red wine

Bipolar disorder

Introduction

Bipolar disorder (BD), or manic-depressive illness, is one of the most common, severe, and persistent mental illnesses. BD is characterized by periods of deep, prolonged, and profound depression that alternates with periods of an excessively elevated or irritable mood known as mania. Between these highs and lows, patients usually experience periods of high functionality and can lead productive lives. Patients with BD typically experience recurrent symptoms for many years. The median time to recurrence of an episode was 87 weeks, with 24% of patients experiencing a recurring episode in 6 months, 36% having an episode in 1 year, and 61% in four years (Baldassano 2009).

BD is 80 to 90% heritable (McMahon 2008). Twin, family, and adoption studies all indicate that BD has a genetic component. First-degree relatives of a person with BD are approximately seven times more likely to develop BD than the rest of the population. The age of onset of BD varies greatly, from childhood to 50 years, with a mean age of approximately 21 years. Most commence between the ages of 15 and 24 years. Some patients diagnosed with recurrent major depression may have bipolar disorder and go on to develop their first manic episode when older than 50 years. However, for most patients, the onset of mania in people older than 50 years should lead to an investigation for medical or neurologic disorders such as cerebrovascular disease (Soreff 2008).

Individuals with BD often experience a variety of other difficulties, including impulsivity, risky behavior (e.g., alcohol abuse, sexual indiscretion, excessive spending), and interpersonal problems. The clinical course is primarily depressive rather than manic. In one longitudinal study, patients were symptomatically ill 47% of the time with depressive (68%), manic (19%), and mixed (13%) symptoms (Hirschfeld 2009). Hypomania may be thought of as a less severe form of mania that does not include psychotic symptoms or lead to major impairment of social or occupational function.

During a depressed phase of BD, patients have a very high rate of suicide and suicide attempts. Approximately 25 to 50% of individuals with BD attempt suicide, and 11% are successful (Soreff 2008). Cardiovascular disease is responsible for many premature deaths in patients with BD, but the highest standardized mortality rate in patients with BD is for suicide. Only lithium has shown efficacy in the long-term prevention of suicide.

Bipolar disorder and comorbidities

Studies have shown that patients with BD are likely to have comorbid psychiatric conditions, which include anxiety, impulse control problems, attention deficit hyperactivity disorder, personality disorder, and eating disorders. The prevalence of comorbid substance abuse is as high as 61%, which is greater than the prevalence of comorbid substance abuse seen with schizophrenia, panic disorder, dysthymia, and unipolar depression.

Common nutrition-related medical comorbidities include alterations in carbohydrate metabolism, lipid metabolism, and weight gain.

The International Society for Bipolar Disorders (ISBD) published consensus guidelines for monitoring patients with BD that includes BMI, waist circumference, glucose levels, and a lipid profile as well as many other aspects of care for these patients (Ng 2009).

Bipolar disorder and nutrition-related behavior—level II evidence
A survey of veterans with a diagnosis of (1) BD (N = 2032), (2) schizophrenia (N = 1895), or (3) no serious mental illness (N = 3065) found that individuals with BD were more likely to report suboptimal eating behavior, including eating fewer than two meals a day and having difficulty obtaining or cooking food. This group was more likely to have gained 10 lb or more in the past 6 months, to have poor exercise habits, and the least likely to report that their health care provider had discussed their eating habits or exercise habits (Kilbourne 2007).

The prevalence of migraine headaches, diabetes, and cardiovascular disease is higher in patients with BD than in the general population. Overweight status in patients with BD is related to a variety of factors, including sedentary lifestyle, poor eating habits, and weight gain associated with psychotropic medications. Similar to other individuals with abdominal obesity, patients with BD are at increased risk for developing metabolic syndrome. In a study of 171 patients with BD, 74% were overweight or obese, and 30% met the criteria for metabolic syndrome (Baldassono 2008).

Genetics

The strongest genetic associations detected in genes of individuals with BD were in biochemical pathways regulated by lithium. The strongest association has been observed within the first intron of diacylglycerol kinase eta (DGKH) gene. DGKH is a key protein in the lithium-sensitive phosphatidyl inositol pathway.

Levels of expression of other genes (oligodendrocyte-myelin genes) appear to be decreased in brain tissue from persons with BD. Oligodendrocytes produce myelin membranes that wrap around and insulate axons to permit the efficient conduction of nerve impulses in the brain. Therefore, loss of myelin is thought to disrupt communication between neurons, leading to some of the thought disturbances observed in BD and related illnesses. Brain imaging studies of persons with BD also show abnormal myelination in several brain regions. Gene expression and neuroimaging studies of persons with schizophrenia and major depression also demonstrate similar findings, indicating that mood disorders and schizophrenia may share some biological underpinnings (Soreff 2008).

Lipids

Omega-3 fatty acids and psychiatry—level III evidence

A review of evidence resulted in a paper approved by three committees of the American Psychiatric Association (the Committee on Research on Psychiatric Treatments, the Council on Research, and the Joint Reference Committee). The reviewers concluded that, although mixed, evidence supported a potential protective effect of omega-3 EFA intake, particularly EPA and DHA, in unipolar and bipolar disorders. There was less evidence of benefit for schizophrenia. Treatment with EPA and DHA were thought to present negligible risks (Freeman 2006).

Bipolar disorder and DHA supplement—level I evidence

A double-blind study of 30 patients with bipolar disorder by Stoll et al. (1999) found DHA levels were significantly lower in severely depressed patients. Provision of a DHA supplement served as an effective mood stabilizer. Study patients consumed 9.6 g/day omega-3 fatty acids (6.2 g/day EPA; 3.4 g/day DHA) for four months. A high ratio of EPA:DHA was desirable (a 2–3:1 ratio is often reported). The dose of omega-3 fatty acid per day must be calculated based on the concentration of omega-3 fatty acid on the supplement label. It is difficult to eat enough dietary fish to get this dose. The placebo group was given olive oil. Stoll et al. (1999) concluded that a deficiency of omega-3 fatty acids may be related to depression and result in cognitive deficit.

DHA can be added to an ongoing present treatment program; there are no known interactions with psychotropic drugs. Stoll reports that dosage can be split between morning and night or given all at night. He advised that the supplement may yield some fishy taste and noted that taking the fish oil with orange juice can reduce the taste. Stoll further noted that omega-3 fatty acids should not be used if a person is taking any type of blood thinners such as high doses of warfarin or aspirin. He also noted that using cod liver or other fish liver oils to achieve high omega-3 doses could result in vitamin A toxicity (Stoll 1999).

Amines

Choline and rapid-cycling bipolar disorder—level I evidence

Some studies have reported dysfunction in high-energy phosphate metabolism in patients with BD. It is possible mitochondrial dysfunction in BD results in not meeting the need for ATP. Since phospholipid synthesis for membrane integrity in the brain consumes 10 to 15% of the total ATP pool, an increased availability of choline may support increased ATP utilization.

Eight lithium-treated patients were assessed using MRI and three professional rating scales at baseline, 2, 3, 5, 8, 10, and 12 weeks of treatment along with a 50-mg choline bitartrate supplement. The supplemented patients had a significant decrease in brain purine levels, which the authors related to the anti-manic effects of adjuvant choline (Lyoo 2003).

The average intake (AI) of choline for adults is 400 to 550 mg/day; the upper limit (UL) is 3000 to 3500 mg/day. It is estimated that on a freely chosen diet, persons consume a total of ~1000 mg/day (Zeisel 2006).

Vitamins

Folic acid–sensitive genetic sites—level II evidence

Folate-sensitive fragile genetic sites were investigated in 40 patients diagnosed with BD. In comparison with a control group, the rate of fragile sites was considerably higher in the BD group ($p < 0.001$) (Demirhan 2009). Patients and relatives of those with BD carrying specific genetic alterations in methylenetetrahydrofolate reductase (MTHR) were shown by Ozbek et al. (2009) to have elevated tHcy, low folate, and vitamin B_{12} compared to levels found in a control group.

Minerals

Magnesium/lithium binding sites—level II evidence

Cellular targets for lithium action involve magnesium-activated enzymes, which are inhibited by lithium. Competition for binding sites has been suggested as a mechanism for the action of lithium (Mota de Freitas 2006).

Supplements

Micronutrient supplementation: A case report—level IV evidence

Frazier et al. (2009) reported a case involving a boy diagnosed with BD Not Otherwise Specified (NOS) at age 6. His diagnosis evolved into obsessive compulsive disorder by age 10 and BD-I with psychotic features by age 12. He received conventional treatment from ages 6 to 12 years. Because vitamins and minerals are important for general physical and mental health, he was then treated for 14 months with a vitamin-mineral supplement, which provided 16 minerals, 4 vitamins, 3 amino acids, and 3 antioxidants, resulting in a *superior outcome*. This included cessation of hallucinations and night terrors, decreased compulsions, sleeping through the night, ability to attend and function in public school, and normalized bowel movements and skin texture, all the while decreasing and ultimately ceasing all psychotropic medications.

ABAB treatment with micronutrient formula—level IV evidence

Rucklidge and Harrison (2010) reported a case of a 21-year-old woman with an 8-year history of psychiatric symptoms who used micronutrient supplements with 1 year of follow-up. She was treated for 8 weeks with a micronutrient formula. Improvements of symptoms and blood levels were normal after treatment and for 8 weeks afterward. When she discontinued the formula, scores on depression, anxiety, and ADHD assessments worsened and the formula was re-introduced. One year later, the patient was in remission from all psychiatric symptoms with no side effects.

Self-reported multivitamin/mineral supplementation—level V evidence

Adults who purchased a multivitamin/mineral supplement online and reported a diagnosis of BD also self-reported change in symptoms through an online survey at baseline, 3 months, and 6 months. The 16 symptoms were those specified in the DSM-IV related to mood. Of 358 individuals whose records were analyzed, 81% reported taking common psychiatric medications. Over the course of the data collection, reported symptom severity decreased and medication use decreased. It was noted that nutrients have been reported to boost medication response. A similar report of children with BD was reported separately.

Although interesting, voluntary self-reported data on diagnosis, medications, and symptoms are often criticized as biased and potentially unreliable. The authors who analyzed the data were not involved in any way with the sale of the product or in contact with the patients whose reports were analyzed. They suggested that objective research regarding micronutrient supplementation for those with BD appears to be warranted (Gately 2009; Rucklidge 2010).

Other

Inositol and bipolar disorder

Inositol, an isomer of glucose found in all living cells, is involved in intracellular signaling, is part of phosphatidyl inositol, and is linked to various neurotransmitter receptors. It has been hypothesized that lithium lowers myo-inositol in several areas of the brain. MRI assessment has shown that patients in manic or depressed states have abnormalities in myo-inositol concentrations, in contrast to patients who are euthymic and healthy controls. Inositol crosses the blood–brain barrier in pharmacological doses.

Inositol supplementation with medication—level II evidence

A small study supplemented usual medications with 12 g inositol for 6 weeks, using D-glucose as a placebo. Six of the 12 of those receiving inositol showed significant improvement in ratings of depression and clinical improvement scales, with no side effects (Chengappa 2000).

Low-inositol diet for low-response patients—level II evidence
A trial of an inositol-deficient diet with 15 patients who had a low response to lithium or valproate concluded the diet had a major effect in reducing the severity of affective disorder symptoms in the first 7 to 14 days of treatment. Lithium inhibits inositol monophosphate, which decreases ability of neurons to generate intermediate compounds. The diet provides ~1 g inositol/day and ~4 g/day are synthesized from glucose. Dietary inositol is not affected by lithium intake, but could account for variability in response to treatment. Restriction of dietary inositol was hypothesized to potentially enhance the effect of lithium (Shaldubina 2006).

*Inositol in bipolar disorder patient with psoriasis:
A case study—level IV evidence*
Kontoangeles et al. (2010) reported a case study of a 62-year-old woman who discontinued lithium treatment due to exacerbation of psoriasis while taking lithium. During 4 years of follow-up, she took 3 g/day of inositol alone. Her mood stabilized and there was remarkable improvement of psoriasis (Kontoangelos 2010).

Myo-inositol in food—level V evidence
In 1980, Clements and Darnell (1980) reported on gas-liquid chromatography analysis for myo-inositol content of 487 foods. Myo-inositol is high in mammalian nerve cells, and it has been hypothesized that increased dietary inositol would improve nerve function in patients with diabetes and those on kidney dialysis. They reported that in a 2500-calorie American diet there was ~900 mg of myo-inositol; 56% was lipid bound. They found it possible to alter the diet to provide from 225 to 1500 mg per day per 1800 calories. Foods high in myo-inositol are fresh fruit (especially cantaloupe and citrus, except lemon), beans, grain (especially oats and bran), and nuts. Leafy vegetables, dairy foods, and meats are low in comparison.

Complementary and alternative treatments—level III evidence
A review of the evidence by Andreescuab et al. (2010) concluded that St. John's Wort and SAMe (S-adenosyl-L-methionine) both have the potential to induce mania and may interact with other medications. Better studies are needed and possible risks understood before these substances are recommended as treatment.

Bipolar disorder and metabolic syndrome—level III evidence
Because premature morbidity and mortality have been observed in persons with BD, a literature review by Taylor and MacQueen (2006) was conducted

for evidence of dysregulation of physiologic systems. Ninety-seven studies met the study criteria and illustrated that BD and metabolic syndrome both exhibit dysregulation in (1) hormonal, (2) immunologic, and (3) autonomic nervous systems. Lifestyle choices and other conditions may predispose individuals for increased risk of diabetes and vascular disease.

Obesity as bipolar disorder?—level II evidence
A study of 400 consecutive individuals who enrolled in a weight management program assessed them for mood, eating, and other disorders. Twenty-nine percent met the criteria for BD. The prevalence of these disorders in the general population is 2 to 4%. The author concluded that perhaps obesity should be thought of as a psychiatric disorder, and enrollees in weight-management programs should be assessed for issues that underlie the weight problems (Kotwal 2005).

Weight gain and lithium—level III evidence
A review of factors affecting those with BD includes (1) the insulin-like effect of lithium, which affects carbohydrate metabolism and stimulates appetite, and (2) the enzyme AMP-protein kinase, which is suspected of mediating weight gain. Sleep disruption, a common occurrence in BD, is associated with altered levels of leptin and ghrelin. Leptin decreases appetite and increases energy expenditure, while ghrelin increases appetite and decreases energy expenditure, either of which can occur in BD. Calories from alcohol, foods high in sucrose, and sweetened beverages may also be a factor in weight changes. These and other lifestyle choices should be included in an assessment of individuals seeking assistance with BD (Simmons-Alling 2008).

Suicide

Introduction

Almost 1 million people worldwide die from suicide annually. In the last 45 years, suicide rates have increased by 60% worldwide. These figures do not include suicide attempts, which are up to 20 times more frequent than completed suicide (http://www.who.int/mental_health/prevention/suicide/suicideprevent/en/).

Traditionally suicide rates have been highest among elderly males, but rates among young people have been increasing and are now among the three leading causes of death among those aged 15 to 44 years in some countries, and the second leading cause of death in the 10 to 24 year age group. The Centers for Disease Control (CDC) reports that in the United States, suicide was the 11th most frequent cause of death in persons age 10 years and older in 2006 (CDC 2010).

Mental disorders (particularly depression and alcohol use disorders) are a major risk factor for suicide in Europe and North America. In Asian countries, impulsiveness plays an important role. Suicide is a complex phenomenon with psychological, social, biological, cultural, and environmental factors involved (WHO http://www.who.int/mental_health/prevention/suicide/suicideprevent/en/).

Diet and mental health in the arctic—level IV evidence

Suicide rates in circumpolar people are among the highest in the world. A hypothesis linking diet and mental health is based on a review of changes in the life circumstances of these people. Chronic diseases such as obesity, diabetes, and cardiovascular disease are on the rise. Rates of seasonal affective disorder (SAD) and anxiety have increased in non-isolated populations. In addition to change in cultural and social circumstances as well as climate/latitude factors, McGrath-Hanna (2003) makes a case for disappearance of the traditional diet as a factor in the mental health of these populations.

Lipids

Cholesterol and suicide in anorexia nervosa

See Chapter 9.

Cholesterol and suicide: A prospective study—level II evidence

Seventy-four adults who sought inpatient or outpatient psychiatric treatment were followed for five years or longer tracking cholesterol levels and suicide or suicide attempts. A plasma cholesterol level of ≤190 mg/dl was defined as the "low cholesterol" group. The only statistically significant difference in the groups was age; the high cholesterol group also had more individuals with hypertension and hypothyroidism. Analysis of data did not support the predicted relationship between lower cholesterol levels and the likelihood of later suicide attempts. The high cholesterol group, if below the median age of 32 years, experienced more severe suicide attempts during the study, which was not explained by the study. Patients with serious suicide attempts had a mean cholesterol level of 213 mg/dl. Other participants had a mean cholesterol level of 209 mg/dl. The 20 participants with severe suicide attempts before joining the study had a mean cholesterol level of 187 mg/dl, qualifying for the "low cholesterol" group (Fiedorowicz 2007).

HDL and suicide attempt in healthy women—level II evidence

A sample of data from the Third National Health and Nutrition Examination Survey (NHANES) was assessed for total cholesterol, HDL, and LDL. For healthy women, an HDL ≤40 mg/dl was associated with increased prevalence of suicide attempts (odds risk = 2.93), although HDL level was not associated with suicidal ideation (Zhang 2005).

Leptin, cholesterol, and suicide attempters—level II evidence
In two studies, Atmaca et al. (2007, 2008) report on cholesterol and leptin levels in violent and nonviolent suicide attempters.

> The patients with schizophrenia had lower serum total cholesterol and leptin levels compared with the controls. Significantly lower total cholesterol and leptin levels were observed in patients who had attempted suicide compared with those who had not. The levels were observed to be low in violent attempters when compared with nonviolent attempters.

The researchers suggest that low serum cholesterol and leptin levels are related to suicidality and violence, two dimensions of suicide attempts (Atmaca et al. 2007, 2008).

Postmortem fatty acids in brain following suicide—level II evidence
Lalovic et al. (2007) reported on 49 fatty acids observed in autopsies of brains of 16 individuals with depression who completed suicide, 23 individuals without depression who completed suicide, and 19 control subjects. The values for all three groups fell into the range of those previously reported for normal subjects. Future studies of enzymes related to lipid metabolism in the brain and carrier proteins in the CNS and brain might give insight into this area. In addition, it was found that when classified into violent or nonviolent suicides the frontal cortex of those who died through violent suicide had lower gray-matter cholesterol than those through nonviolent suicides.

5-HIAA, cholesterol, and suicide attempts—level II evidence
An investigation regarding the link between serum cholesterol and CSF 5-HIAA (a serotonin metabolite) in humans revealed a significant association, which was not associated with violence/nonviolence, with suicide attempts or survival of attempts (Asellus 2010).

Review of evidence—level III evidence
Fiedorowicz and Haynes (2010), in a review of evidence in 2010, listed the psychiatric syndromes associated with low cholesterol as anorexia nervosa, bipolar disorder, borderline personality disorder, major depressive disorder, and seasonal affective disorder, along with behaviors of suicide, suicide attempts, and violence. The neurobiologic effects of

low cholesterol were summarized as (1) improving membrane stability, (2) reducing membrane permeability, (3) possible reduction of serotonin transporter activity, and (4) involvement in synapse formation and myelin production. Low cholesterol and altered metabolic function may increase impulsivity, which may be linked to violent behavior, including suicide.

The researchers concluded that (1) current evidence does not support considering low serum cholesterol a risk factor for suicide, (2) links between low cholesterol and suicidality could be correlational, not causal, and (3) studies of lipid-lowering drugs have not indicated increased depression and suicide in healthy populations. They suggest practitioners actively assess and monitor hyperlipidemia as well as changes in behavior and mental status after beginning a lipid-lowering agent (Fiedorowicz and Haynes 2010).*

Another study including patients with a history of suicide attempts, patients without a history of suicide attempts, and healthy controls found odds ratios of 1.8 mg/dL to 2.0 mg/dL for low cholesterol in all groups (Perez-Rodriquez 2008).

Other

Food insufficiency and suicidal ideation—level III evidence
Data from the National Health and Nutrition Examination Survey III (NHANES III) assessed depressive disorders and suicidal symptoms for 15- and 16-year-old adolescents. Adolescents were classified as "food insufficient" if a family respondent reported that the family sometimes or often did not have enough to eat. Almost 5% of 15- to 16-year-old adolescents reported that they had attempted suicide, and 38.8% reported at least one suicidal symptom. Food-insufficient adolescents were significantly more likely to have had dysthymia, thoughts of death, a desire to die, and to have attempted suicide. Alaimo (2002) concluded that there is a strong association between food insufficiency and depressive disorder and suicidal symptoms in U.S. adolescents. (See also Chapter 10—Quality of Life, Wellbeing, and Stress.)

Conclusions

1. Symptoms occurring in major depression, the depressive phases of bipolar disorder, atypical depression, and suicide may directly affect nutritional intake as well as indirectly affect food procurement and preparation.

---

* Reprinted with permission from "Cholesterol, mood, and vascular health: Untangling the relationship." *Current Psychiatry Online* July 2010; 9 (7). Published by Current Psychiatry.

2. Populations with higher intakes of omega-3 fatty acids have lower rates of depression and treatment with omega-3 fatty acids of individuals with depression may reduce symptoms.
3. Phospholipids are seen as the potential link between DNA and environmental factors such as diet and nutrient intake.
4. Amino acids tryptophan and tyrosine, as precursors of neurotransmitters, can affect the signals/communication within the central nervous system. Diet composition may affect the availability of these amino acids.
5. Vitamins (biotin, B_{12}, C, folic acid) and minerals (chromium, iron, lead, magnesium, selenium, and zinc) are known to affect those with mood disorders. Effects may be related to deficiency or excesses.
6. Factors such as vegetarian diets, food availability/sufficiency, and use of supplements can affect mood and mental status.
7. Assessment and discussion of eating habits and dietary intake of patients with mood disorders is likely to uncover factors that may be compromising physical and mental health.
8. Research has resulted in conflicting results regarding low serum cholesterol levels in those who are depressed, or who have attempted or completed violent or nonviolent suicide. Brain composition at autopsy reports that the frontal cortex in brains of violent, successful suicides contained lower gray matter cholesterol than those who completed nonviolent suicide.
9. The relationship of omega-3 and omega-6 fatty acids to mood may extend to nonclinical variation in affect and not only to overt psychopathology.

References

Alaimo, K, CM Olson, and EA Frongillo. Family food insufficiency, but not low family income, is positively associated with dysthymia and suicide symptoms in adolescents. *J Nutr* 2002; 132(4):719–725.

American Psychiatric Association. *Diagnostic and Statistical Manual of Mental Disorder-IV.*, 317–320. 4th ed. Washington DC: American Psychiatric Association, 1994.

Andreescuab, C, H Benoit, H Mulsantac, and J E Emanuela. Complementary and alternative medicine in the treatment of bipolar disorder – A review of the evidence. *Affect Disord* 2008; 110(1):16–26.

Asellus, P, P Nordstrom, and J Jokinen. Cholesterol and CSF 5-HIAA in attempted suicide. *J Affect Disord* 2010; 125(1–3):388–392.

Assies, J, F Pouwer, A Lok, et al. Plasma and erytrocyte fatty acid patterns in patients with recurrent depression: A matched case-control study. *PLoS One* 2010; 5(5):e10635. Accessed September 15, 2010.

Atmacaa, M, M Kuloglu, E Tezcan, and B Ustundag. Serum leptin and cholesterol values in violent and non-violent suicide attempters. *Psychiatry Res* 2008; 158(1):87–91. Epub 2007 Dec 26.

Atmacaa, M, M Kuloglu, E Tezcan, and B Ustundag. Serum leptin and cholesterol levels in schizophrenic patients with and without suicide attempts. *Acta Psychiatr Scand* 2003; 108(3):208–14.

Baldassano, CF. Promoting wellness in patients with bipolar disorder: Strategies to move beyond maintaining stability and minimizing adverse events in effective long-term management. *Maintaining wellness in patients with bipolar disorder: Moving beyond efficacy to effectiveness.* Meeting of the American Academy of Clinical Psychiatrists. 2009. http://www.currentpsychiatry.com/pdf/Supp/SupplCP1010_BD_2.pdf. Published October 2009.

Beezhold, BL, CS Johnston, and DR Daigle. Vegetarian diets associated with healthy mood states: a cross-sectional study in Seventh Day Adventist adults. *Nutr Journal* 2010; 9:26. http://nutrj.com/content/9/1/26.

Benton, D. Selenium intake, mood and other aspects of psychological functioning. *Nutr Neurosci* (2002) 5(6);363–374.

Bryden, KS, RC Peveler, A Stein, et al. Clinical and psychological course of diabetes from adolescence to young adulthood: A longitudinal cohort study. *Diabetes Care* 2001; 24(9):1536–1540.

Center for Disease Control. http://www.cdc.gov/violenceprevention/suicide/statistics/index.html accessed 7/30/10.

Cizza, G, P Gold, G Chrousos, and P Ravn. Depression: A major, unrecognized risk factor for osteoporosis? *Trends in Endocrin & Metab* 2001; 12(5):198–203.

Chengappa, KN, J Levine, S Gershon, et al. Inositol as an add-on treatment for bipolar depression. *Bipolar Disord* 2000; 2(1):47–55.

Clements, RS. Jr and B Darnell. Myo-inositol content of common foods: Development of a high-myo-inositol diet. *Am J Clin Nutr* 1980; 33:1954–1967.

Conklin, SM, SJ Manuck, JK. Yao, et al. Serum omega-3 fatty acids are associated with depressive symptoms and neuroticism. *Psychosomatic Med.* 2007; 69:932–934.

Coppen, A and C Bolander-Gouaille. Treatment of depression: Time to consider folic acid and vitamin B^{12}. *J Psychopharmacology* 2005; 19(1):59–65.

Davidson, JRT, K Abraham, KM Connor, and MN McLeod. Effectiveness of chromium in atypical depression: A placebo-controlled trial. *Biol Psychiatry* 2003; 53:261–264.

Demirhan, O, D Tastemir, and Y Sertdemir. The expression of folate sensitive fragile sites in patients with bipolar disorder. *Yonsei Med J* 2009; 50(1):137–141.

Docherty, JP, D Sack, M Roffman, et al. A double-blind, placebo-controlled, exploratory trial of chromium picolinate in atypical depression: Effect on carbohydrate craving. *J Psychiatric Prac* 2005; 11(5):302–314.

Dunkley, EJC, GK Isbister, D Sibbitt, et al. The Hunter Serotonin Toxicity Criteria: Simple and accurate diagnostic decision rules for serotonin toxicity. *QJM* 2003; 96:635–642.

Eastmond, DA, JT Macgregor, and RS Slesinski. Trivalent chromium: Assessing the genotoxic risk of an essential trace element and widely used human and animal nutritional supplement. *Crit Rev Toxicol* 2008; 38(3):173–190.

Edwards, R, M Peet, J Shay, and D Horrobin. Omega-3 polyunsaturated fatty acid levels in the diet and in red blood cell membranes of depressed patients. *J Affect Disorder* 1998; 48(2–3):149–155.

Evans-Olders, R, S Eintracht, and LJ Hoffer. Metabolic origin of hypovitaminosis C in acutely hospitalized patients. *Nutrition* 2010; 26(11–12):1070–1074.

Fiedorowicz, JG, and WH Coryell. Cholesterol and suicide attempts: A prospective study of depressed inpatients. *Psychiatry Res* 2007; 152(1):11–20.

Fiedorowicz, JG, and WG Haynes. Cholesterol, mood, and vascular health: Untangling the relationship. *Current Psychiatry Online* July 2010; 9 (7). http://www.currentpsychiatry.com/article_pages.asp?aid=8766

Frazier EA, MA Fristad, and LE Arnold. Multinutrient supplement as treatment: literature review and case report of a 12-year-old boy with bipolar disorder. *J Child Adol Psychopharmacol*. 2009; 19(4):453–60.

Freeman, M, JR Hibbeln, KL Wisner, et al. Omega-3 fatty acids: evidence basis for treatment and future research in psychiatry. *Clin Psychiatry* 2006; 67(12):1954–1967.

Gaby, AR. L-tryptophan is back. *Original Internist* 2010; 17(1):47(2).

Gan, R, S Eintracht, and LJ Hoffer. Vitamin C deficiency in a university teaching hospital. *Fed Amer Soc Exper Biol J* 2007; 21:A104.

Ganji, V, C Milone, MM Cody, et al. Serum vitamin D concentrations are related to depression in young adult U.S. population: The Third National Health and Nutrition Examination survey. *Int'l Arch Med* 2010; 3:29.

Gately, D and BJ Kaplan. Database analysis of adults with bipolar disorder consuming a miocronutrient formula. *Clin Med Psychiatry* 2009; 2:3–16.

Ginsberg, Lawrence D. 2010 meeting of the Amer Psychiatric Assoc. http://medscape.com/viewarticle/722744. (accessed 1/10/12)

Gosney, MA, MF Hammond, A Shenkin, and S Allsup. Effect of micronutrient supplementation on mood in nursing home residents. *Gerontology* 2008; 54(5):292–299.

Haller, J. Vitamins and Brain Function. *Nutritional Neuroscience*. ed. HR Lieberman, RB Kanarek, and C Prasad, 211. CRC Press: Taylor & Francis Group, Boca Raton, FL 2005.

Hirschfeld, R. Making efficacious choices: The integration of pharmacotherapy and nonpharmacologic approaches to the treatment of patients with bipolar disorder. *Maintaining wellness in patients with bipolar disorder: Moving beyond efficacy to effectiveness*. Meeting of the Amer Acad Clin Psychiatrists. 2009. http://www.currentpsychiatry.com/pdf/Supp/SupplCP1010_BD_2.pdf. Published October 2009.

http://www.intarchmed.com/content/3/1/29

http://www.who.int/mental_health/prevention/suicide/suicideprevent/en/ accessed 8/5/11.

Hoogendijk, WJG, P Lips, MG Dik, et al. Low blood levels of vitamin D may be associated with depression in older adults. *Arch Gen Psychiatry* 2008; 65(5):508–512.

Hudson, C, SP Hudson, T Hecht, and J MacKenzie. Protein source tryptophan versus pharmaceutical grade tryptophan as an efficacious treatment for chronic insomnia. *Nutr Neuroscience* 2005; 8(2):121–127. http://report.nih.gov/NIHfactsheets/ViewFactSheet.aspx?csid=48. accessed 8/14/2011.

Jacka, Felice N, Julie A Pasco, et al. Association of western and traditional diets with depression and anxiety in women. *Am J Psychiatry* 2010; 167(3):305–311.

Jarvis, C. Mental Health Assessment. In *Physical Examination and Health Assessment*. 101–124. Philadelphia: W.B. Saunders Co. 1992.

Jazayeri S, M Tehrani-Doost, SA Keshavarz, et al. Comparison of therapeutic effects of omega-3 fatty acid eicosapentaenoic acid and fluoxetine, separately and in combination, in major depressive disorder. *Aust N Z J Psychiatry*, 2008; 42(3):192–198.

Jones, WO and BD Nidus. Biotin and hiccups in chronic dialysis patients. *J Renal Nutr* 1991; 1(2):80–83.

Kay, LK. Supplements for serotonin-related ailments. *Today's Dietitian* 2004; 6(7):44–45.

Kessler RC, PA Berglund, O Demler, et al. Lifetime prevalence and age-of-onset distributions of DSM-IV disorders in the National Comorbidity Survey Replication (NCS-R). *Arch Gen Psychiatry* 2005; 62(6):593–602. http://www.nimh.nih.gov/health/publications/the-numbers-count-mental-disorders-in-america/index.shtml

Kilbourne, AM, DL Rofey, JF McCarthy, et al. Nutrition and exercise behavior among patients with bipolar disorder. *Bipolar Disord* 2007; 9(5):443–452.

Kinsman, RA and J Hood. Some behavioral effects of ascorbic acid deficiency. *Am J Clin Nutr* 1971; 24:455–464.

Kontoangelos, K, N Valdakis, I Zervas, et al. Administration of inositol to a patient with bipolar disorder and psoriasis: A case report. *Cases Journal* 2010; 3(1):69.

Kornsteiner, M, I Singer, and I Elmadfa. Very low n-3 long-chain polyunsaturated fatty acids status in Austrian vegetarians and vegans. *Ann Nutr Metab* 2008; 52:37–47.

Kotwal, R. Patients in weight management clinics more likely to have bipolar disorder than general population. Sixth International Conference on Bipolar Disorder (ICBD) Pittsburgh. Abstract 125. Medscape, June 2005.

Kuczmarski, MF, AC Sees, L Hotchkiss, et al. Higher Healthy Eating Index-2005 scores associated with reduced symptoms of depression in an urban population: Findings from the Healthy Aging in Neighborhoods of Diversity Across the Life Span (HANDLS) Study. *J Amer Diet Assoc* 2010; 110(3):383–389.

Lalovic, A, E Levy, L Canetti, et al. Fatty acid composition in postmortem brains of people who completed suicide. *J Psychiatry Neurosci* 2007; 32(5):363–370.

Lalovic A, E Levy, G Luheshi, et al. Cholesterol content in brains of suicide completers. *Int J Neuropsychophamacol* 2007; 10(2):159–166.

Levenson, CW. Zinc: The new antidepressant? *Nutr Rev* 2006; 64(1):39–42.

Levina, A, and PA Lay. Chemical properties and toxicity of chromium (III) nutritional supplements. *Chem Res Toxicol* 2008; 21(3):563–571.

Litzinger, MHJ, J Takeshita, and M Litzinger. SSRIs and serotonin syndrome. *US Pharm.* 2008; 33(11):29–37. http://uspharmacist.com/content/d/featured_articles/c/11467/ accessed 10/19/2010.

Lyoo, IK, CM Demopulos, F Hirashima, et al. Oral choline decreases brain purine levels in lithium-treated subjects with rapid-cycling bipolar disorder: A double-blind trial using proton and lithium magnetic resonance spectroscopy. *Bipolar Disord* 2003; 5(4):300–306.

Ma, B, and M Taylor. Omega-3 fatty acids and depression in noninstitutionalized older women. *Top Clin Nutr* 2004; 19(2):117–129.

Maes, M, A Christopher, J Delanghe, et al. Lowered omega-3 polyunsaturated fatty acids in serum phospholipids and cholesterol esters of depressed patients. *Psychiatry Res* 1999; 85:275–291.

Maes, M, R Smith, A Christopher, et al. Lower serum high-density lipoprotein cholesterol (HDL-C) in major depression and in depressed men with serious suicidal attempts: Relationship with immune-inflammatory markers. *Acta Psychiatrica Scandanavica* 1997; 95(3):212–221.

Martins, J. EPA but not DHA appears to be responsible for the efficacy of omega-3 long chain polyunsaturated fatty acid supplementation in depression: evidence from a meta-analysis of randomized controlled trials. *J Am Coll Nutr*. 2009; 28(5):525–542.

McGrath-Hanna, NK, DM Greene, RJ Tavenier, and A Bult-Ito. Diet and mental health in the arctic: Is diet an important risk factor for mental health in circumpolar peoples? – A review. *International J Circumpolar Health* 2003; 62(3):228–241.

McLeod, MN and RN Golden. Chromium treatment of depression. *Int J Neuropsychopharmacology* 2000; 3:311–314.

McMahon, F. Genetics. Advances in Bipolar Disorder Detection and Management. *APA 2008: New Findings in Adult and Pediatric Bipolar Disorder*. Medscape Perspectives on the American Psychiatric Association 161st Annual APA Meeting. http://www.medscape.com/viewarticle/577369?src=mp&spon=17&uac=39259FT. Published July 15, 2008.

Morgan, RE, LA Palinkas, EL Barrett-Connor, and DL Wingard. Plasma cholesterol and depressive symptoms in older men. *Lancet* 1993; 341:75–79.

Morris, MS, M Fava, PF Jacques, et al. Depression and folate status in the U.S. population. *Psychotherapy and Psychosomatics* 2003; 72:80–87.

Mota de Freitas, D, MM Castro, and CF Geraldes. Is competition between Li+ and Mg2+ the underlying theme in the proposed mechanism for the pharmacological action of lithium salts in bipolar disorder? *Accounts Chem Res* 2006; 39(4):283–291.

Ng, F, OK Mammen, I Wilting, et al. The International Society for Bipolar Disorders (ISBD) consensus guidelines for the safety monitoring of bipolar disorder treatments. *Bipolar Disord* 2009; 11:559–595.

Owen, AJ, MJ Batterham, YC Probst, et al. Low plasma vitamin E levels in major depression: Diet or disease? *Eur J Clin Nutr* 2005; 59(2):304–306.

Ozbek, Z, CI Kuckkali, E Ozkok, et al. Effect of the methylenetetrahydrofolate reductase gene polymorphisms on homocysteine, folate and vitamin B^{12} in patients with bipolar disorder and relatives. *Prog Neuropsychopharmacol Biol Psychiatry* 2008; 32(5):1331–1337.

Peet, M. International variations in the outcome of schizophrenia and the prevalence of depression in relation to national dietary practices: An ecological analysis. *Br J Psychiatry* 2004; 184:404–408.

Peet, M and DF Horrobin. A dose-ranging study of the effects of ethyl-eicosapentaenoate in patients with ongoing depression despite apparently adequate treatment with standard drugs. *Arch Gen Psychiatry* 2002; 59:913–919.

Perez-Rodriquez, MM, E Baca-Garcia, C. Diaz-Sastre, et al. Low Serum cholesterol may be associated with suicide attempt history. *J Clin Psychiatry* 2008; 69(12):1920–1927.

Prator, BC. Serotonin syndrome. *J Neurosci Nurs* 2006; 38(2):102–105.

Ramos, MI, LH Allen, MN Haan, et al. Plasma folate concentrations are associated with depressive symptoms in elderly Latina women despite folic acid fortification. *Am J Clin Nutr* 2004; 80(4):1024–1028.

Rayman, M, A Thompson, M Warren-Perry, et al. Impact of selenium on mood and quality of life: A randomized, controlled trial. *Biol Psychiatry* 2006; 15(2):147–154.

Rogers, PJ, and HM Lloyd. Nutrition and mental performance. *Proc Nutr Soc, Inst Food Res in Great Britain* 1994; 53:443–456.

Rogers, PJ, KM Appleton, D Kessler, et al. No effect of n-3 long-chain polyunsaturated fatty acid (EPA and DHA) supplementation on depressed mood and cognitive function: A randomized controlled trial. *Br J Nutr* 2008; 99(2): 421–431.

Rucklidge, JJ, D Gately, and BJ Kaplan. Database analysis of children and adolescents with bipolar disorder consuming a micronutrient formula. *BMC Psychiatry* 2010; 10:74.

Rucklidge, JJ and R Harrison. Successful treatment of bipolar disorder II and ADHD with a micronutrient formula: A case study. *CNS Spectr* 2010; 15(5):289–295.

Rude, RK and ME Shils. Magnesium. In *Modern Nutrition in Health & Disease*. ed. M. Shils, M Shike, A C Ross, B Caballero, and RJ Cousins. 223–248. Lippincott Williams & Wilkins, New York 2005.

Roy-Byrne, P, DA Gorelick, and SR. Marder. Unusual dietary habits in a patient with schizotypal personality disorder: Interaction of nutritional status and psychopathology. *J Psychiatric Treatment and Eval* 1983; 5:67–69.

Rondanelli, M, A Giacosa, A Opizzi, et al. Effect of omega-3 fatty acid supplementation on depressive symptoms and on health-related quality of life in the treatment of elderly women with depression: A double-blind, placebo-controlled, randomized clinical trial. *J Amer College Nutr* 2010; 29(1):55–64.

Sachdev, PS, RA Parslow, O Lux, et al. Relationship of homocysteine, folic acid and vitamin B^{12} with depression in a middle-aged community sample. *Psychological Med* 2005; 35(4):529–538.

Sanders TA. DHA status of vegetarians. *Prostaglandins Leukot Essent Fatty Acids* 2009; 81:137–141.

Scheinfeld, NS and SB Freilich. *Biotin Deficiency*. http://emedicine.medscape.com/article/984803-overview accessed 1/27/2011.

Shaldubina, A, Z Stahl, M Furzpan, et al. Inositol deficiency diet and lithium effects. *Bipolar Disord* 2006; 8(2):152–159.

Shaw, K., J Turner, and C Del Mar. Tryptophan and 5-Hydroxytryptophan for depression. Article No.: CD003198. A *Cochrane Review* Abstract, 2003. http://www.cochrane.org/reviews/en/ab003198.html; reviewed 2008.

Smith, AD, Y-I Kim, and H Refsum. Is folic acid good for everyone? *Am J Clin Nutr* 2008; 87:517–33.

Smith, KA, C Williams, and PJ Cowen. Impaired regulation of brain serotonin function during dieting in women recovered from depression. *Br J Psychiatry* 2000; 176:72–75.

Soreff, S and LA McInnes. Bipolar affective disorder. http://emedicine.medscape.com/article/286342-overview. Updated: Sep 18, 2008.

Steegmans, PH., AW Hoes, AA Bak, et al. Higher prevalence of depressive symptoms in middle-aged men with low serum cholesterol levels. *Psychosomatic Med* 2000; 62:205–211.

Stoll, AL, WE Severus, MP Freeman, et al. Omega-3 fatty acids in bipolar disorder: A preliminary double-blind, placebo-controlled trial. *Arch Gen Psychiatry* 1999; 56(5):407–412.

Su, K-P. Biological Mechanism of antidepressant effect of omega-3 fatty acids: How does fish oil act as a "Mind-Body Interface", Table 1. Overlapping of symptoms of acute sickness behavioral associated with IFN-α therapy and the somatic symptoms in MDD. *Neurosignals* 2009; 17:144–152.

Taylor, MJ, S Carney, J Geddes, and G Goodwin. Folate for depressive disorders. Article. No:CD003390. DOI: 10.1002/14651858. *Cochrane Database System Review*. 2003; (2). www.cochrane.org/reviews/en/ab003390.html

Taylor, V and G MacQueen. Associations between bipolar disorder and metabolic syndrome: A review. *J Clin Psychiatry* 2006; 67(7):1034–41.

Tiemeier, H, HR van Tuijl, A Hofman, et al. Plasma fatty acid composition and depression are associated in the elderly: The Rotterdam Study. *Am J Clin Nutr* 2003; 78(1):40–46.

Tiemeier, H, HR van Tuijl, A Hofman, et al. Vitamin B^{12}, folate and homocysteine in depression: The Rotterdam Study. *Am J Psychiatry* 2002; 159(12):2099–2101.

Torres, SJ, C Nowson, and A Worsley. Dietary electrolytes are elated to mood. *Br J Nutr* 2008; 100:1038–1045.

Uranga, RM. and JN Keller. Diet and age interactions with regards to cholesterol regulation and brain pathogenesis. *Curr Gerontol Geriatr Res*. 2010; article 2196883.

World Health Organization. http://www.who.int/mental_health/prevention/suicide/suicideprevent/en/ accessed 8/15/11.

Wilson, JD. Vitamin deficiency and excess. *Harrison's Principles of Internal Medicine*, ed. KJ Isselbacher, E Braunwald, J. Wilson, JB Martin, AS Fauci, and DL Kasper. 472–480. McGraw-Hill Inc., San Francisco 1994.

Yatzidis, H, D Koutisicos, B Agroyannis, et al. Biotin in the management of uremic neurologic disorders. *Nephron* 1984; 36:183–186.

Young, SN. Amino acids, brain metabolism, mood and behavior. In *Nutr Neurosci*. 136–137. CRC Press. Tylor & Francis Group. New York. 2005.

Ziesel, SH and MD Niculescu. Choline and phosphatidylcholine. *Modern Nutrition and Health and Disease*. 10th edition. p. 526. ed. ME Shils, M Shike, AC Ross, et al. Lippincott, Williams & Wilkins. Philadelphia, 2006.

Zhang, J, RD McDeown, JR Hussey, et al. Low HDL cholesterol is associated with suicide attempt among young healthy women: The Third National Health and Nutrition Examination Survey. *J Affect Disord* 2005; 89(103):25–33.

Zhang M, L Robitaille, S Eintracht and J Hoffer. Vitamin C provision improves mood in acutely hospitalized patients. *Nutrition* 2011; 27(5):520–533.

chapter eight

Schizophrenia

Introduction

Schizophrenia is a disorder featuring psychotic symptoms such as delusions or hallucinations, with or without insight into their pathological nature. Symptoms may also include disorganized speech or catatonic behavior (DSM-IV 1994).

Schizophrenia occurs in approximately 1% of the population nationally and around the world. According to the National Institutes of Mental Health, approximately 2.4 million American adults have schizophrenia in a given year (http://report.nih.gov/NIHfactsheets). Schizophrenia is 80% heritable, occurring in 45 to 50% of monozygotic twins and 15 to 17% of dizygotic twins. Symptoms commonly begin to appear in males during their late teens and early twenties and appear in females in their twenties or early thirties.

Symptoms of schizophrenia may be categorized as positive, negative, or cognitive. Positive symptoms such as command auditory or visual hallucinations may tell the individual to do or not do something. The action may be trivial but also may be harmful to the individual experiencing the hallucination or to others. Positive symptoms may wax and wane over time and usually become less severe as people age.

Negative symptoms imply the absence of something. Negative symptoms may include a lack of social contacts, lack of initiative, socially awkward behavior, loss of energy or motivation, as well as decreased emotional displays of affect combined with discomfort when interacting with other people. Negative symptoms tend to become more prominent over time.

Cognitive symptoms of schizophrenia involve deficits in memory, decision-making, and problem solving, although these symptoms are not as severe as those seen in persons with Alzheimer's dementia. Together, negative and cognitive symptoms are associated with persistent disability commonly seen in individuals with schizophrenia (Coyle 2006; Lindenmayer 2006).

Cognitive domains include:
Executive domain (altered abstraction, mental flexibility)
Memory domain (verbal, spatial, facial)
Intellectual domain (language, spatial)

Niacin skin flush test for diagnosis of schizophrenia—level II evidence

A noninvasive niacin skin flush test used to indicate impaired arachidonic acid-related signal transduction in schizophrenia has been validated. During a 20-minute observation time, the response was significantly different in patients with schizophrenia at a $p < 0.00002$. Authors reported a sensitivity of 90% and a specificity of 75% (Puri 2001).

Nutrition-related comorbidities

Studies have shown that 40—90% of afflicted patients were not being treated for comorbidities of obesity, hypertension, hyperlipidemia, and diabetes (Smith 2005). These conditions associated with schizophrenia are related to quality of diet, specific nutrients and their metabolites, weight status, psychotropic medications, or metabolic factors linked to genetic factors. Also implicated are lifestyle choices such as smoking, exercise, and use of alcohol and drugs.

Patients with schizophrenia who smoke were found to have lower erythrocyte DHA and EPA compared with nonsmokers. Smoking status needs to be taken into account when studying essential fatty acids (EFA) in this population (Hibbeln 2003).

A retrospective analysis of inpatient and outpatient insurance claims data in Iowa showed that patients with schizophrenia (n = 1074) vs. controls (n = 726,262) were more likely than controls to have chronic conditions such as hypothyroidism, chronic obstructive pulmonary disease, diabetes with complications, hepatitis C, fluid/electrolyte disorders, and nicotine abuse (Carney 2006).

It has been noted that schizophrenia and obesity were correlated before the availability of atypical antipsychotic medications. One explanation for the increased risk of obesity, metabolic syndrome, and diabetes is a tendency to store fat in abdominal areas subcutaneously or in inner visceral locations. A computed tomography study showed that with equal total body fat and equal subcutaneous fat, there was three times the amount of visceral fat stored by patients with schizophrenia. In this respect, patients with schizophrenia who were not taking antipsychotic medications were no different from those who were taking antipsychotics (Meyer 2005). Ryan reports an increased incidence of impaired glucose tolerance in first episode, drug naïve patients diagnosed with schizophrenia (Ryan 2003).

It is recommended that if an individual gains 5% from baseline weight during routine monitoring, that individual should be assessed and referred to a weight management program (Meyer 2005). It has been demonstrated that programs of moderate exercise and behavior modification in food selection are successful with this population, and significant health gains can be achieved with 2 to 15 lb of weight loss.

> For a person weighing 130 lb (59 kg), 5% is 6.5 lb (2.95 kg).
> For a person weighing 200 lb (90.9 kg), 5% is 10 lb (4.55 kg).

Schizophrenia and oxidative stress

Neuronal membranes contain a high proportion of polyunsaturated fatty acid (PUFA) and are a site of oxidative stress. Production of reactive oxygen or decreased antioxidant protection occurs in schizophrenic patients. Findings suggest that multiple neurotransmitter systems may be faulty (Fendri 2006).

> Oxidative stress: an imbalance between the generation of reactive oxygen species and antioxidant defense capacity of the body.

Lipids

AA and DHA with different diets—level II evidence

Decreased EFA content has been observed in cell membranes of various tissues from patients with schizophrenia, including neural cell membranes. A comparison of red blood cell (RBC) PUFA in unmedicated patients and a control group on diets from two cultures (Malaysian and Indian) showed no difference in arachidonic acid (AA) between any of the groups. Levels of DHA were significantly increased in both cultural groups of unmedicated patients. Diet-related differences in DHA between populations from India and Malaysia were much greater than differences between patients with schizophrenia and controls. Researchers suggested that RBC levels of AA and DHA might be reflective of abnormal metabolism of phospholipids and fatty acids rather than directly related to schizophrenia (Peet 2004).

EPA = placebo—level II evidence

Eighty-seven patients with schizophrenia, while continuing on medications, were given 3 g/day of ethyl-EPA or a placebo for 16 weeks. Improvement was no greater in those treated with EPA than with subjects on medication treatment only (Fenton 2001).

Psychopathology and EPA/DHA, vitamins E and C—level II evidence

Thirty-three patients with schizophrenia treated with a mixture of EPA/DHA (180:120 mg), 400 IU vitamin E, and 500 mg vitamin C morning and evening for 4 months showed levels of EFA in red cell membranes were significantly increased while there was a significant decrease in psychopathology. Although biochemical levels returned to pretreatment levels after a 4-month washout period, the clinical improvement was significantly retained (Arvindakshan 2003).

Omega-3 fatty acids and antipsychotic properties: A Cochrane review—level III evidence

A review of five short, small studies suggested that enriched omega-3 oil might have some antipsychotic properties. Omega-3 supplementation of medicated and nonmedicated patients showed an improved mental state. Individuals already on antipsychotic medication had a greater improvement. Some studies showed no clear effects and there was not a clear dose response (Joy 2003).

Omega-3 fatty acids decreased progression to psychosis—level I evidence

Patients at high risk for developing psychosis (mild, transient, or family history of psychosis) were assigned either to a placebo group that received coconut oil or to a group that received 700 mg EPA and 480 mg DHA for 12 weeks. Twenty-seven percent of the placebo group had progressed to experiencing psychosis; 4.9% of the treatment group progressed to psychosis (Amminger 2010).

Remission of symptoms with EPA: Case study—level IV evidence

A closely observed case study was reported of an individual diagnosed with schizophrenia 3 years previously, but who had remained unmedicated, although receiving regular care along with 2 g/day of EPA. His negative and positive symptoms decreased in 6 months and continued in remission with a regular schedule of psychiatric care before, during, and following the start of EPA. Positive symptom scores went from 46 to 7; negative symptom scores went from 16 to 3 (Puri 1998). Reduced turnover of neuronal phospholipid was demonstrated by MRI spectroscopy (Puri 2000). Later observations by the authors postulated that outcomes may have been the result of the inhibition of a specific phospholipase enzyme (Richardson 2000).

Variation in study results—level IV evidence

Possible explanations for presently unexplainable results in studies of EPA and schizophrenia include (1) neuroleptic drugs may block the response to EPA, (2) dosage may be too high or too low, (3) patients may change their dietary intake of kind or amount of fats, and (4) the design of the study can influence detection of significant change (Horrobin 2003).

Phospholipids

All membranes are high in phospholipids. Nerve dendrites and synapses are 80% lipid by weight. All neurotransmitters have to cross those membranes during nerve signaling. These compounds link the EFAs from the environment (diet) and the genetic control of enzymes that control EFAs in the body. Growing data suggests that optimal control of schizophrenia, bipolar disorder, and severe depression lies in understanding phospholipid metabolism in neuronal membranes.

> Phospholipid refers to a fatty acid with a phosphorous-oxygen duo, plus other molecules, such as inositol, choline, or ethanolamine. Lecithin is one example of a phospholipid.

Unique genetic combinations result in removal of EPA, DHA, AA, and GLA from phospholipid structure. An intake ratio of EPA to DHA of 3:1 decreases the overactivity of one of the phospholipase enzymes that breaks down phospholipids structures (Jones 2001).

Amino acids and protein

In a case-control study, a majority of patients with schizophrenia had elevations of methionine and a smaller subgroup had elevated homocysteine. This could be related to folate levels and single-carbon metabolism (Regland 2005).

One-carbon metabolism and gluten sensitivity—level II evidence

Intolerance to the protein gluten, which manifests as celiac disease, was found to have approximately the same prevalence as schizophrenia. From a group of 1401 patients with schizophrenia, 23.1% had moderate to high levels of IgA-AGA, an antibody indicating sensitivity to gliadin, compared with 3.1% of the control group of 900 (Cascella 2011).

Vitamins

Folate and negative symptoms—level II evidence

A study of 91 outpatients with schizophrenia found that folate concentration was inversely related to the Scale for Assessment of Negative

Symptoms total score. Findings could reflect low dietary intake, cigarette smoking, alteration in neurotransmitter synthesis, or decreased enzyme activity (Goff 2005).

Low folate levels associated with fourfold to sevenfold risk of schizophrenia—level II evidence

In a comparison of 35 patients with schizophrenia to 104 unrelated controls, levels of vitamins B_6 and B_{12} and homocysteine (Hcy) did not differ between the groups. Those with folate below the tenth percentile of controls were associated with an approximate four- to sevenfold risk of having schizophrenia. Elevated Hcy levels and mutation of the gene 677C > T (the gene for MTHF reductase enzyme) were not associated with increased risk (Muntiewerff 2003).

Vitamin D and mental illness—level II evidence

Patients with depression, schizophrenia, or alcohol addiction (25 to 35 per group) have been compared with healthy controls for vitamin D status. Vitamin D levels were significantly lower in all three patient groups compared to the control group (Schneider 2000).

A greater than tenfold variation in prevalence of schizophrenia was found at different geographic sites, generally increasing with latitude. A meta-analysis of 49 studies indicated that environmental factors such as those contributing to high infant mortality and those associated with living at higher latitudes, both of which influence vitamin D level, seem related to the incidence of schizophrenia. Those who live at higher latitudes also have (1) low fish consumption (<5 kg/person/year), (2) dark skin, and (3) less access to prenatal and infant care (Kinney 2009). These findings were combined with observations of vulnerability to immune dysfunction beginning around puberty in individuals who develop schizophrenia. Authors propose these factors support a unifying hypothesis of schizophrenia, "which suggests theoretical bridges between different lines of evidence on schizophrenia and offers explanations for many puzzling findings about schizophrenia" (Kinney 2010). (See also Chapter 1.)

Minerals

Zinc and copper in schizophrenic males—level II evidence

In an investigation of zinc and copper nutrition, blood levels of hospitalized criminal and noncriminal individuals with schizophrenia were compared. Mean plasma zinc values were significantly lower in criminal subjects when compared to noncriminal subjects (Table 8.1) (Tokdemir 2003).

Table 8.1 Ratios of Zinc and Copper in Criminals with Schizophrenia and Noncriminals with Schizophrenia

	Mean Plasma Zn µ/dL	Mean Plasma Cu µ/dL
Criminals	68 ± 1.55	104 ± 1.8
Noncriminals	81 ± 2.73	93 ± 2.92

Source: Tokdemir et al. 2003.

Magnesium deficiency

Magnesium is necessary for (1) ATP metabolism, (2) activity of enzymes using vitamins B_1 and B_6, and (3) the biosynthesis of serotonin, GABA, and melatonin. Magnesium-deficient patients may have depression, agitation, confusion, and disorientation. Psychotic behavior has been observed in 50% of patients with hypomagnesemia, which also coincides with extrapyramidal symptoms.

Magnesium supplementation—level II evidence

In a group of 20 institutionalized individuals with chronic schizophrenia, five were found to be magnesium deficient without any other electrolyte abnormalities. A supplement of 600 mg magnesium per day as MgCl or a magnesium amino acid combination did not normalize magnesium levels in 3 of the 4 who took the supplement. Kanofsky (1991) questioned the absorption of magnesium in these patients. In healthy subjects, magnesium is 50 to 70% absorbed.

Minerals (Mg, Zn, Cu) and antipsychotic medications—level II evidence

Minerals in plasma and erythrocytes of newly hospitalized patients with schizophrenia (N = 56) were compared with a control group (N = 20). Zinc and magnesium levels differed in the two groups but copper levels did not differ. Treatment with haloperidol and risperidone improved erythrocyte magnesium levels and plasma zinc levels. The researchers concluded that (1) the ratio of plasma copper to erythrocyte magnesium and (2) the ratio of plasma copper to plasma zinc are two important markers of acute paranoid schizophrenia. They believe that increase of magnesium cell level is important for the antipsychotic action of haloperidol and risperidone (Nechifor 2004).

Supplements

Vitamin C, oxidative stress, and outcome of schizophrenia—level II evidence

In examining oxidative damage and change associated with drug treatment in a group of 48 patients with schizophrenia and a group of 40 healthy volunteers, serum superoxide dismutase (SOD) and malondialdehye (MDA) were higher in patients with schizophrenia, indicating oxidative stress. Plasma ascorbic acid was lower in patients with schizophrenia, indicating less capacity for protecting against oxidative stress. Ascorbic acid increased while SOD and MDA decreased with treatment using an atypical antipsychotic and supplemental ascorbic acid of 500 mg/day. Scores on the Brief Psychiatric Rating Scale also improved significantly ($p < 0.05$) (Dakhale 2004, 2005).

> Superoxide dismutase (SOD) acts as both an antioxidant and an anti-inflammatory agent.
> Serum malondialdehye (MDA) is a marker for decomposition of unsaturated lipids as a by-product of AA metabolism.

Nutritional status and response to supplements—level II evidence

In a study of 61 patients with schizophrenia, patients were found to have high red cell saturated fatty acids and low monounsaturated fatty acids and PUFA. Those that did not take supplements had low vitamin B_{12} and high Hcy. Two patients exhibited EFA deficiency, and seven showed signs of marginal DHA deficiency.

It was interesting that none of the nutritionally deficient patients were predicted by clinicians to have poor diets. The authors, however, felt that since supplements of B vitamins, soybean oil, and fish oils corrected the low biochemical levels that this confirmed that the likely cause of abnormal blood levels was inadequate dietary intake (Kemperman 2006).

Megavitamin therapy—level II evidence

In 1999, a trial of megavitamin therapy was conducted with 22 schizophrenic patients in Australia. An established orthomolecular psychiatrist known to the researchers prescribed a regimen of 18 tablets of vitamins (including vitamins A, B_6, B_{12}, C, and E; thiamin; riboflavin; niacin; and folic acid) based on blood levels of vitamins and allergy testing. A control group (n = 7) was compared with the experimental group during the 5-month trial. The control group regimen included 17 placebo pills and 25 mg of vitamin C.

A brief symptom inventory (BSI) was completed by the patients monthly and a monthly behavior disturbance inventory (BDI) was completed by a family member. Although patient and family reports were consistent, the reports on symptoms and behavior indicated only weak statistical trends. The authors commented that the BSI was not a measure of core schizophrenia symptomology, but rather indicated anxiety, depression, and interpersonal sensitivity.

Serum levels of thiamin, pyridoxine, folic acid, and vitamins B_{12} and A showed a statistically significant increase with treatment (Tables 8.2 and 8.3).

Table 8.2 Changes Observed in Subjects Who Took Recommended Supplements

Vitamin	Average Experimental Dose	Experimental Serum Levels: Start→End	Experimental Serum Levels: Change
A	6000 iu	568 → 879[a]	↑ 311
Thiamin (B_1)	1345 mg	58.4 → 87.4[b]	↑ 29
Niacin (B_2)	320 mg	66.8 → 68.8	↑ 2.0
Pyridoxine (B_6)	6223 mg	57.4 → 86.7[b]	↑ 29.3
C	2822 mg	0.81 → 1.03	↑ 0.22
E	204 iu	8.7 → 8.2	↓ 0.5
B_{12}	25 mg	296.7 → 1356[b]	↑ 1059.3
Folic acid	5.2 mg	8.1 → 46.2[b]	↑ 38.1

[a] t test change significant at $p = 0.05$.
[b] t test change significant at $p = 0.001$.

Table 8.3 Changes Observed in Control Subjects

Vitamin	Controls Serum Levels: Start → End	Controls Serum Levels: Changes
A	540 → 528	↓ 12
Thiamin (B_1)	59.5 → 54.7	↓ 4.8
Niacin (B_3)	65.2 → 66.7	↓ 1.5
Pyridoxine (B_6)	50.3 → 18.1[a]	↓ 32.2
C	0.87 → 0.45	↓ 0.42
E	9.5 → 8.1	↓ 1.4
B_{12}	280 → 264.2	↓ 15.8
Folic acid	7.0 → −7.3	↑ 0.3

[a] change: t test significant $< p = 0.001$.

The authors commented that it would be erroneous to conclude the therapy is not effective; that is, such a conclusion would be a type 2, false negative, error.

Control subjects' measures of nutritional status went generally downward over the term of the study (Vaughan and McConaghy 1999).

Other

Tetrahydrobiopterin (BH4) and schizophrenia—level II evidence

Tetrahydrobiopterin (BH4) is a cofactor involved in maintaining the availability of neurotransmitters dopamine, noradrenaline, and serotonin.

A group of 154 patients with schizophrenia was compared with 37 healthy controls for levels of fasting plasma biopterin (a measure of BH4). After controlling for gender, age, ethnicity, neuroleptic use and dose, phenylalanine-to-protein ratio, and plasma phenylalanine levels (which stimulate BH4 synthesis), it was found that plasma total biopterins were 34% lower in the patient group. Further study of the BH4 system in schizophrenia seems warranted (Richardson 2005).

Water intoxication/dilutional hyponatremia—level IV evidence

Consuming more water than can be excreted results in water intoxication, also known as dilutional hyponatremia. In the mentally ill, the condition may be termed psychogenic polydipsia. It is said to be most common in the schizophrenia population (Townsend 1994), but is also described in case studies of those with bipolar disorder (Duraiswamy 2011) and bulimia (Steckler 1995) as well as athletes, young children, and the elderly.

Symptoms such as seizures, psychosis, confusion, disorientation, and inappropriate behavior may be misinterpreted and delay diagnosis and treatment. Other symptoms include nausea and vomiting. Plasma levels of sodium (Na) <120 mmol/L are of concern (N = 132–144 mmol/L); symptoms are often associated with concentrations <110 mmol/L and severe symptoms occur at levels of 90 to 105 mmol/L. Untreated hyponatremia can lead to coma and death (Farrell 2003).

Development of hyponatremia can be gradual or abrupt. Individuals can become chronically imbalanced at relatively low levels of sodium. Gradual correction of hyponatremia is important; 8 to 10 mmol/L over 24 hours is recommended. Sudden large changes in sodium or fluid in nerve cells can result in pontin myelinolysis or extrapontin myelinolysis. Treatment may also include altered intake of fluid or sodium as well as close monitoring of weight changes.

> Central pontine myelinolysis: a neurological condition caused by severe damage of the myelin sheath of nerve cells and brain stem. Swelling of cells secondary to fluid shifts is problematic in the brain cells because the skull does not allow for expansion that other cells may have.

Schizophrenia and calorie needs—level II evidence

A comparison of calorie needs using common formulas for energy expenditure was compared with calorie needs based on individualized indirect calorimetry using a deuterium dilution. Two formulas (Harris-Benedict and Schofield) were found to overestimate calorie needs in the eight male schizophrenics taking clozapine by about 16%, or an average of 280 calories per day. If the estimations were not corrected, the extra 280 calories per day might result in approximately 2.5 kg (5.5 lb) weight gain over 10 weeks (Sharpe 2005).

Schizophrenia and diet history—level II evidence

Assessment of 4-day dietary intake histories of 88 patients diagnosed with schizophrenia or schizoaffective disorder found that the patient group consumed significantly fewer calories, carbohydrates, protein, total fat, saturated fat, monounsaturated fatty acids, PUFA, fiber, folate, sodium, and alcohol than a matched group from the National Health and Nutrition Examination Survey (NHANES) population and consumed more caffeine. This suggests that the higher proportion of obesity in the schizophrenic group is related to factors such as medication side effects and reduced physical activity than to intake (Henderson 2006).

A 24-hour diet recall of 146 outpatients with schizophrenia was compared to U.S. population standards. Saturated fatty acid and PUFA intake was significantly higher in patients with schizophrenia than in controls. No differences were found with regard to dietary intake of gamma-linolenic acid (GLA), eicosapentaenoic acid (EPA), and docosahexaenoic acid (DHA). Intake of vitamins A, C, and E were not different between the two groups. The observed cell membrane deficits in PUFA and EFA content do not appear to derive from decreased dietary supply. It was suggested that abnormal membrane phospholipid metabolism might be causative (Strassnig 2005).

Schizophrenia and diet pattern—level II evidence

Using a food frequency questionnaire and the Food Guide Pyramid, the eating patterns of 30 hospitalized patients with schizophrenia were compared to 30 matched subjects in a control group. Female patients consumed less milk and dairy products, fresh vegetables, fruits, chicken, and nuts compared to members of the control groups. Full-fat cream and carbonated beverages were included in their intakes. BMI levels were not significantly different, although percent body fat was significantly higher in patients compared to controls.

Males reported use of more full-fat cream, hydrogenated fats, and less red meat and nuts than male controls. BMI levels of patients were significantly higher in male patients, but percent body fat did not differ significantly (Amani 2007).

Prevention of weight gain following start of antipsychotic medications—level II evidence

Sixty-one first-episode, drug-naïve patients with schizophrenia who participated in eight flexible behavioral intervention modules were randomly assigned to a medication (risperidone, olanzapine, or haloperidol) and randomly assigned to the intervention modules or to a control group that received usual care. The intervention modules included behavioral interventions, nutrition, and exercise. Outcomes were BMI and weight, taken at baseline and weekly for 3 months. Eleven patients in the intervention group gained >7% of baseline weight (~4.1 kg [9 lb]), while 26 of the control group increased weight by >7% (~6.9 kg [15 lb]). Clinically meaningful weight gain was defined as a gain of 7% above baseline (Alvarez-Jimenez 2006).

Programs for management of weight gain induced by antipsychotic medications—level II evidence

Weight gain following start of antipsychotic medications is a common reason for patients to discontinue medication. Weight gains of 25 to 60 lb over the course of several years have been reported. Many patients are willing, able, and successful when weight management programs are offered. A medical record review of 35 patients indicated that most frequently used interventions included regular visits with a dietitian, a self-directed diet, and a stated treatment goal of weight loss. Patients in the group reviewed sustained a loss of approximately half their initial weight gain (O'Keefe 2003).

During a 12-week program in a partial hospitalization program, 30 patients in a study group lost an average of 2.7 kg (6 lb) (2.7% of body weight) compared to 15 patients in a control group who gained an average of 2.9 kg (6.4 lb) (3.1% of body weight). Those in the intervention group reported improvements in hunger rating and nutrition knowledge, and an increase in days and minutes of exercise per week. The program included two group sessions and one 15-minute individual session per week. Topics for these groups included healthful weight management techniques, meal planning, label reading, food shopping and preparation, portion control, and healthy snacking. Behavioral management techniques such as developing a slower eating style and differentiating emotional from physiological hunger were also included in the program (Vreeland 2003).

Of an initial group of 31 patients on antipsychotic medications, 20 completed a 52-week program that included behavioral interventions, nutrition, and exercise. Improvements of weight, BMI, hemoglobin A_{1c}, blood pressure, exercise level, and nutrition knowledge were seen in the program group. Patients who did not receive the program continued to gain weight (Menza 2004).

A treatment program modeled after the Diabetes Prevention Project using a cognitive behavioral therapy (CBT) treatment group was compared to a control group. The CBT group lost an average of 5.4 lb (range: 1 to 20 lb) or 2.9% of body weight. The control group lost 1.3 lb or 0.6% of body weight (Weber 2006).

The "Acute Solutions for Wellness" outpatient program in Ireland was offered to 47 selected inpatients. The program consisted of eight modules offered in 4 weeks and included topics of health living, physical activity, the food pyramid, recommended food servings, fat and salt in the diet, healthy and unhealthy eating habits, high fiber diet, and controlling hunger. A workbook was provided as well as group instruction by a trained nurse. Weight loss or maintenance was seen in 70% of the patients. An assessment 15 to 24 days following the program showed 14 patients had gained weight; the remainder either maintained their previous loss or continued to lose weight (Bushe 2008).

Conclusions

1. Studies show mixed results regarding whether individuals with schizophrenia have eating patterns significantly different from those without schizophrenia.
2. Weight gain and disorders related to overweight status (hypertension, hyperlipidemia, and diabetes) are common in patients with schizophrenia. This was documented before the widespread use of current antipsychotic medications, although evidence indicates that these drugs are associated with weight gain and comorbidity. Weight gain is a common reason for noncompliance with medication prescriptions.
3. Individuals with schizophrenia have been observed to have differences in metabolism (for example, a different pattern of body fat storage) and nutrient levels (vitamin C, folate, EFA, phospholipids, magnesium, and copper-to-zinc ratios) compared to those without schizophrenia.
4. Using laboratory energy expenditure methods, common formulas for determining calorie needs for those with schizophrenia have been shown to overestimate energy needs by several hundred calories per day.
5. Studies illustrate that those with schizophrenia are interested in changing habits and can be successful in multidisciplinary weight management programs. Successful programs address eating habits, nutrition, activity levels, and other aspects of weight management, as well as inclusion of follow-up. CBT has also been demonstrated to be helpful.

6. Lifestyle habits such as smoking and use of alcohol and drugs are higher in populations of people with schizophrenia, which often affect weight, nutrition, and comorbidity.

References

Alvarez-Jimenez, M, C Gonzalez-Blanch, JL Vasquez-Barquero, et al. Attenuation of antipsychotic-induced weight gain with early behavioral intervention in drug-naive first episode psychosis patients: A randomized controlled trial. *J Clin Psychiatry* 2006; 67(8):1253–1260.

Amani, R. Is dietary pattern of schizophrenia patients different from healthy subjects?. *BMC Psychiatry* 2007; 9:15.

American Psychiatric Association. *Diagnostic and Statistical Manual of Mental Disorder- IV*, 4th ed. 273. Washington DC: American Psychiatric Association, 1994.

Amminger, GP, MR Schafer, K. Papageorgiou, et al. Long-chain omega-3 fatty acids for indicated prevention of psychotic disorders: A randomized, placebo controlled trial. *Arch Gen Psych*; 2010; 67(2):146–154.

Arvindakshan, M, M Ghate, PK.Ranjekar, et al. Supplementation with a combination of omega-3 fatty acids and antioxidants (vitamins E and C) improves the outcome of schizophrenia. *Schizophr Res* 2003; 62(3):195–204.

Bushe, CJ, D McNamara, C Haley, et al. Weight management in a cohort of Irish in-patients with serious mental illness (SMI) using a modular behavioral programme. A preliminary service evaluation. *BMC Psychiatry* 2008; 8:76.

Carney, CP, L Jones and RE Woolson. Medical comorbidities in women and men with schizophrenia: A population-based controlled study. *J Gen Intern Med* 2006; 21(11):1133–1137.

Cascella, NG, D Kryszak, B Bhatti, et al. Prevalence of celiac disease and gluten sensitivity in the United States Clinical Antipsychotic Trials of Intervention Effectiveness (CATIE) study populations. *Schizophr Bull* 2011; 37(1):94–100.

Coyle, J. Treating the negative symptoms of schizophrenia: An expert interview. *Medscape* Psychiatry & Mental Health, 2006;11(2). http://www.medscape.com/viewarticle/546990

Dakhale, GN, Khanzode SD, Khanzode SS, et al. Oxidative damage and schizophrenia: The potential benefit by atypical antipsychotics. *Neuropsychobiology* 2004; 49(4):205–209.

Dakhale, GN, SD Khanzode, SS Khanzode, and A. Supplementation of vitamin C with atypical antipsychotics reduces oxidative stress and improves the outcome of schizophrenia. Psychopharmacology (Berl). 2005; 182(4):494–498.

Duraiswamy, K, NP Rao, G Venkatasubramanian G, et al. Psychogenic polydipsia in bipolar affective disorder—a case report. Gen Hosp Psychiatry. 2011; 33(1):84.

Farrell, DJ and L Bower. Fatal water intoxication. *J Clin Pathol*. 2003; 56(10): 803–804.

Fendri, C, A Mechri, G Khiari, et al. Oxidative stress involvement in schizophrenia pathophysiology: A review. javascript:AL_get(this, 'jour', 'Encephale.'); Encephale.2006; 32(2 Pt 1):244–252. English translation of original French abstract. http://www.ncbi.nlm.nih.gov/pubmed/16910626 accessed 9/30/2010.

Fenton, WS, F Dickerson, J Boronow, et al. A placebo-controlled trial of omega-3 fatty acid (ethyl eicosapentaenoic acid) supplementation for residual symptoms and cognitive impairment in schizophrenia. *Am J Psychiatry* 2001; 158:2071–2074.

Goff, DC, T Bottiglieri, E V Shih, et al. Folate, homocysteine, and negative symptoms in schizophrenia. *Am J Psychiatry* 2005; 162(7):1387–1388.

Henderson, D, C Borba, T Daley, et al. Dietary intake profile of patients with schizophrenia. *Ann Clin Psychiatry* 2006; 18(2):99–105.

Hibbeln, JR, KK Makino, CE Martin, et al. Smoking, gender, and dietary influence on erythrocyte essential fatty acid composition among patients with schizophrenia or schizoaffective disorder. *Biol Psychiatry* 2003; 53(5):431–441.

Horrobin, DF. Omega-3 Fatty Acid for Schizophrenia. Letter to the editor. *Am J Psychiatry* 2003; 160:188–189.

http://report.nih.gov/NIHfactsheets/ViewFactSheet.aspx?csid=67 (accessed 1/10/12)

Jones, JW. and M Sidwell. Essential fatty acids and treatment of psychiatric diseases. *Original Internist Inc.*2001; 8:5.

Joy, CB., R Mumby-Croft, and L A Joy. Polyunsaturated fatty acid supplementation for schizophrenia. Art. No.: CD001257. DOI: 10.1002/14651858.CD001257. pub2. A Cochrane Review Abstract. The Cochrane Collection (2003 and 2006). http://www.cochrane.org/reviews/en/ab001257.html.

Kanofsky, JD. Magnesium deficiency in chronic schizophrenia. *Int J Neurosci* 1991; 1:87–90.

Kemperman, RF, M Veurink, and T Van der Wal. Low essential fatty acid and B-vitamin status in a subgroup of patients with schizophrenia and its response to dietary supplementation. *Prostaglandins Leukotrienes Essential Fatty Acids* 2006; 74(2):75–85.

Kinney, D, K Hintza, EM Shearera, et al. A unifying hypothesis of schizophrenia: Abnormal immune system development may help explain roles of prenatal hazards, post-pubertal onset, stress, genes, climate, infections, and brain dysfunction. *Med Hypotheses* 2010; 75(3):555–563.

Kinney, DK, P Teixeira, D Hsu, et al. Relation of schizophrenia prevalence to latitude, climate, fish consumption, infant mortality, and skin color: A role for prenatal vitamin D deficiency and infections? *Schizophr Bull* 2009; 35(3):582–595.

Lindenmayer, JP and A Khan. *The American Psychiatric Publishing Textbook of Schizophrenia* ed. JA Lieberman, T.S. Stroup and D.O. Perkins. American Psychiatric Publishing, Washington DC 2006: 187–221.

Menza, M. B Vreeland, S Minsky, et al. Managing atypical antipsychotic-associated weight gain: 12-month data on a multimodal weight control program. *J Clin Psychiatry* 2004; (65(4):471–477.

Meyer, JM. Schizophrenia and the Metabolic Syndrome: An expert interview. *Medscape,* Psychiatry and Mental Health 2005; 8(1). http://www.medscape.com/viewarticle/506136

Muntiewerff, JW, N van der Put, T Eskes, et al. Homocysteine metabolism and B-vitamins in schizophrenic patients: Low plasma folate as possible independent risk factor for schizophrenia. *Psychiatry Res;* 2003; 121;1(1):1–9.

Nechifor, M, C Vaideanu, I Palamaru, et al. The influence of some antipychotics on erythrocyte magnesium and plasma magnesium, calcium, copper and zinc in patients with paranoid schizophrenia. *J Amer Coll Nutr* 2004; 23(5): 549S–551S.

O'Keefe, CD, DL Noordsu, TB Liss, and H. Weiss. Reversal of antipsychotic weight gain. *J.Clin Psychiatry* 2003; 64(8):907–912.

Peet, M. S Shah, K Selvam, and CN Ramchand. Polyunsaturated fatty acid levels in red cell membranes of unmedicated schizophrenic patients. *World J Biol Psychiatry* 2004; 5(2):92–99.

Puri, BK, T Easton, I Das, et al. The niacin skin flush test in schizophrenia: A replication study. *Int J Clin Pract* 2001; 55(6):368–370.

Puri, BK. and AJ Richardson. Sustained remission of positive and negative symptoms of schizophrenia following treatment with eicosapentaenoic acid. *Arch Gen Psychiatry* 1998; 55:188–189.

Puri, BK, AJ Richardson, DF Horrobin, et al. Eicosapentanenoic acid treatment in schizophrenia associated with symptom remission, normalization of blood fatty acids, reduced neuronal membrane phospholipid turnover and structural brain changes. *Int'l J Clin Pract* 2000; 54(1):57–63.

Regland, B. Schizophrenia and single-carbon metabolism. *Prog Neuropsychopharmacol Biol Psychiatry* 2005; 29(7):1124–1132.

Richardson, A.J., T.Easton, and BK Puri. Red cell and plasma fatty acid changes accompanying symptom remission in a patient with schizophrenia treated with eicosapentaenoic acid. *Eur Neuropsychpharmacol* 2000 10(3):189–193.

Richardson, MA, LL Read, T Clelland, et al. Evidence for a tetrahydropiopterin deficit in schizophrenia. *Neuropsychobiology* 2005; 52(4):190–201.

Ryan, MCM, P Collins, and JH Thakore. Impaired fasting glucose tolerance in first-episode, drug-naïve patients with schizophrenia. *Am J Psychiatry* 2003; 160(2): 284–289.

Schneider, B, B Weber, A Frensch, et al. Vitamin D in schizophrenia, major depression, and alcoholism. *J Neural Transm* 2000; 107(7):839–842.

Sharpe, J-K, NM Byrne, TJ Stedman, and AP Hills. Resting energy expenditure is lower than predicted in people taking atypical antipsychotic medication. *J Amer Diet Assoc* 2005; 105(4):612–615.

Smith, M. APA: Schizophrenia Patients Go Untreated for Comorbidities. MedPage report from Toronto, May 24, 2006. http://www.medpagetoday.com/Psychiatry/2005APAMeeting/tb/3385

Steckler, TL. Central pontine myelinolysis in a patient with bulimia. *South Med J* 1995; 88(8):858–859.

Strassnig, M, JS Brar, and R Ganguli. Dietary fatty acid and antioxidant intake in community-dwelling patients suffering from schizophrenia. *Schizophr Res* 2005; 76:343–351.

Tokdemir, M, SA Polat, Y Acik, et al. Blood zinc and copper concentrations in criminal noncriminal schizophrenic men. *Arch Androl* 2003; 49(5):365–368.

Townsend, R. General information regarding water intoxication. Drug Abuse Education Information and Research 1994. http://www.erowid.org/chemicals/other/water_info1.shtml

Vaughan, K and N McConaghy. Megavitamin and dietary treatment in schizophrenia: A randomized, controlled trial." *Aust New Zealand J Psychiatry* 1999; 33:84–88.

Vreeland, B, S Minsky, M Menza, et al. A program for managing weight gain associated with atypical antipsychotics. *Psychiatr Serv* 2003; 54:1155–1157.

Weber, M and K Wyne. A cognitive/behavioral group intervention for weight loss in patients treated with atypical antipsychotics. *Schizophr Res* 2006; 83(1):95–101.

chapter nine

Starvation, eating disorders, craving, dieting, and bariatric surgery

Introduction

Changing food intake as a conscious, purposeful decision is experienced differently than changes in food intake being imposed by a life circumstance without much personal control. A chosen course of action leads to more acceptance of the mental and physical changes that accompany a decrease in dietary intake. While the individual's attitude and perspective are quite different, the metabolic changes are the same.

Starvation may be defined as (1) total food energy deprivation, (2) any prolonged submaintenance food intake, or (3) the physiologic state that results when food intake is chronically inadequate. Starvation may be imposed from outside circumstances such as food unavailability, lack of purchasing power, social conditions, or disease, or by willful control by the individual who is starving.

Fasting is often defined as total denial of all food energy. It is also thought of as the metabolic condition of any person after an overnight sleep. It may also be defined as any diet that is restricted to only a few foods, such as a juice fast (Hoffer 1994). A prolonged fast lasts several days or longer. Fasting is generally a purposeful decision by the fasting individual. It may be part of a religious ritual, a means of showing protest, required as preparation for surgery, or a part of a program intended to improve health.

Dieting most often refers to a conscious change in dietary intake, usually for a stated purpose such as weight loss, or prevention or treatment of a health condition. Dieting may be a change in type or quantity of foods eaten. It may also involve changing the time and place of eating or other behaviors.

Malnutrition is defined as any disorder of nutrition. Malnutrition may result from unbalanced or insufficient diet or from defective absorption, assimilation, or utilization of foods. Traditionally, malnutrition is classified in four categories: obesity, marasmus (energy, protein, vitamin, and mineral deprivation), kwashiorkor (protein deprivation), and marasmus-kwashiorkor mixed. Any deficit in calories or protein may result in a deficit of vitamins and minerals as well.

High-calorie malnutrition is caused by adequate calories with insufficient vitamins and minerals. This is most likely when (1) intake is high in low-nutrient calories such as sugar or alcohol, or (2) foods are low in vitamins and minerals in proportion to the calories they provide. In purposeful food or calorie restriction, it is possible to obtain many vitamins and minerals in supplements instead of food, although supplements may not provide a complete and balanced array of necessary nutrients.

Conditions such as anorexia nervosa (AN) result in starvation. Bariatric/gastric bypass surgery, some cancers, and acquired immune deficiency syndrome (AIDS) can produce states of inadequate nutrition that fit the definition of starvation or malnutrition. Consequences of starvation are not 100% predictable because of individual variation. Some individuals have a larger initial storage of nutrients, some have greater nutritional needs than others, and others have altered or less efficient absorption and metabolism.

The term *orthorexia* is used to describe obsession with a perfect diet or a fixation on righteous eating. The diet followed is restrictive, is often based on a philosophy, and is believed to have a redemptive quality. Orthorexia is often accompanied by a morally superior attitude, and those who obsess over every meal may end up socially isolating themselves. Followers of a specific philosophy may band together in communities, rather than exhibit the isolation often seen in AN. The practices become a problem when the behavior hinders the person's ability to take part in everyday society and spills over into areas of functioning. The role of the nutritionist or other professional is to recognize when the desire for healthful eating crosses the line of becoming an obsession and to challenge underlying distorted belief systems (Mathieu 2005). For the original essay on orthorexia by Steve Bratman, MD, see http://www.orthorexia.com/?page_id=6. He describes his personal involvement with "kitchen spirituality" and with patients he has observed with what he coined "orthorexia."

Questions for self-assessment for orthorexia and comments of skeptics from the field of nutrition may be found in an article on www.WebMD.com titled "Good Diets Gone Bad" (http://www.webmd.com/mental-health/anorexia-nervosa/news/20001117/orthorexia-good-diets-gone-bad?page=2.)

Starvation

Mental changes during starvation: The historic work of Ancel Keys, Josef Brozek, and Austin Henschel—level II evidence

In the 1940s, Ancel Keys, Josef Brozek, and Austin Henschel conducted a scientific study of starvation of volunteer conscientious objectors who expressed altruistic and patriotic reasons for wanting to participate in this study. The study was a government-supported project to obtain scientific data for making decisions when planning aid to Europe following World War II. It was expected that refeeding of populations that had been exposed to famine would be needed, and it was unknown at that time what would be required for rehabilitation from starvation. The study was conducted at the University of Minnesota, in a setting for optimal monitoring and observation by a multidisciplinary team.

The 24 male participants were provided 1600 calories per day (approximately two-thirds of the calories needed) including 50 g of protein and 30 g of fat with carbohydrate to make up the balance of the calories. Foods were selected to supply vitamins and minerals adequate to meet their calculated needs. They were served two meals a day consisting of a rotation of three different daily menus composed of foods predicted to be commonly available in Europe. The starvation period lasted six months. Volunteers lost approximately 25% of their body weight.

Rehabilitation was monitored after the period of starvation, and some experimental groups received a multivitamin supplement. During both time periods, participants were monitored for physiological, radiological, biochemical, and psychological changes. The authors commented that "the bond between the physiological status of the organism and the 'psyche' is closer than is sometimes realized. The dominance of the 'body' becomes prominent under severe physical stress" (Keys 1950).

Emotional and personality changes

Observations were recorded regarding behavior, personality, and intelligence. Common changes were measured in magnitude and direction by psychological tests over time, chiefly the Minnesota Multiphasic Psychological Inventory (MMPI). During the starvation phase, the men were generally more serious and less happy than during the control period. Common changes included depression, irritability, and lack of will to utilize unaffected intellectual talents. Depression was partly related to inability to sustain mental or physical effort and feeling inefficient. Individuals became frustrated due to the discrepancy between what they wanted to do and what they were able to do. Apathy grew out of repeated failure. Although the scientists warned against a simple explanation that did not include other variables, a linear relationship was found between depression scores and the level of calorie supplementation during the rehabilitation phase. Rehabilitation groups who received a higher initial calorie allotment grew less depressed more quickly than did those whose calorie allotment was raised more gradually.

During the restrictive period, participants became less emotionally stable. Irritability increased to the point that it was an individual and group problem. They were aware of the irritability but not entirely able to control outbursts of temper, sulking, and aggression. It was reported that at times it was a real effort for the men to maintain socially acceptable behavior. In comparing themselves to prestarvation personal characteristics, volunteers described themselves as lacking in self-discipline and self-control, indecisive, restless, sensitive to noise, unable to concentrate, and markedly nervous. Increased nail biting and smoking were felt to be related to the increased nervousness. Humor and high spirits between the participants present at the start of the experiment gradually disappeared. What little humor remained at the end of the starvation period was described as mostly sarcastic.

Tests of personality changes found statistically significant increases in scores of social introversion, depression, and emotional instability. A statistically significant decrease in measures of social leadership, self-confidence, and freedom from nervous tension was found. Changes in MMPI scores showed a significant rise in depression, hypochondriasis, and hysteria, which were regarded as a diffuse psychoneurosis (defined as the neurotic triad). During rehabilitation, improved scores tended to parallel the caloric intake, restoring subjects to the control prestarvation test levels within one year. The authors commented, "Because the changes were produced by restriction of the diet and reversed by means of controlled nutritional rehabilitation, we may speak of an experimental neurosis" (Keys 1950, p. 871).

Partial completions

Although there was an absence of a tendency to develop aggressive, antisocial, or "character neuroses," four subjects were not able to carry out the experimental routine. Three of these did show significant elevations in the MMPI's Psychopathic Deviation scale. After being given opportunities to try again after eating off the program diet and based on subject's diaries, interviews, and clinical impressions, a few subjects were released from the study.

One subject had a strong desire to stay on the experimental routine but could not. His conflict over his desire to leave the study and desire to save face precipitated a borderline psychotic episode. A second subject could not come to terms with his extra eating and lack of weight loss, which were not part of his view of himself as a "perfect person." He was able to dissociate himself from reality in "mental blackouts while eating" and rationalize his behavior as psychological protection. A third subject was summarized as having latent personality weaknesses that were amplified and brought to the surface by the stress of the starvation regimen. A fourth subject had personality deterioration ending in two attempts at self-mutilation (successful contrived accidents aiming at severing his fingers). He felt his stated "philosophy of pushing himself mentally and physically was breaking down." He felt his character was shaken by temptation and when he collapsed from weakness on a treadmill test it was an acute emotional upset because he felt he had failed to live up to the standards he had set for himself. He wrote in his diary that the conflict regarding his ability was so intense he wondered if he was losing his mind. He described being "closer to violence than he had ever had before" because he "wanted to be at breakfast at Mother's" instead of in the program. Changes in his psychological testing indicated someone who was in stress so severe he was not able to handle it. At one point before being released from the study, he had a short admission to the university psychiatric hospital unit (Keys 1950, p. 895).

One subject was accepted into the experiment with a history of cyclothymic disorder (cyclic periods of manic, then depressive, experiences). Interviewers felt his strength of character, insight, and a record of accomplishment would allow him to be an acceptable subject. He did experience some mood swings during the study, at one time fearing he was going crazy. However, he completed the program and ended with a feeling of being personally strengthened by the ordeal. He gained a lot of satisfaction from his success and was proud of his accomplishment. At the end of the rehabilitation period, his MMPI scores on the psychoneurotic scales were as good as, or better than, his prestarvation levels (Keys 1950, p. 880).

No change in intellect

The clinical impressions of the investigative team and quantitative tests indicated intellectual capacity was unchanged. Subjects could think clearly and talk intelligently, although with decreasing speed. Subjects complained of inability to concentrate, difficulty in developing thoughts, and impaired judgment, and that their general alertness and comprehension had declined. Investigators concluded that the subject's estimate of decreased intellectual ability must be attributed to physical and emotional factors. Narrowing interests, apathy, and lack of initiative led the men to conclude they had suffered actual loss of intellect (Keys 1950, p. 859).

Behavioral changes related to stress of starvation

The authors stated that, in general terms, actual behavior might be considered partly a result of basic personality, past experience, present status, and the particular circumstances at any given time. Being observable and reportable, behavior around food, eating, and other parts of the experiment was reported in detail. Subjects became preoccupied with thoughts of food as the period of deprivation lengthened. They became possessive of their food, guarding their place in line while waiting for meals. They used increased spices and salt. They became isolative because interacting with others required too much energy. They ate silently and deliberately, sometimes toying with their food to make it seem to go further. Many saved some food to eat later, creating personal rituals. Some made their meals last two hours. Although most were well educated and refined, they routinely licked their dishes at the end of a meal. Food dislikes disappeared, and the taste appeal of the three rather repetitious menus increased. They became intolerant of any spoilage or wasting of food by nonsubjects in the dining room. Food, recipes, cookbooks, and menus became the topic of frequent conversation, reading, and daydreams. Some even changed their career plans to consider going into grocery, restaurant, or agriculture businesses. A few, however, expressed disgust with the "animal attitude" toward food. Gum chewing and coffee drinking ultimately had to be limited due to extreme use. Coffee was limited to nine cups a day. When some men chewed up to 40 packs of gum per day, the limit of 2 packs a day was established.

Although personal appearance and grooming deteriorated, bathing was not neglected. It was a source of pleasure, providing a means of getting warm and relieving aches, pains, and fatigue. Loss of 25% of body weight reduces the insulation of fat layers beneath the skin, leading to loss of body heat and feeling cold (Keys 1950).

Resting metabolic rate (RMR) decreases within days of starvation. This is related to the reduction of conversion of inactive thyroid hormone (T-4) to active form (T-3) within a few days or even hours of starting a starvation diet.

Another effect of starvation is the reduction of messenger RNA (mRNA) and a decrease in synthesis of protein. The amount of mRNA may vary with the level of immediate food intake. Protein synthesis can drop by 30 to 40% or more. Protein breakdown also decreases in similar amounts, so following adaptation to starvation a person can survive with a loss of only 3 to 4 g of body protein per day. Adequate levels of potassium, phosphorous, zinc, and magnesium are especially important in supporting protein sparing (Hoffer 1994).

Eating disorders

Comparison of starvation of conscientious objectors to patients with eating disorders

Individuals with AN demonstrate some of the same behaviors observed in the study by Keys et al. Individuals with AN are less able to concentrate, focus their attention, or attend to what is going on. This can potentially decrease their ability to benefit from treatment, unless treatment includes consumption of adequate nutrients. Their willful drive to not eat, to lose weight, and to be in control of their food intake prevails over feelings of hunger and fatigue as well as desire for the pleasures of taste. Compensatory behaviors may include (1) eating pepper, mustard, and other spicy condiments but without the food usually accompanying the condiments, (2) mealtime rituals, and (3) spending time preparing appealing food for others to eat.

Individuals suffering from bulimia nervosa (BN), often at a normal weight for height, have been known to lose up to 50% of their body weight in several months through dieting and purging. This is double the weight loss of the individuals in the starvation study.

In the study by Keys (1950), with a food intake of about two-thirds of their caloric needs (about 1600 calories/day), male volunteers lost 70% of their body fat while also losing 24% of the active lean tissue, even though the protein intake was the amount recommended as adequate for adults (0.75 g/kg of body weight) (Keys 1950). Protein and energy intake each influence the efficient use of the other. Body weight is lost rapidly at first. In Keys' study, weight had reached a plateau by 24 weeks. This is a lifesaving adaptation. Basal metabolic rate (BMR) was reduced by 40%. Response to starvation is also related to biochemical individuality (Williams 1956).

An example: If an individual's energy needs are about 2000 calories per day, a normal-weight adult would have 60 to 70 days of fuel stored in the form of fat. An adult has about 12 kg (26.4 lb) of protein in the body. Loss of 50% of lean tissue is incompatible with survival, even though this protein could theoretically supply 2 weeks of energy.

Potentially individuals with eating disorders may display some of the adaptation to starvation such as reduction of blood pressure, core temperature, and heart rate. Fluid accumulation results in swelling within the skin and subcutaneous tissues. Other features of human starvation include anemia, altered heart mass and function, decreased pulmonary mechanical function, diminished response to stimuli to breathe, altered gut anatomy, mildly impaired absorptive function, impaired healing, altered metabolism of many drugs, and immunodeficiency. Catecholamine turnover and action is also decreased (Torun 1994). Catecholamine levels affect mental status.

Stability of weight and successful accommodation preserves homeostasis as long as a new condition is not superimposed. An individual may stay in a state of fragile accommodation indefinitely. Maintenance in this state requires less-than-critical depletion of lean tissue, weight stability, normal plasma albumin (a blood protein), and normal lymphocyte (white cell) count. If fever and rapid heartbeat occur with a rise in serum and urinary urea levels, this indicates a lack of ability to preserve the fragile accommodation to starvation. Food restriction too severe for accommodation results in persistent weight loss until death occurs. The mortality rate of those with eating disorders is 3 to 5% (Crow 2009). Death is often associated with a BMI of 13 or below (Hoffer 1994).

Eating disorders: Nutritional consequences

Nutritional consequences accompany the psychopathology of both AN and BN. An individual with BN may lose and regain one-third to one-half of his or her body weight, often appearing "normal" to friends and family. Regaining weight quickly often replaces protein tissue with fat deposits, to the detriment of overall health. The process of inducing vomiting and abusing laxatives (up to 50 laxatives per day is not unknown) may alter function of the gastrointestinal tract, decreasing absorption and rate of peristalsis. This may induce medical emergencies such as low potassium and loss of chloride through stomach acid (as well as tooth erosion requiring dental caps). Physical discomfort during reestablishment of regular peristalsis and bowel function is often quite worrisome to these individuals.

An individual restricting both amount and types of food, such as occurs with AN, can become malnourished in macronutrients of carbohydrate, protein, and fat as well as micronutrients of vitamins and minerals. Subclinical levels of malnutrition and biochemical changes are followed by visible signs of deficiency. Documentation of observable physical signs of altered nutritional status, known as "nutrient-based lesions" or "cutaneous lesions," by professionals from around the world describes and compares how these psychiatric conditions produce as many as 40 such lesions (Table 9.1) (Strumia 2005; Glorio 2000). The critical value for the frequent occurrence of nutrient-based lesions is a BMI of ≤16 (Hediger 2000).

Vitamin and mineral deficiency diseases are possible after a few weeks or months of restriction. Water-soluble vitamins may be depleted in a matter of weeks. Vitamin B_6 (pyridoxine), vitamin C (ascorbic acid), and vitamin B_2 (riboflavin) may all become depleted in 2 to 6 weeks. Vitamin B_1 (thiamin) may become depleted in 1 to 2 weeks. Fat-soluble vitamins may be stored, so generally take longer for depletion depending on intake before a deficient dietary intake begins. Vitamin E (alpha-tocopherol) may take 6 to 12 months for depletion. Vitamin A may become depleted in 1 to 2 years (Haller 2005).

Table 9.1 Selected Lesions and Frequencies Observed in Patients with Eating Disorders

Lesion	Frequency observed %
Xerosis	58.3%
Nail changes	45.8
Cheilitis	41.6
Gingivitis	33.3
Carotenoderma	20.8
Hyperpigmentation	12.5
Seborrheic dermatitis	8.3

Source: Strumia 2005.

Observations of plasma, enzyme functions, and skin—level II evidence
Phillipp and colleagues (1988) documented normal but highly variable plasma levels of some vitamins in a group being treated for eating disorders. The group consisted of 24 patients with BN and 6 patients with AN. Vitamin C was reduced in several patients. Very high, nearly toxic, levels of retinol (vitamin A) were detected in 4 of the 30 patients. These patients reported hair loss, dryness, and fissures of the lips and dermatitis. Vitamin-dependent enzymes related to vitamins B_1, B_2, and B_6 were low in activity, indicating probable subclinical functional deficiencies. Prolonged prothrombin times in 5 patients were possibly related to low vitamin K intake or reduced production in the intestine. Dietary intakes reported by group members and analyzed by computer indicated a reduced intake of vitamins A, E, C, B_1, B_2, B_6, B_{12}, and folic acid. Phillipp recommended monitoring the vitamin status of eating disorder patients on a regular basis (Phillipp 1988).

Tracking nutritional changes in 61 patients with AN revealed with re-nourishment red blood cell folate and zinc increased, but did not reach normal levels. They recommend supplementation with these two nutrients during treatment (Castro 2004).

Pellagra in anorexia nervosa—level II evidence
Pellagra is the deficiency disease caused by insufficient amounts of niacin and tryptophan. Nonspecific symptoms include dermatological, neuropsychiatric, and gastrointestinal symptoms. Classic symptoms are called "the 3 Ds": diarrhea, dermatitis, and dementia. Most commonly observed in AN are erythema on sun-exposed areas, glossitis, and stomatitis. Neuropsychiatric and cognitive symptoms include dementia preceded by fatigue, nervousness, irritability, depression, confusion, apathy, and memory impairment.

Diagnosis can be confirmed with a 24-hour urine test for niacin metabolites and 5-hydroxy-indole-acetic acid. A total of <1.5 mg per 24 hours of N-methylnicotinamide and pyridine indicates severe niacin deficiency. Treatment of adults is 100 mg nicotinamide every 6 hours for several days until acute symptoms resolve and 50 mg every 8 to 12 hours until all skin lesions heal (Roman 2006). Prousky (2003) reports cutaneous symptoms resolve within 24 to 48 hours using a dose of 150 to 500 mg of niacin. This compares to the 20 to 35 mg/day Reference Daily Intake for healthy adults and teens.

Neuroimaging indicates change in brain
Neuroimaging techniques and their interpretation are increasing our understanding of normal processes in the control of food intake, hunger, and satiety. AN is associated with an enlargement of cerebral spinal fluid spaces, which generally recovers with refeeding. However, some cortical areas fail to correct when weight is restored. Functional changes in AN

reverse with weight recovery. However, reduction of 5-HT$_{2A}$ (serotonin) receptor binding remains after long-term weight restoration.

Studies of BN provide evidence of brain atrophy in the absence of significant weight loss but potentially related to chronic dietary restriction. Thalamic and hypothalamic serotonin transporters are less available in BN. Transporter availability continues to decrease with a longer duration of illness. Thus, BN is associated with changes in function and structure of the brain despite being at a normal weight (Stomatakis 2003).

Marsh et al. (2009) illustrated with fMRI altered patterns of brain activity in women with BN. More severe cases were related to less self-regulation, associated with acute tryptophan depletion, which can translate to an increase in impulsive behaviors.

Binge-eating and genes

A study of several gene mutations and the relationship to binge eating found 100% of carriers with the melanocortin 4 receptor (MC4R) mutation reported binge eating, as compared with 14.2% of obese subjects without mutations (P < 0.001) and none of the normal-weight subjects without mutations (Branson 2003).

Twin study suggesting genetic link for eating disorders—level II evidence
Twin and family studies suggest that genetic variations contribute to the development of BN and AN. In one study, probands who met modified criteria for (1) BN or (2) BN with a history of AN (BAN) were assessed using trained raters, self-report assessments, and DNA samples. A total of 365 patient/relative pairs were analyzed. (see Table 9.2.)

Proband: Individuals identified independently of their relatives.

Examples of findings in genetic studies include specific genes on several chromosomes that have been linked to (1) drive-for-thinness and obsessionality in AN and (2) susceptibility for self-induced vomiting have been associated with chromosome 10p (Devlin 2002).

Lipids

Cholesterol and suicide in anorexia nervosa
Researchers led by Angela Favaro evaluated serum cholesterol and nutritional status before refeeding in a sample of seventy-four patients with anorexia nervosa. Subjects who reported previous suicide attempts, impulsive self-injurious behavior, or current suicidal ideation showed significantly lower cholesterol levels than subjects without suicidality. Decreased cholesterol levels were negatively correlated with the severity

Table 9.2 Evidence of Genetic Links in Eating Disorders

Number of Relatives of Individuals with BN	Percentage of Relatives of Individuals with BN	Diagnosis of a Relative with BN or AN
N = 62	34.8	Diagnosed as having BN
N = 49	27.5	Diagnosed as having BN with history of AN
N = 35	19.7	Diagnosed as having AN
N = 32	18.0	Diagnosed as having an eating disorder not otherwise specified
Relatives of individuals with bulimia and a history of AN		
N = 42	22.5	Diagnosed as having BN
N = 67	35.8	Diagnosed as having BN with history of AN
N = 48	25.7	Diagnosed as having AN
N = 30	10.6	Diagnosed as having an eating disorder not otherwise specified

Source: Kaye et al. 2004.

of depressive symptoms in all patients with the exception of those with recurrent binge eating. The relationships between cholesterol levels and suicidal behavior and ideation did not seem to be affected by other nutritional and metabolic factors considered in this study (Favaro 2004).

Anorexia nervosa and polyunsaturated fatty acids—level V evidence Epidemiological evidence, reports of symptoms, comorbidity, and consequences suggest that polyunsaturated fatty acid and phospholipid abnormalities are significant in AN. Dietary restriction is hypothesized to cause essential fatty acid deficiencies and polyunsaturated fatty acid abnormalities, which might contribute to the physical and mental symptoms and the maintenance of the disorder (Ayton 2004).

See also Chapter 7 "Mood Disorders: Depression, Bipolar Disorder, and Suicide."

Vitamins

Scurvy in anorexia nervosa: A case study—level IV evidence
A case of scurvy in a 46-year-old female with anorexia nervosa was reported by Christopher, Tammaro, and Wing (2002). The woman's total diet consisted of small portions of boiled chicken and white rice. She was 5'5" in height and weighed 85 lb (N = 113 to 137 lb). She had lost 22 lb (10 kg) in the previous

3 years. Two weeks before admission to the hospital for her skeleton-like appearance, the patient noticed easy bruising and a nonitching rash. The examining physician observed (1) papules surrounding the base of each hair follicle from ankle to knee (petechiae), (2) ecchymoses (large areas of under-the-skin bleeding), and (3) lanugo (fine, downy-like hair) on the arms. Wasting at the temples; small, red spots on the hard palate; and periodontal disease were also observed. Other conditions included kidney stones, general malnutrition, and anemia. They commented that scurvy develops after 1 to 3 months of deficient ascorbic acid intake, the severity being related to the length of time without vitamin C. In spite of resistance, the patient was provided with ascorbic acid via vitamin supplements and food. The ecchymoses and petechiae resolved over 7 days. Perifollicular papules persisted.

Minerals

Magnesium and anorexia nervosa—level II evidence

One study found 60% of patients with AN had low serum magnesium at admission and up to 3 weeks into the refeeding program. Magnesium deficiency can cause weakness, constipation, seizures, and arrhythmias. The investigators recommended a minimum of measuring serum magnesium at admission and weekly for the first 3 weeks of refeeding (Birmingham 2004).

Other

BMI: Hospitalization and death—level II evidence

Studies report mean BMI at admission to hospital for intense treatment as 12.5 ± 0.9 kg/m^2 (Gentile 2008), and 12.09 and 13.2 kg/m^2 (Thiels 2008). In a long-term study, patients with a BMI over 11.5 had an average standardized mortality ratio (SMR) of about 7 and those with BMI lower than 11.5 had SMR above 30 (Rosling 2011). During the 1992–1993 famine in Somalia, some individuals with a BMI <12 survived, but only with special care (Collins 1995).

Food craving

Craving is often defined as a consuming desire or yearning. Ninety-seven percent of women and 68% of men report episodes of food cravings (Yanovski 2003). Core features of food craving are related to (1) strength of craving, (2) difficulty resisting eating, (3) feeling anxious when the craved food was unavailable, and (4) a change in speed of food consumption (Gendall 1997).

Food deprivation does not appear to be a necessary condition for food cravings to occur. Rather, food cravings are closely associated with mood;

mood may be an antecedent and a consequence of craving. Food cravers had higher ratings of boredom and anxiety during the day; dysphoric mood was prominent prior to the cravings themselves (Hill 1991). The data of 108 women studied over 2 years suggest that women have a stable core of foods for which they experience cravings, relatively independent of estradiol levels, BMI, or degree of dietary restraint (Rodin 1991).

Craving for chocolate—level II evidence

Much has been written concerning chocolate and its possible effect on mood and the "addiction" to chocolate. In a study by Hill and Heaton-Brown (1994), 25 healthy women recorded food cravings over 5 weeks. The average number of cravings recorded was just under two per week. Craving for chocolate amounted to 49% of all the food cravings.

In another study, 20 self-identified chocolate "addicts" rated depression, guilt, and craving higher than control subjects did before eating chocolate and feelings of contentment and relaxation lower than control subjects felt. However, eating chocolate resulted in increased feelings of guilt in the "addicts" and no significant changes in feeling depressed or relaxed. Correlational data showed little evidence of a relation between addiction to chocolate or the pharmacological (e.g., xanthine-based) effects of chocolate and the liking for chocolate (Macdiamid 1995).

Craving carbohydrate—level II evidence

In a study of self-classified individuals, a nonsignificant correlation existed between protein cravers' ratings of craving intensity and mood. A significant positive correlation existed between both male and female carbohydrate cravers' ratings of craving intensity and almost all mood scales assessed. The correlation between craving intensity and mood existed predominately with individuals who craved sweet, carbohydrate-rich foods (Christensen 2001).

Craving with calorie restriction—level II evidence

It is commonly believed that dieting and the restriction of specific types of foods produces cravings for these foods. In a clinical trial, Harvey and associates (1993) studied changes in self-reported cravings by 93 obese subjects with Type 2 diabetes who were randomly assigned to either (1) a balanced, low-calorie diet (LCD) of 1000 to 1200 kcal/day with all foods allowed in moderation or (2) a program that included a 12-week period of a very low calorie diet (VLCD) of 400 kcal/day with only lean meat, fish, or fowl allowed. There were significant decreases in cravings for all types of foods over the 20 weeks of the study for both the VLCD and the LCD conditions.

There was no evidence to support the belief that restricting intake of certain foods leads to increased craving for these foods or that the magnitude of weight loss is related to food cravings (Harvey 1993).

Brain function and craving—level II evidence

Writing that dysfunction of the prefrontal cortex is implicated in craving for drugs and food, Uher et al. (2005) reported that a 10 Hz repetitive transcranial magnetic stimulation to the left dorsolateral prefrontal cortex inhibits the development of craving.

Dieting: Fasting and restricting food intake

Physical effects of dieting—level III evidence

Conventional weight reduction diets cause an adaptation no different from that occurring during starving of nonobese individuals except increased protein sparing has been shown in weight loss by obese individuals. Metabolism during prolonged fasting results in low-insulin status after the body's limited carbohydrate stores are depleted. Insulin deficiency and ketosis are prevented by an intake of carbohydrate of about 130 to 150 g/day. Fasting of 5 weeks duration reduced glucose uptake by the brain by 50%, or about 40 g/day. Positron studies have shown in fully conscious humans a dramatic decrease of glucose use in all brain regions after 3 weeks of fasting. One to three days of fasting causes protein turnover from muscle proteins. Protein destruction decreases in 7 to 10 days. In 3 weeks, protein loss/nitrogen excretion is even lower—one-third to one-half the initial rate. One week of fasting causes a loss of about 75 g (about 2.5 oz.) of protein, used to make about 45 g of glucose per day. During this week, the brain gradually switches to ketone oxidation for energy (Hoffer 1994).

Psychological effects of dieting—level II evidence

For many years, Polivy and Herman studied those who restrain or do not restrain their eating and the effects on weight, eating behavior, and psychological characteristics. While not producing weight loss that is maintainable, chronic dieting and deprivation of pleasurable foods results in predictable cognitive, emotional, and behavioral changes. Changes include future gorging, guilt, preoccupation with food, decreased inhibition of urges to eat, stronger responses to emotion-inducing situations, more anxiety and neurosis, narcissism, higher stress levels, and lower general well-being (Herman 1975). Polivy suggests that individuals should be informed about the probability of these consequences if they begin food restriction. She suggests that establishing a healthful lifestyle, incorporating favorite foods in moderate quantities and gradual increments in physical activity rather than counterproductive "restraint," will lead to long-term maintenance of health (Polivy 1996).

In a review of weight loss therapy, including the psychological ramifications, Johnstone (2007) states subject's feelings of well-being, quality of life, libido, energy or fatigue, and anxiety or depression are important in dieting. Decrease in psychological performance during weight loss may be related to fatigue. Faster weight loss produces greater fatigue. Pre-fasting levels of fatigue did not resume until 2 weeks of ad libitum feeding after fasting.

Health at Every Size (HAES)—level IV evidence

In a review of research findings, Bacon and Aphramor (2011) point out that the Health at Every Size (HAES) approach is associated with statistically significant and clinically relevant improvements including mood, self esteem, body image, and eating behaviors. The HAES approach includes increased awareness of the body's response to food, which increases internal regulation of eating. This approach is similar to intuitive or mindful eating. This approach did not worsen health conditions.

A measure of intuitive eating—level V evidence

An assessment tool for intuitive eating was developed and psychometrically evaluated by Tylka (2006). Intuitive eating is characterized by eating based on physiological hunger and satiety cues rather than situational and emotional cues and is associated with psychological well-being. Her evaluation found that Intuitive Eating Scores (IES) were negatively related to eating disorder symptomatology, body dissatisfaction, poor interoceptive awareness, pressure for thinness, internalization of the thin idea and body mass, and positively related to indices of well-being.

Bariatric surgery

There are four types of bariatric surgery, also known as gastric bypass surgery: (1) adjustable gastric band (AGB), (2) Roux-en-Y gastric bypass (RYGB), (3) biliopancreatic diversion with a duodenal switch (BPD-DS), and (4) vertical sleeve gastrectomy (VSG). In general, these procedures generally (1) bypass most of the stomach and small intestine and reroute food intake to the lower intestine (Roux-en-Y gastroplasty), or (2) section off a small part of the stomach, but allow the contents of that section to go through the small intestine and large intestine for absorption (vertical banded gastroplasty).

The four types of surgery yield different nutritional consequences related to the size of the stomach and the accommodations made involving the small and large intestine and the availability of digestive enzymes. (See also http://win.niddk.nih.gov/publications/gastric.htm Accessed 8/23/2010.)

The simplest type of gastric surgery is laparoscopic gastric banding. It is reported to be reversible in contrast to some of the other types of surgery. Lanthaler and colleagues (2010) report that 53% of patients in their 9-year follow-up had at least one complication requiring reoperation. In an earlier, separate report, researchers comment, "Of our patients, 73% would not agree to gastric banding again" (Lanthaler 2009). Refinements of these procedures continue to be developed.

Psychological aspects of severe obesity: Early reports—level III evidence

As early as 1992, Stunkard and Wadden (1992) reviewed the literature regarding (1) the burden experienced by those who are severely obese and (2) the results of bariatric surgery. At that time, they concluded there is no single personality type that describes the severely obese and that "epidemiological and clinical studies refute the popular notion that overweight persons as a group are emotionally disturbed." They commented this population does not report greater levels of general psychopathology than average weight-control subjects and in light of the discrimination and prejudice from others they experience, "it is truly surprising that severely obese persons show no greater disturbances than persons of average weight ... a remarkable testimony to the resilience of the human spirit." Examples of weight-specific problems include lack of confidence due to inability to maintain a weight loss, isolation from others not able to understand their frustration, and humiliation for not fitting into theater or airplane seats.

Candidates for bariatric surgery and psychiatric diagnoses—level II evidence

In 2004, evaluations of 90 patients to determine appropriateness for bariatric surgery were reviewed by Sarwar and colleagues (2004). The Weight and Lifestyle Inventory, the Questionnaire on Eating and Weight Patterns, and the Beck Depression Inventory-II were used in addition to evaluation by a psychologist. Almost two-thirds of the subjects had a psychiatric diagnosis; major depressive disorder was most common. More than half of these were receiving psychiatric treatment at the time. Three patients were not recommended for surgery. Thirty-one percent were recommended for additional psychiatric or nutritional counseling.

Previous maltreatment and bariatric surgery candidates—level II evidence

Studies of various types of maltreatment show that candidates for bariatric surgery report two to three times greater maltreatment than normative samples. Emotional abuse and emotional neglect are associated with higher rates of depression and lower self-esteem than sexual abuse and other physical abuse. As reported by Grilo and associates (2005), the incidence of male and female maltreatment was not significantly different. Reported maltreatment was not significantly associated with BMI, binge eating, or eating disorder feature, preoperatively or at 12-month follow-up. Additional controlled research may increase understanding on this lack of significant association.

Literature reports demonstrate that self-esteem, positive emotions and eating behavior all improve dramatically after gastric bypass surgery. Stamina, mobility, mood, self-confidence, body image, and feelings of well-being all increase following weight loss from bariatric surgery.

Bariatric surgery and eating disorders—level II evidence

Evaluation by structured interview of 174 candidates for bariatric surgery found 13.8% had a diagnosis of an eating disorder, 22% had a history of affective disorder, and 15.5% had a history of anxiety disorder (Rosenberger 2006).

Psychopathology pre- and postbariatric surgery—level II evidence

In a national study of 1027 individuals described as morbidly obese who were seeking surgical help for weight loss, Maddi and colleagues (1997, 2001) gathered data including MMPI assessment presurgery, and 6 months and 1 year postsurgery, personal interviews, family history, and a medical examination. Medical comorbidity was low except for hypertension. They found an unusually high level of psychopathology compared to the normal population, consisting of depressive disorder, anxiety, and somatization features. This was linked to family of origin substance abuse and abuse of the individual including sexual abuse (N = 124), physical abuse (N = 119), and emotional abuse (N = 103). The authors commented that symptoms, coping skills, and long-standing habits could not be assumed to disappear with surgery or weight loss. Therefore, psychological support counseling is advised following bariatric surgery for the morbidly obese.

Obesity is defined as a BMI of ≥30.
Extreme or morbid obesity is defined as a BMI ≥40.

This work provides evidence that psychopathology declines following bariatric surgery. The unusually high levels of psychopathology before surgery may be a joint function of the factors producing the morbid obesity and a reaction to the obesity itself. That psychopathology declines following surgery to levels expected in the general population indicates that the patients were becoming more positive about their lives.

Long-term effects—level II evidence

To investigate the long-term effects of surgically induced weight loss on the psychological functioning of 62 morbidly obese patients, researchers led by Van Gemert (1998) compared preoperative and postoperative psychometric test results in a cross-sectional study. Three psychometric tests were administered before and after surgery. Follow-up occurred after 4 to 8 years. Surgical treatment resulted in a weight loss of between 50 to 150 lb.

The psychometric test results before surgery demonstrated somatization (conversion of mental experiences or states into bodily symptoms), depression, denial of emotional stress, social incompetence, and an indifferent attitude toward certain aspects of interpersonal behavior. All psychopathology, except for somatization, disappeared after surgical treatment. The level of self-esteem greatly improved. The psychopathology before surgery was almost totally reversed after sustained, surgically induced weight loss. This suggests that the preoperative psychological disturbances are the result, rather than the cause, of morbid obesity (Van Gemert 1998).

Binge eating before and after bariatric surgery—level II evidence

Fifty morbidly obese patients, classified as binge or non-binge-eaters, were assessed before and after gastric bypass surgery using the Beck Depression Inventory (BDI), the Three-Factor Eating Questionnaire (TFEQ), and the Eating Disorder Examination (EDE). Although the two groups differed markedly before operation, they were largely indistinguishable 4 months afterward. All binge eating had ceased and mood had improved markedly. Restraint in eating increased; hunger and disinhibition decreased. Eating concern, shape concern, and weight concern scores dropped. The overall findings indicate that gastric bypass surgery had a positive short-term impact on nonbinge and binge eaters alike (Kalarchian 1999).

A follow-up study of 99 patients at 2 years and 7 years found that BMI remains below presurgery levels, but significant weight gain was reported by a subgroup of 46% who reported recurrent loss of control over eating (Kalarchian 2002). Switching to "grazing" to replace binge eating is also reported as a postsurgical risk factor (Saunders 2004).

Postsurgical avoidance of food/eating—level II evidence

Case studies have been reported of individuals who do not fit the criteria of AN, but have pathological eating habits related to an intense fear of regaining weight. Segal (2004) proposes a new diagnosis of Post-Surgical Eating Avoidance Disorder (PSEAD). After ruling out anorexia or bulimia, simple phobias, and organic causes, criteria for the proposed diagnosis include (but are not limited to) (1) alteration in nutritional tests not in line with the surgical technique lasting for ≥ 2 months, (2) denial, and (3) perception that more weight loss would yield positive results, despite evidence to the contrary. Food avoidance has been associated with development of Wernicke–Korsakoff syndrome (Fandino 2005).

Anxiety, depression, and quality of life following bariatric surgery—level II evidence

A report by Kruseman et al. (2010) described 80 of 141 patients who completed the final visit after being followed for 8 years after bariatric surgery. Neither BMI nor body composition at baseline were predictive of successful weight loss at 8 years. A younger age at the time of surgery was associated with a loss of $\geq 50\%$ of excess weight. Screening scores for depression and anxiety were unchanged between baseline and last visit. There was a 27% prevalence rate for depression and a 46% prevalence rate for anxiety.

Present weight minus Desired weight = Excess weight

At baseline, patients scored high on most Eating Disorder Inventory-II (EDI-II) scores compared to reference values. At last visit, paired comparisons showed significantly improved scores for four scales: drive for thinness, bulimia, body dissatisfaction, and ascetism. Results from the systematic interview conducted during the last visit indicated that 51% of patients (N = 41) suffered from irregular episodes of binge eating or night eating syndrome during the previous month. A higher number of psychological consultations seemed to predict success. Therefore, the authors conclude, it seems the usual screening for eating disorders before surgery should be followed by periodic follow-up screenings, as these disorders can occur any time afterward. It remains unclear whether the operation could trigger these disorders, and more research is needed in this field. Eighty-five percent of participants were satisfied with results and would undergo it again. Data provides support for investigator's recommendation to widen the definition of success to include problematic

Table 9.3 Suicide Rates per 10,000 Individuals

	U.S. Suicide Rate	Rate after Bariatric Surgery
Women	0.7	5.2
Men	2.4	13.7

Source: Tindle et al. 2010.

eating behaviors and psychological well-being as well as excess weight loss (Kruseman et al. 2010) (See Table 9.3).

Impact of bariatric surgery on health status—level II evidence

Comparing 22 patients in a 4-year follow-up with 39 members of a control group waiting for surgery, surgical patients indicated they had more negative experiences with eating, but reported subjective improvement in health status. Qualitative data indicated (1) improved energy and self-esteem, (2) a fundamental shift in their relationship with food, which included food having a reduced role in their lives, and (3) a feeling of being more in control of their food intake (Ogden 2005).

A 2- to 3-year follow-up of 100 patients who had laparoscopic gastric bypass surgery found 57% adhered to vitamin-mineral supplement recommendations. Blood chemistries detected low ferritin, hematocrit, thiamin, and vitamin D. There was an absence of depression; 32% used antidepressant medications. There was a perceived benefit of 83 ± 18 (on a scale of 0 to 100) from having had the surgery (Welch 2011). A 6-year follow-up found significantly increased employment and existence of a partnership as well as improvement in anxiety, depression, and health-related quality of life (Nickel 2007).

Reappearance of symptoms after nine years—level II evidence

A 9-year follow-up of 462 gastric bypass patients tracked weight and psychological test results. Weight loss in the group averaged 103 lb (36% of the presurgery weight) and was largely maintained over the years of follow-up. Results showed that preoperative psychological issues were significantly reduced in the first 2 years following surgery. During these first 2 years, medical and mental health clinic support was available. Results of testing for anxiety, depression, well-being, and self-control indicated a return to presurgery levels after 2 years. Improvements in vitality and general health ratings did not revert to presurgery levels (Waters 1991). Commenting on Waters' work, Greenburg (2003) noted three suicides and two deaths from alcohol abuse occurred in a group followed for mental health ($N = 157$).

Data from a 10-year period in Pennsylvania indicated the rate of suicide following bariatric surgery substantially exceeded the U.S. suicide death rate. Seventy percent occurred within the first 3 years after surgery (Tindle 2010).

Long-term emotional support appears to be an essential ingredient for successful bariatric surgery.

Factors influencing unfavorable outcomes of bariatric surgery—level IV evidence

Although the mental health of patients improved following bariatric surgery, the benefits may be transient. Negative personality profiles, detrimental eating patterns, and negative body image may persist. Noncompliance with nutritional regimens occurs in almost two-thirds of cases. Nearly one-third of individuals having this surgery have a history of substance abuse disorder (Song 2008).

Psychological support

Some think surgery may be the most effective therapeutic option for weight reduction in carefully selected patients with morbid obesity resistant to conventional treatment. However, some agree "surgical treatment is not the solution but an important precondition for successful management of morbid obesity," wrote Kinzl (2002). A majority of the obese individuals expressed interest in psychological support postoperatively, but only a minority of this patient group (about one-quarter) ultimately enlisted psychological support on a regular or irregular basis according to one study. Some specific psychological areas have proven to be particularly important, including (1) change of self-esteem as a consequence of weight loss, (2) problems in adopting new eating behaviors along with the risk for developing a new eating disorder behavior, and (3) development of problem-solving skills (Kinzl 2002).

Cognitive behavioral therapy following bariatric surgery—level II evidence

Cognitive behavioral therapy (CBT) designed for the special needs of postoperative bariatric surgery patients has received positive feedback from patients. The group addresses both eating-related changes and psychological issues. Measures of success include psychosocial factors, eating behaviors, and weight loss (Saunders 2004).

Ashton and colleagues (2009) report on preoperative CBT for reducing binge eating. Four sessions of CBT reduced the number of weekly binge-eating episodes in data from 243 surgery candidates.

Bariatric Quality of Life Index—level V evidence

A condition-specific, 30-item instrument to evaluate the quality of life following bariatric surgery was created and validated with 133 patients by Weiner and associates (2005). It was tested at baseline, 1, 6, and 12 months following surgery. It was validated against the Gastrointestinal Quality of Life Index and the Bariatric Analysis and Reporting Outcome System and was reported ready for clinical use.

Nutritional effects of bariatric surgery with psychological implications—level II evidence

Bariatric surgical procedures are designed to produce reduced intake, various degrees of malabsorption of nutrients, or both. Micronutrient deficiencies, especially those involved in erythropoiesis (generation of red blood cells) and bone metabolism, are common to nearly all bariatric surgery patients. The patients consistently responded to supplementation. The pattern of deficiency suggests that absorption of micronutrients is more dependent upon the functioning of the gut as a whole than on the capacity of any single segment of the gut (Cannizzo 1998).

Pre- and postoperative nutrition—level II evidence

Assessment of nutritional status before bariatric surgery and 12 months following Roux-en-Y surgery indicates numerous candidates are poorly nourished before bariatric surgery (Table 9.4).

Appropriate supplements appear to improve nutritional status for many in spite of the compromised absorption following the surgery. Investigators recommend use of supplements and follow-up assessment using serum values at short intervals for early correction of deficiencies.

Vitamins

Vitamin B_{12}—level II evidence

Maleskey (2005) reports on B_{12} deficiency following Roux-en-Y bypass surgery. Subclinical deficiency of serum B_{12} levels (200 to 300 pg/ml) (N = 200 to 1100 pg/ml) can cause neuropsychiatric or neurological abnormalities such as paresthesia, sensory loss, ataxia, and dementia without anemia or macrocytosis. The amount of vitamin B_{12} in most oral multivitamin supplements is not adequate to maintain optimal long-term serum vitamin B_{12} levels in these patients. Monthly injections are often recommended for individuals with serum vitamin B_{12} levels of <300 pg/ml.

Chapter nine: Eating disorders, craving, dieting, and bariatric surgery 175

Table 9.4 Nutritional Status Prior to Bariatric Surgery

Investigator and Nutrients	Presurgery	Postsurgery
Bavaresco 2008 (n = 48)		1 year—intake
Calorie intake		770–1035 cal/day
Protein intake	15.6% were deficient	0.5 g/kg/day
Albumin	12.2%"	8.9% were deficient
Iron	3.4%"	14.6%"
Calcium		16.7%"
Madan 2006 (n = 100)		1 year—serum, labs
Vitamin A	11% had low levels	17% had low levels
B_{12}	13%	3%
Vitamin D-25	40%	21%
Zinc	30%	36%
Iron	16%	6%
Ferritin	9%	3%
Selenium	58%	3%
Folate	6%	11%
Dias 2006 (n = 40)		Every 4 months for 1 year—diet recall
Calories		529–866 cal/day
Iron		Inadequate
Zinc		Inadequate

Source: Bavarasco et al. 2008; Madan et al. 2006; and Dias et al. 2006.

Thiamin: A case study—level IV evidence

A group of researchers led by Loh et al. (2004) report a case of a patient who, two months after laparoscopic bariatric surgery, developed Wernicke's encephalopathy (WE). The clinical symptoms of inattentiveness, ataxia, and ophthalmoplegia were described. An initial MRI demonstrated characteristic injuries of WE. Repeat imaging showed resolution after 4 months of thiamin supplementation. The patient had normal gait but persistent memory deficits.

The researchers commented that even with early recognition and aggressive therapy, acute WE commonly results in permanent disability due to the irreversible cytotoxic effects on specific regions of the brain. They recommend prevention of the predictable time course of WE following bariatric surgery by administration of parenteral thiamin beginning at 6 weeks postoperatively in malnourished patients.

Thiamin levels before bariatric surgery—level II evidence

A review of medical records of 303 patients indicated that 15.5% had significant thiamin deficiency prior to having bariatric surgery (Table 9.5).

Table 9.5 Thiamin Levels before Bariatric Surgery

Males	3.2 µg/ml or 0.032 µg/dl
Females	2.4 µg/ml or 0.024 µg/dl
Normal levels* 200 µg/ml or 0–2.0 µg /dl	

Source: Carrodeguas et al. 2005.

* *Harrison's Principles of Internal Medicine*, Volume I, 16th ed., A-9, 2005. Mc Graw-Hill Medical Publishing Division, New York.

Excluded from the review were records of patients who had been taking a vitamin supplement, had a history of frequent alcohol consumption, a history of any malabsorptive disease, or previous gastric bypass surgery (Carrodeguas 2005).

Minerals

Calcium—level II evidence

A report of 3- to 6-year follow-up of 243 patients with various types of bariatric surgery found low and normal vitamin D levels were associated with elevated parathyroid levels suggestive of inherent calcium malabsorption following gastric bypass surgery (Johnson 2006).

Iron—level II evidence

Of 174 patients followed for 2 years, 15 to 38% developed low ferritin (a measure of iron nutritional status), depending on the type of surgery. Three patients had low serum albumin and were hospitalized and fed parenterally for three weeks (Skroubis 2002).

Supplements

Standard supplementation may not be adequate—level II evidence

Follow-up of 137 patients who had Roux-en-Y surgery and who had prescriptions for a standard multivitamin supplement designed for this population was reported by Gastyear and colleagues (2008). Follow-ups were conducted at 3, 6, 9, 12, 18, and 24 months to determine need for addition supplements. At 3 months, 34% needed at least one additional specific supplement; at 6 months, 59% needed additional supplementation; and at 24 months, 98% needed additional supplementation.

Other

Nutritional monitoring following bariatric surgery—level IV evidence

Monitoring of nutritional status postbariatric surgery at 3, 6, and 12 months, and then annually is recommended by Breiter (2006). Monitoring should include biochemical assessment of thiamin, folic acid, vitamin B_{12}, vitamin D, calcium, and iron (See Table 9.6).

Managing the diet following bariatric surgery

Habits such as using food as a coping mechanism for stress, emotional issues, and boredom, eating as the primary aspect of social relationships, or the presence of an eating disorder need to be addressed for long-term success at maintaining loss of weight. Making dietary changes is often emotionally difficult. The eating routine after bariatric surgery requires a high level of motivation and perhaps a sense of possible serious, immediate consequences if not followed. Presurgery consultation and education along with postsurgery follow-up is recommended.

An eating protocol following surgery generally includes the following: intake beginning with 1 oz. of noncarbonated, clear liquid every 15 minutes. Sugar is omitted for patients whose surgery bypassed the small intestine. From Day 2 up to one month after surgery, high-protein, full-fat liquids are consumed in five to six small meals of 2 to 3 oz. each. When tolerated, this transitions to pureed foods. From Week 4 to Week 8 after surgery, the diet transitions to five to six meals of 2 oz. of semisolid or soft, low-fat foods. Protein supplements are added as needed.

From Week 8, regular-textured, low-fat foods are consumed in five meals of 2 to 3 oz. each. Maintenance includes attention to use of lean meats, vegetables, fruits, low-fat dairy, and whole grain foods. Fluid intake to maintain hydration is necessary. A vitamin/mineral supplement and protein supplement, as prescribed, are part of maintaining adequate nutritional status. A long-term schedule of three to five meals of 4 to 6 oz. of food is common.

---

1 oz. = about two standard tablespoons of food

Possible food intolerances include tough meat; fresh, fluffy bread; rice; pasta; fibrous foods such as celery, mushrooms, citrus membranes, and fruit skins; fried foods; and foods with added sugars (Marcus 2005).

Stressful consequences of bariatric surgery—level III evidence
In addition to psychological consequences of a history of obesity, recovery from surgery, extensive adjustment of lifestyle, and possible nutrient deficiencies that may alter mental status, other consequences of

Table 9.6 Recommendations for Monitoring Nutrient Deficiencies Following Bariatric Surgery

Nutrient	Frequency of Monitoring	Biochemical Monitor	Reference Range	Supplement Intervention
Thiamin (vitamin B$_1$)	3, 6, and 12 months after surgery, then annually	Serum thiamin	1.6–4.0 µg/dl	50–100 mg IV or IM
Folic acid/folate	3, 6, and 12 months after surgery, then annually	Serum folic acid	225–600 ng/mL deficiency indicated at <4 ng/ml or <9 mmol/L	1 mg/day
Vitamin B$_{12}$	3, 6, and 12 months after surgery, then annually	CBC— mean corpuscular volume Serum B$_{12}$ or MMA + homocysteine levels	>100 fL indicates deficiency >400 mg/dl	350 µg oral/day or 1000 µg sublingual/day or 2000 µg/month IM
Vitamin D	3, 6, and 12 months after surgery, then annually	Plasma levels of 25-hydroxy vitamin D*	25–40 ng/ml <20 ng/ml indicates deficiency	400 IU/day
Calcium	3, 6, and 12 months after surgery, then annually	Parathyroid hormone Serum ionized calcium***	65 pg/ml 4.4–5.3 mg/dl	2 g/day of calcium as calcium citrate**
Iron	3, 6, and 12 months after surgery, then annually	Serum ferritin Total iron-binding capacity Serum iron	30–300 ng/ml 3.0–8.5 µg/ml 250–450 µg/dl F - 60–140 µg/dl M - 75–150 µg/dl	Deficiency may need treatment by oral or IV iron, then 300 mg ferrous sulfate three times per day, along with ascorbic acid

Source: Breiter et al. 2006.

* Calcitrol or 1,25 (OH) dihydroxyvitamin D is not as valuable an indicator due to influences of calcium, phosphorus, and parathyroid hormones.
** Calcium carbonate is not as readily absorbed due to low levels of acid present.
*** Serum ionized calcium may or may not be affected.

bariatric surgery may cause embarrassment. These include hair loss for 3 to 12 months, unusual body odors, presence of excess skin, and up to 10 foul-smelling bowel movements per day (secondary to expected fat malabsorption) (Fujioka 2005).

Conclusions

1. Restricting energy/calorie intake, even if accompanied by adequate intake of vitamins and minerals, affects mental functioning and the brain, whether it is from starvation, dieting, fasting, or AN. Personality, mood, emotional stability, and behavior are affected. Intellect appears not to be affected.
2. Mental changes produced by experimental restriction of the diet and reversed by means of controlled nutritional rehabilitation, are termed "experimental neurosis."
3. In the absence of sufficient vitamins and minerals, deficiencies that may affect mental status may arise, even if calories are adequate or excessive. These deficiencies may cause physical, cutaneous, nutrient-based lesions that are corrected by treating with the deficient nutrient. Assessment of nutritional status needs to be more broadly defined than assessment of only weight, BMI, body composition, or intake of macronutrients.
4. A genetic link for eating disorders appears likely. Neuroimaging of individuals with AN or BN indicates changes in the brain may or may not normalize following treatment.
5. Craving is not a reliable indicator of a lack or deficiency of chocolate, caffeine, or carbohydrates. These substances do affect mood.
6. Self-esteem, positive emotions, and eating behavior all improve dramatically after gastric bypass surgery. Stamina, mobility, mood, self-confidence, body image, and feelings of well-being also increase following weight loss from bariatric surgery.
7. Long-term follow-up after bariatric surgery indicates that psychological problems may reappear and nutritional deficiencies are likely to develop. Individuals often do not follow recommendations for psychological and nutritional support and monitoring after bariatric surgery.
8. "Success" following bariatric surgery should include not only loss of weight, but also maintenance of nutritional status, eating behaviors, and psychological well-being.
9. Resources regarding bariatric surgery:
   Suggestions for the Pre-Surgical Psychological Assessment of Bariatric Surgery Candidate (Lemont 2004). The American Society for Bariatric Surgery. Available online. http://www.asbs.org/html/pdf/PsychPreSurgicalAssessment.pdf.

Bariatric Nutrition: Suggestions for the Surgical Weight Loss Patient (Aills 2008). Reprints: American Society for Metabolic and Bariatric Surgery, 100 SW 75th Street, Suite 201, Gainesville, FL 32607. E-mail: info@asmbs.org.

References

Aills, L, J Blankenship, C Buffington, et al. *Bariatric Nutrition: Suggestions for the Surgical Weight Loss Patient*. American Society for Metabolic and Bariatric Surgery. Allied Health Sciences Section Ad Hoc Nutrition Committee. 2008.

Ashton, K, M Drerup, A Windover, and L Heinberg. Brief, four session group CBT reduces binge eating behaviors among bariatric surgery candidates. *Surg Obes Related Diseases* 2009; 5(2):257–262.

Ayton, Agnes K. Dietary polyunsaturated fatty acids and anorexia nervosa: Is there a link? *Nutr Neuroscience* 2004; 7(1):1–12.

Bacon, L and L Aphramor. Weight science: Evaluating the evidence for a paradigm shift. *Nutr J* 2011; 10:9. http://www.nutritionj.com/content/10/1/9

Bavaresco, M., S Paganini, TP Lima, et al. Nutritional course of patients submitted to bariatric surgery. *Obes Surg* 2008; 20(6):716–721.

Birmingham, CL, D Laird, JH Puddicombe, and J Hlynsky. Hypomagnesmia during refeeding in anorexia nervosa. *J Eat Weight Dis* 2004; 9(3):236–237.

Branson, R, N Potoczna, JG Kral, et al. Binge eating as a major phenotype of melanocortin 4 receptor gene mutations. *N Engl J Med*. 2003; 348(12):1096–103.

Breiter, AM. and L Greiman Potential nutrient deficiencies following gastric bypass surgery. *Weight Management Dietetic Practice Group Newsletter*. (a practice group of the American Dietetic Association) 2006; 4(1):15–17.

Cannizzo, F, Jr. and JG, Kral. Obesity surgery: A model of programmed undernutrition. *Curr Opin Clin Metab Care* 1998; 1(4):363–368.

Carrodeguas, L, O Kaider–Person, S Szomstein, et al. Preoperative thiamin deficiency in obese population undergoing laparoscopic bariatric surgery. *Surg Obes Rela Dis* 2005; 1(6):517–522.

Castro, J, R Deulofeu, A Gila, et al. Persistence of nutritional deficiencies after short-term weight recovery in adolescents with anorexia nervosa. *Int'l J Eat Dis* 2004; 35(2):169–178.

Christensen, L and L. Pettijohn. Mood and carbohydrate cravings. *Appetite*. 2001; 36(2):137–145.

Christopher, K, D Tammaro, and EJ Wing. Early scurvy complicating anorexia nervosa. *So Med J* 2002; 95(9):1065–1066.

Collins, S. The limit of human adaptation to starvation. *Nature Med* 1995; 1(8):810–814.

Crow, S.J., CB Peterson, SA Swanson, et al. Increased mortality in bulimia nervosa and other eating disorders. *Am J Psychiatry* 2009; 166:1342–1346.

Dias, MC, AG Ribeiro, VM Scabim, et al. Dietary intake of female bariatric patients after anti-obesity gastroplasty. *Clinics* 2006; 61(2):93–98.

Devlin B, SA Bacanu, KL Klump, et al. Linkage analysis of anorexia nervosa incorporating behavioral covariates. *Hum Mol Genet*, 2002; 11(6):689–696.

Fandino, JN., AK Benchimol, LN Fandino, et al. Eating avoidance disorder and Wernicke-Korsakoff syndrome following gastric bypass: An under-diagnosed association. *Obes Surg* 2005; 18(8):1207–1210.

Favaro, Angela, Lorenza Caregaro, Lorenza Di Pascoli, Francesca Brambilla, and Paolo Santonastaso. "Total serum cholesterol and suicidality in anorexia nervosa." *Psychosomatic Med* 2004; 66:548–552.

Fujioka, K. Follow-up of nutritional and metabolic problems after bariatric surgery. *Diabetes Care* 2005; 28(2):481–484.

Gastygear, C, M Suter, RC Gaillard, and V Guisti. Nutritional deficiencies after Roux-en-Y gastric bypass for morbid obesity often cannot be prevented by standard multivitamin supplementation. *Am J Clin Nutr* 2008; 87(5):1128–1133.

Gendall, K. A., P. R. Joyce, and P. F. Sullivan. Impact of definition on prevalence of food cravings in a random sample of young women. *Appetite* 1997; 28(1):63–72.

Gentile MG, GM Manna, R Ciceri, and E Rodeschini. Efficacy of inpatient treatment in severely malnourished anorexia nervosa patients. *Eat Weight Disord*. 2008; 13(4):191–7.

Glorio, R, M Allevato, A De Pablo, et al. Prevalence of cutaneous manifestations in 200 patients with eating disorders. *International J Derm* 2000; 39(5):348–353.

Greenberg, I, Psychological aspects of bariatric surgery. *Nutr in Clin Pract* 2003; 18:14–130.

Grilo, CM, RM Masheb, M Brody, et al. Childhood maltreatment in extremely obese male and female bariatric surgery candidates. *Obes Res* 2005; 13(1):123–130.

Haller, J, Vitamins and Brain Function. *Nutritional Neuroscience*. ed. HR Lieberman, RB Kanarek, and C Prasad. 229. CRC Press: Taylor & Francis Group, Boca Raton, FL 2005.

Harvey, J, RR Wing, and M Mullen. Effects on food cravings of a very low calorie diet or a balanced, low calorie diet. *Appetite* 1993; 21(2):105–115.

Hediger, C, B Rost, and P. Itin. Cutaneous manifestations in anorexia nervosa. *Schweizerische Medizinische Wochenschrift* 2000; 130(16):565–575.

Herman, CP and J Polivy. Anxiety, restraint and eating behavior. *J Abnormal Psychology* 84(1975):666–672.

Hill, AJ and L Heaton-Brown. The experience of food cravings: A prospective investigation in health women. *J Psychosomatic Res* 1994; 38(8):801–814.

Hill, A J, CF Weaver, and JE Blundell. Food craving, dietary restraint and mood. *Appetite* 1991; 17(3):187–197.

Hoffer, L. John. Starvation. In *Modern Nutrition in Health and Disease*. ed. ME Shils, JA Olson, and M Shike. 939. Lea & Febiger, Philadelphia 1994.

Hoffer, L. John. Starvation. *Modern Nutrition in Health and Disease*. ed. ME Shils, JA Olson, and M Shike. 927–242. Lea & Febiger, Philadelphia 1994.

Johnson, JM, JW Maher, EJ DeMaria, et al. The long-term effects of gastric bypass on vitamin D metabolism. *Ann Surg* 2006; 243(5):701–704.

Johnstone, A. Fasting–The ultimate diet? *Obes Rev* 2007; 8(3):211–222.

Kalarchian, MA, MD Marcus, GT Wilson, et al. Effects of bariatric surgery on binge eating and related psychopathology. *Eating and Weight Dis* 1999; 4(1):1–5.

Kalarchian, M, MD Marcus, EW Labouvic, et al. Binge eating among gastric patients at long-term follow-up. *Obes Surg* 2002; 12(2):270–275.

Kaye WH, B Devlin, N Barbarich, et al. Genetic analysis of bulimia nervosa: Methods and sample description. *Int'l J Eat Disord*, May 2004; 35(4):556–70.

Keys, A, J Brozek, A Henschel, et al. *The Biology of Human Starvation*, Minneapolis: University of Minnesota Press, 1950; 767–905. Quote p. 767.

Ibid. 871.

Ibid. 895.

Ibid. 880–904.

Ibid. 859–863.

Kinzl, JF, E Trefal, M Fiala, and W Biebl. Psychotherapeutic treatment of morbidly obese patients after gastric banding. *Obes Surg* 2002; 12(2):292–294.

Kruseman, Maaike, Anik Leimgruber, Flavia Zumbach, and Alain Golay. Dietary, weight, and psychological changes among patients with obesity, 8 years after gastric bypass. *J Am Diet Assoc* 2010; 110:527–534.

Lanthaler, M, F Algner, J Kinzl, et al. Long-term results and complications following adjustable gastric banding. *Obes Surg* 2010; 8:1078–1085.

Lanthaler M, S Strasser, F Aigner, et al. Weight loss and quality of life after gastric band removal or deflation. *Obes Surg* 2009; 19(10):1401–1408.

LeMont D, MK Moorehead, MS Parish, et al. Suggestions for the pre-surgical psychological assessment of bariatric surgery candidate. The American Society for Bariatric Surgery. 2004. Available online. http://www.asbs.org/html/pdf/PsychPreSurgicalAssessment.pdf accessed 8/23/2010.

Loh Y, WD Watson, A Verma A, et al. Acute Wernicke's Encephalopathy following bariatric surgery: Clinical course and MRI correlation. *Obes Surg* 2004; 14(1):129–132.

Macdiarmid, JI and MM Hetherington. Mood modulation by food: An exploration of affect and cravings in chocolate addicts. *Brit J Clin Psychology* 1995; 34 (Part 1):129–138.

Madan, SK, WS Orth, DS Tichansky and CA Ternovits. Vitamin and trace mineral levels after laparoscopic bypass. *Obes Surg* 2006; 16(5):603–606.

Maddi, SR, D Khoshaba, M Persico, et al. Psychosocial correlates of psychopathology in a national sample of the morbidly obese. *Obes Surg* 1997; 7:397–404.

Maddi, SR, S Ross, R Fox, et al. Reduction in psychopathology following bariatric surgery for morbid obesity. *Obes Surg* 2001; 11(6):680–685.

Maleskey, GE. "Subclinical B_{12} deficiency in post Roux-en-Y gastric bypass patients." *J Amer Diet Assoc* 2005; 105(8):29.

Marcus, E. Bariatric Surgery: The role of the RD in patient assessment and management. *SCAN* (Newsletter of Sports, Cardiovascular and Wellness Nutritionists, a practice group of the American Dietetic Association) 2005; 24(2):18–20.

Marsh, R, JE Steinglass, AJ Gerber, et al. Deficient activity in the neural systems that mediate self-regulatory control in bulimia nervosa. *Arch Gen Psychiatry* 2009 Jan; 66(1):51–63.

Mathieu, Jennifer. What is orthorexia? *J Amer Diet Assoc* 2005; 105(10):1510–1512.

Nickel, MK., TH Loew, and E Bachler. Change in mental symptoms in extreme obesity patients after gastric banding, Part II: Six-year follow up. *Int'l J Psychiatry in Med* 2007; 37(1):69–79.

Ogden, J, C Clement, S Aylwin, and A Patel. Exploring the impact of obesity surgery on patients' health status: A quantitative and qualitative study. *Obes Surg* 2005; 15(2):266–272.

Phillipp, E, K-M Pirke, M Seidl, et al. Vitamin status in patients with anorexia nervosa and bulimia nervosa. *International J Eat Dis* 1988; 8(2):209–218.

Polivy, Janet. Psychological consequences of food restriction. *J Amer Diet Assoc* 1996; 96(6):589–592.

Prousky, JE. Pellagra may be a rare secondary complication of anorexia nervosa: A systematic review of the literature. *Altern Med Rev* 2003; 8(2):180–185.

Rodin, J, J. Mancuso, J. Granger, and E. Nelbach. Food cravings in relation to body mass index, restraint and estradiol levels: A repeated measures study in healthy women. *Appetite* 1991; 17(3):177–185.

Roman, G. "Nutritional Disorders of the Nervous System". *Modern Nutrition and Health and Disease.* 10th ed 1371. ed ME Shils, M Shike, AC Ross, B Caballero, RJ Cousins. Lippincott, Williams & Wilkins. Philadelphia, 2006.

Rosenberger, PH, KE Henderson, and CM Grilo. Psychiatric disorder comorbidity and association with eating disorders in bariatric surgery patients: A cross-sectional study using structured interview-based diagnosis. *J Clin Psychiatry* 2006; 67(7):1080–1085

Rosling, AM, P Sparén, C Norring, and A-L von Knorring. Mortality of eating disorders: A follow-up study of treatment in a specialist unit 1974–2000. *Int'l J Eat Dis* 2011; 44:304–310.

Sarwar, DB, NI Cohn, LM Gibbons, et al. Psychiatric diagnoses and psychiatric treatment among bariatric surgery candidates. *Obes Surg* 2004; 14(9):1148–1156.

Saunders, R. Grazing: A high-risk behavior. *Obes Surg* 2004; 14(1):98–102.

Saunders, R. Post-surgery group therapy for gastric bypass patients. *Obes Surg* 2004; 14(8):1128–1131.

Segal, A, KD Kinoshita, and M Larino. Post-surgical refusal to eat: Anorexia nervosa, bulimia nervosa or a new eating disorder? A case series. *Obes Surg* 2004; 14(3):353–360.

Song, A and MH Fernstrom. Nutritional and psychological considerations after bariatric surgery. *Aesthet Surg J* 2008; 28(2):195–199.

Stamatakis, EA and MM Hetherington. Neuroimaging in eating disorders. *Nutri Neuroscience* 2003; 6(6):325–334. see also www.maney.co.uk/journals/nns and www.ingentaconnect.com/content/maney/nns

Strumia, R. Dermatologic signs in patients with eating disorders. *Am J Clin Derm* (2005); 6(3):165–173.

Stunkard, AJ and TA Wadden. Psychological aspects of severe obesity. *Am J Clin Nutr* 1992; 55:524S–532S.

Thiels, C. Forced treatment of patients with anorexia. *Curr Opin Psychiatry.* 2008; 21(5):495–498.

Tindle, HA, B Omalu, A Courcoulas, et al. Risk of suicide after long-term follow-up from bariatric surgery. *Am J Med* 2010; 123(11):1036–1042.

Torun, B and F Chew. Protein-Energy Malnutrition In *Modern Nutrition in Health and Disease* ed. ME Shils, JA Olson, and M Shike. 950–976. Lea & Febiger, Philadelphia 1994. Williams, R. *Biochemical Individuality.* 232. John Wiley & Sons, New York 1956.

Tylka, TL. Development and psychometric evaluation of a measure of intuitive eating. *J Couns Psychol* 2006; 52(2):226–240.

Uher, R, D Yoganathan, A Mogg, et al. Effect of left prefrontal repetitive transcranial magnetic stimulation on food craving. *Biol Psychiatry* 2005; 58(10):840–842.

Van Gemert, WG, RM Severeijns, JW Greve, et al. Psychological functioning of morbidly obese patients after surgical treatment. *Int'l J Obes Rel Metab Dis* 1998; 22(5):393–398.

Waters, GS, WJ Pories, MS Swanson, et al. Long-term studies of mental health after the Greenville gastric bypass operation for morbid obesity. *Amer J Surg* 1991; 161:154–158.

Weiner, S, S Sauerland, M Fein, et al. The Bariatric Quality of Life index: A measure of well-being in obesity surgery patients. *Obes Surg.* 2005; 15(4):538–545.

Welch, G, C Wesolowski, S Zagarins, et al. Evaluation of clinical outcomes for gastric bypass surgery: Results from a comprehensive follow-up study. *Obes Surg* 2011; 21(1):18–28.

Yanovski, S Sugar and fat: Cravings and aversions. Symposium: Sugar and fat- From genes to culture. *J Nutr* 2003; 133(3):835S–837S. http://jn.nutrition.org/content/133/3/835S.full.pdf+html

chapter ten

Quality of life, well-being, and stress

Quality of life

The World Health Organization defines quality of life (QOL) as a "complete state of physical, mental and social well-being and not merely the absence of disease or infirmity" (World Health Organization 2006). It has been described as "enjoying life," "being happy and satisfied with life," and "being able to do what you want to do when you want to do it."

Health-related quality of life (HRQOL) assessments are applicable to many populations and often have established national norms. Indices for assessing QOL for patients with specific conditions have also been validated. These include, but are not limited to, Impact of Weight on Quality of Life (IWQOL), Obesity Related Well-Being (ORWELL) (deZwaan 2002), the Gastrointestinal Quality of Life Index, the Bariatric Quality of Life Index, the Pediatric Quality of Life Inventory, the Skindex-Teen (for adolescents with skin disease), and an Oral Health-Related Quality of Life assessment tool.

Core dimensions of HRQOL in clinical research as described by Kral and colleagues (1992) included (1) physical complaints/well-being, (2) psychological distress/well-being, (3) functional status, (4) role functioning, (5) social functioning/well-being, and (6) health/perceived QOL. They concluded the improvement in QOL for those who had lost weight following bariatric surgery justifies surgically treating severely obese patients.

Nutritional quality of life (NQOL)

Nutrition has two types of impact on QOL writes Daniela Schlettwein-Gsell (1992) from the Institute of Experimental Gerontology in Basel, Switzerland. First, the preventive effect of nutrition is greatest in early childhood and gradually diminishes with age. Second, the effects of pleasure and enjoyment offered by food, the social aspects of meals, and the impact of eating on self-esteem all increase with age. Food for prevention is based on scientific calculation and experiments. Food for enjoyment is chosen by psychological mechanisms, which have their roots in childhood. QOL, as it can be related to nutrition, depends on the balance between these two (Schlettwein-Gsell 1992).

An initial step in the definition of NQOL by Barr and Schumacher (2003) utilized focus groups, surveys, and a consensus process. Six aspects of NQOL, along with selected relevant descriptors include:

1. Food impact (enough food to be satisfied, affordability, taste, and enjoyment)
2. Self-image (pleased, confused, related to food eaten)
3. Psychological factors (reward, happy, guilty, frustrated)
4. Social/interpersonal (stress regarding food, someone to talk to)
5. Physical (food-related condition caused problems sleeping, using the bathroom)
6. Self-efficacy (knew what to eat, felt confident) (Barr and Schumacher 2003)

Barriers to using knowledge—level II evidence

During in-depth interviews, individuals with serious mental illness described healthy eating accurately. Accomplishing healthy eating included internal barriers (such as negative perceptions, eating for comfort, decreased taste, and satiation) and external barriers (such as inconvenience, availability, social pressures, and medication side effects). Recommendations for caregivers included individualized education, opportunities to prepare and taste healthy foods, inquiry and discussion about eating in response to emotions, and the impact of medications (Barre 2011).

Well-being and mental energy

Well-being has been defined as a state of health, happiness, comfort, and prosperity. Two kinds of psychological well-being are (1) hedonic, short-term pleasures elicited by the senses and (2) eudaimonic, longer-term positive states such as feeling gratification or a sense of meaning. Six concepts included in eudaimonic well-being are autonomy, environmental mastery, self-acceptance, positive relations with others, personal growth, and purpose in life (Huppert 2004).

A related concept, "mental energy," may be included in a sense of well-being. The concept of mental energy was defined during workshops convened by the International Life Sciences Institute in 2004. The group defined mental energy as the ability to perform mental tasks, the intensity of feelings of energy/fatigue, and the motivation to accomplish mental and physical tasks. The model included the concepts of cognition, mood of energy, and motivation as core dimensions of mental energy. It has also been defined as "a three-dimensional construct consisting of mood (transient feelings about the presence of fatigue or energy), motivation (determination and enthusiasm) and cognition (sustained attention and vigilance)" (Gorby 2010). The group noted that variables that can influence mental energy include, among others, genetics, nutrition, pain, drugs, disease states, and sleep (International Life Sciences Institute 2006).

Feelings of mental energy and mood influence behavior. For example, a lack of energy may influence a person to avoid physical or mental work. A person may try to increase energy by eating, drinking, taking dietary supplements or drugs, or sleeping.

Methods that can be used to assess mental energy include tests of cognitive performance, vigilance and reaction time when making choices, mood questionnaires, electrophysiological techniques, brain scanning technologies, and ambulatory monitoring (Leiberman 2006, 2007). Measures of mental energy may be influenced by self-awareness, literacy, and faking answers (depending on the stakes) dependent on measurement outcome (O'Connor 2006).

Vitamins

Thiamin supplement—level II evidence

One hundred twenty females, 99 of which had normal transketolase levels (a measure of thiamin status) to begin, took a 50-mg supplement of thiamin for two months. Before and after scores on the Profile of Mood States (a self-reported measure of mood) were compared with a group who did not take thiamin. Subjects who took thiamin more than doubled their scores on clear-headedness and mood as well as increasing their quickness of reaction time. The placebo group showed no changes in these areas. Reported improvement included feelings of confidence, composure, and elation, although these changes were not statistically significant (Benton 1997).

Annual large-dose vitamin D and mental well-being—level I evidence

In a large group of elderly women given oral doses of 500,000 IU of vitamin D annually for 3 to 5 years, there was not a statistically significant correlation between vitamin D status and mental well-being. There was an inconsistent pattern observed for improved scores of anxiety or depression in a subgroup of those receiving the supplement (Sanders 2011).

Supplements

Healthy male volunteers—level I evidence

A study of 80 healthy male volunteers who took a moderate level multivitamin/mineral supplement for 28 days assessed the volunteers for anxiety and measures of perceived stress. Somatic complaints and dietary intake were assessed as well. The diet questionnaire revealed that 51% of the treatment group and 29% of the placebo group ate fresh fruit or vegetables less than daily and approximately half of both groups ate only white bread.

Relative to the placebo group, the treatment group had consistent and statistically significant reductions in anxiety and perceived stress. The treatment group rated themselves as less tired, better able to concentrate, and registered significantly fewer somatic symptoms on day 28. Physical symptoms reported were watery eyes, upset stomach, congested nose, headache, feverishness, ringing in the ears, eye strain, sore throat, persistent cough, shortness of breath, chills, malaise, sneezing, muscle aches, neck ache, and nasal discharge. As well as improving psychological well-being, the researchers believed the supplement protected participants against minor illness during the study (Carroll 2000).

Supplements, cognition, and multitasking—level I evidence

Nine weeks' supplementation with a multivitamin/mineral resulted in improved scores of mood and fatigue, increase in accuracy and speed in mathematical processing, and increase in altered response times during a Stroop test for 216 participants. A subgroup of participants also had significant reductions in homocysteine levels (Haskell 2010).

Other

Foods, nutrients, and alcohol related to anxiety, depression, and insomnia—level II evidence

A study of reported food and nutrient consumption and mental well-being of 27,111 males in Helsinki found that reported mood varied with food intake. Three symptoms addressed mental well-being: reports of anxiety, depression, and insomnia. The subjects were questioned annually for 3 years, at which times they were asked about their symptoms for the past 3 months. Men who reported anxiety, depression, and insomnia at baseline and all three annual visits were included in the analysis. Males with a history of previous mental disorders were excluded. Reported intake of fish, milk, meat, vegetables, margarine, coffee, and alcohol were evaluated as well as total calorie intake. These dietary factors were compared with males who did not report any of the symptoms that indicated a decreased state of mental well-being.

Findings showed intake of energy was 1 to 3% higher and intake of alcohol was 30 to 33% higher in men with symptoms than in those without symptoms. Men reporting symptoms obtained 6% of their total energy intake from alcohol; those without symptoms obtained 4% of their total energy intake from alcohol. In men with anxiety, omega-6 fatty acid intake was 7% greater than the intake of those without anxiety. Intake of omega-3 fatty acids or fish was not associated with anxiety or depressed mood. Body mass index was lower in subjects with any of the symptoms, despite a higher caloric intake. Tryptophan intake showed no association with mental well-being in this population. Intake of vitamin D and folic acid exceeded the daily recommendation and showed no association with mental well-being.

The researchers acknowledged limitations of the study related to data being reported rather than observed and that the cross-sectional method disallows assumption of causal association. They noted that the relationship between nutrition and mental well-being is a new area of study and that more attention needs to be paid to the intake of nutrients in patients suffering from depression, anxiety, and insomnia and the mechanisms of any action (Hakkarainen 2004).

It was noted in an editorial letter from Pierre Astorg that the lack of association in the above study may have been related to the research protocol providing (1) vitamin E, (2) beta-carotene, (3) both E and beta-carotene, or (4) a placebo. Since systemic oxidant stress decreases blood omega-3 fatty acids and the subjects were smokers, the evidence of association with depression could have been influenced by the supplements. Astorg also mentioned that dietary alpha-linolenic acid (ALA) is converted to eicosapentaenoic acid (EPA) at a very low rate and also questioned whether the study asked about participant's use of fish oil supplements and not merely fish intake (Astorg 2005).

Stress

The original definition of stress by Hans Selye was "the non-specific response of the body to any demand for change" (Selye 1984). Laboratory animals exposed to stresses such as blaring light, deafening noise, extremes of heat or cold, or perpetual frustration exhibited pathologic changes such as stomach ulcerations, shrinkage of lymphoid tissue, and enlargement of the adrenals. Persistent stress caused these animals to develop diseases similar to human heart attacks, stroke, kidney disease, and rheumatoid arthritis. Contrary to the prevailing belief that most diseases were caused by specific but different pathogens, Selye proposed that different insults could cause the same disease, not only in animals, but in humans as well. Stress may potentially affect the immune system, hormones, brain neurotransmitters, prostaglandins, enzyme systems, and metabolic activities.

This phenomenon is also known as general adaptation syndrome (GAS) or the fight or flight response. One definition used today defines stress as a condition or feeling experienced when a person perceives that demands exceed the personal and social resources the individual is able to mobilize. Eustress is stress with a positive connotation, such as winning a prize.

A biopsychosocial model includes three components of stress: an external element (the environment), an internal element (the physiological and biochemical response), and the interaction between these two factors. The interaction includes the cognitive processes that result from the interaction.

Stress may include some combination of anger or irritability, anxiety, and depression. Acute stress may be related to stomach, gut, and bowel problems such as heartburn, acid stomach, flatulence, diarrhea, constipation, and irritable bowel syndrome. Chronic stress may be related to loneliness, poverty, bereavement, depression, and frustration. The worst aspect of chronic stress is that people get used to it but the effects continue. Chronic stress may result in suicide, violence, or physical disease.

The physical response to stress includes a rise in blood glucose to furnish quickly available fuel for muscle cells, decrease in blood flow to the gastrointestinal (GI) tract, and increase of blood flow to muscles in the arms and legs.

Oxidative stress

Oxidative stress refers to the effect of greater than normal production of free radical molecules in the mitochondria during energy metabolism and the potential damage to lipids in cell membranes, DNA, enzymes, neurotransmitters, and structural proteins by these high-energy free radicals. Polyunsaturated phospholipids in neuronal membranes are particularly vulnerable to damage by free radicals. It is hypothesized that antioxidants such as vitamins E and C might prevent this damage.

Ethane as a biomarker for lipid peroxidation—level I evidence

Exhaled ethane is a specific biomarker for cerebral omega-3 polyunsaturated fatty acid peroxidation in humans. The cerebral source of ethane is the docosahexaenoic acid (DHA) component of membrane phospholipids. Patients with schizophrenia have been shown to have increased free radical-mediated damage and cerebral lipid peroxidation (Puri 2008).

Oxidative stress in psychiatric disorders—level V evidence

Research regarding oxidative stress in numerous chronic diseases prompted Tsaluchidu and colleagues (2008) to investigate whether there was published evidence of oxidative stress in psychiatric disorders. Their extensive search of psychiatric disorders led them to conclude the majority of psychiatric disorders are associated with increased oxidative stress. The authors concluded those with disorders in which there might be increased lipid oxidation might benefit from fatty acid supplementation.

Amino acids

Lysine and anxiety—level I evidence

Wheat is low in the essential amino acid lysine. Prolonged dietary lysine inadequacy in animals increases stress-induced anxiety. A 3-month, randomized, double-blind study tested whether lysine fortification of wheat reduces anxiety and stress response in family members whose diet staple was wheat. Lysine-fortification reduced plasma cortisol response in females, as well as sympathetic arousal in males as measured by skin conductance. These results suggest that diet might have the capacity to reduce stress-induced anxiety and possibly improve communal QOL and mental health (Smirga 2004).

Carbohydrate

Awareness of hypoglycemia and stress—level II evidence

A group of scientists concluded that acute stress (such as giving a speech) during hypoglycemia reduces symptom awareness and the ability to detect hypoglycemia, even at evoked levels of plasma glucose below 2.8 mmol/L (about 30 mg/dl) (Pohl 1998).

Vitamins

B_6 and bereavement stress—level II evidence

Pyridoxine is involved in 100 metabolic reactions, some involving the neurotransmitters serotonin and gamma aminobutyric acid (GABA). For example, pyridoxine is a cofactor for 5-hydroxytryptophan decarboxylase, an enzyme in the biosynthesis pathway of serotonin.

A study that tested plasma pyridoxine (vitamin B_6) status as a predictor of overall psychological distress and mood states following bereavement provided measures for 75 HIV-1-positive subjects and 58 HIV-1-negative homosexual males.

Pyridoxine deficiency status, determined with a bioassay of erythrocyte aspartate aminotransferase activity, found pyridoxine deficiency was a significant predictor of increased overall psychological distress in this model. A post hoc analysis of specific mood states showed pyridoxine deficiency status was significantly associated with increases in depression, fatigue, and confused mood levels, but not with those of anxiety, anger, or vigor. The authors suggested that adequate pyridoxine status might be necessary to avert psychological distress in the setting of bereavement (Baldewicz et al. 1998).

Other

Stress and gastrointestinal symptoms

GI illness and symptoms have the potential for compromising nutritional status through altered absorption of nutrients or change of food choices and nutrient intake.

Irritable bowel syndrome (IBS), Crohn's disease, and other GI diseases can be quite disturbing to individuals experiencing them and influence the individual's physical, emotional, and social QOL. Anxiety states and depression are frequent in patients with IBS. Fright and depression reduce acid secretion and reduce gastric motor activity. In comparison, anger and resentment accelerate acid secretion and increase GI motor activity. Symptom severity has been related to chronic stressors such as divorce, serious illness, lawsuits, and housing difficulty (Wilhelmsen 2000).

Baseline psychological stress has been linked to having persistent GI symptoms and seeking health care for them over time. Short-term psychological stress levels do not appear important for explaining changes in symptoms. Researchers recommended clinicians consider psychological factors in the treatment of patients with IBS (Koloski 2003).

A study of the placebo effect, in which participants were informed that they were receiving a placebo, resulted in improvement in the Symptom Severity Scale and the Adequate Relief Scale in 3 weeks. The researchers commented that placebo treatment of conditions such as depression, anxiety, and chronic pain, which have subjective outcomes, are often effective (Kaptchuk 2010).

Assessment of a random population of adults found that subjects with functional gastrointestinal disorders (FGID), IBS, functional bloating, constipation, diarrhea, and unspecified functional bowel disorder have a higher risk of psychological illness than somatic illness. Subjects with psychological illness have a higher risk of severe GI symptoms than controls. Patients with FGID suffer more severe GI symptoms that non-patients do (Alander 2005).

Functional refers to a disease that involves symptoms without tissue damage. Functional disease may be iatrogenic, idiopathic (no known cause), or psychogenic.

Psychological distress and inadequate dietary intake in immigrant women—level II evidence

Data regarding daily nutrient intake were compared with quartile scores measuring psychological distress in Vietnamese women intermarried and living in Korea. Subjects with the highest stress scores were more likely to skip breakfast and to change their dietary habits after living in Korea than those in groups with low stress scores. Those with the highest stress scores were less likely to consume milk or dairy products, eat regular meals, or have balanced diets and have lower consumptions of energy, carbohydrate, protein, fat, calcium, zinc, thiamin, riboflavin, and folate. The prevalence of being underweight (BMI <18.5) increased from the lowest to highest quartiles of psychological distress scores (Hwang 2010).

Vitamin D hypovitaminosis and chronic pain—level II evidence
Chronic pain decreases QOL and sense of well-being. Investigators have linked pain from several conditions to a deficiency of vitamin D, although deficiency is not defined uniformly. Whether the link is causal is still in question, but worth further investigation. In a group of patients with diabetic neuropathy, 81% had vitamin D levels <30 ng/ml (Soderstrom 2011).

Vitamin D and mental well-being
A nonclinical group of women ≥70 years old who were given an annual dose of 500,000 IU vitamin D for 3 to 5 years were also measured on three psychological or health tests. Serum vitamin D levels were 41% higher after receiving the supplement, but there was no difference in mental well-being between the supplemented group and those who were on depressive medication at baseline (Sanders 2011). (See also Chapter 3 "Aggression, Anger, Hostility, and Violence.")

Conclusions

1. Historically, two aspects of NQOL are described as (1) the preventive aspect, which is greatest in childhood and (2) the pleasure aspect, which increases with age. New knowledge describes additional aspects of the role of nutrients and nutritional status in NQOL.
2. Stress may refer to psychological/emotional/social stress, physical stress, or biochemical oxidative stress.
3. Social or emotional stress is observationally related to GI symptoms.
4. Psychiatric disorders are associated with oxidative stress, which may affect lipids in cell membranes, DNA, enzymes, and neurotransmitters.
5. Supplements of thiamin or multivitamins/minerals have been observationally linked to improvement in mood, perceived stress, and anxiety.
6. Anti-oxidant vitamins such as C and E may counter oxidative stress and the effects on polyunsaturated lipids.

References

Alander, T, K Svardsubb, S-E Johansson, and L Agreus. Psychological illness is commonly associated with functional gastrointestinal disorders and is important to consider during patient consultation: A population-based study. *BMC Med* 2005; 3(1):8. see also http://www.biomedcentral.com/1741-7015/3/8

Astorg, P. Omega-3 fatty acids and depression. Letter to the editor, *Amer J Psychiatry* 2005; 162:402.

Baldewicz, T, K Goodkin, DJ Feaster, et al. Plasma pyridoxine deficiency is related to increased psychological distress in recently bereaved homosexual men. *Psychosomatic Med* 1998; 60(3):297–308.

Barr, J and G Schumacher. Using focus groups to determine what constitutes quality of life in clients receiving medical nutrition therapy: First steps in the development of a nutrition quality-of-life survey. *J Amer Diet Assoc*. 2003; 103:844–851.

Barre, LK, JC Ferron, KE Davis, and R Whiteley. Healthy eating in persons with serious mental illnesses: Understanding and barriers. *Psychiatr Rehabil J* 2011; 34(4):304–310.

Benton, D, R Griffiths, and J Haller. Thiamine supplementation improves mood and cognitive functioning. *Psychopharmacology* 1997; 129:66–71. de Zwaan, M, JE Mitchell, LM Howell, et al. Two measures of health-related quality of life in morbid obesity. *Obes Res*. 2002; 10(11):1143–1151.

Carroll, D, C Ring, M Suter, and G Willemsen. The effects of an oral multivitamin combination with calcium, magnesium, and zinc on psychological well-being in healthy young male volunteers: A double-blind placebo-controlled trial. *Psychopharmacology* 2000; 150:220–225.

Hakkarainen, R, T Partonen, J Haukka, et al. Food and nutrient intake in relation to mental wellbeing. *Nutr J* 2004; 3:14. see also http://www.biomedcentral.com

Haskell, CF, B Robertson, E Jones, et al. Effects of a multi-vitamin/mineral supplement on cognitive function and fatigue during extended multi-tasking. *Hum Psychopharmacology* 2010; 25(6):448–61.

Huppert, FA and N Baylis. Well-being: Towards an integration of psychology, neurobiology and social science. *Phil Trans R Soc Lond. B*. 2004; 359:1447–1451.

Hwang, J-Y, SE Lee, SH Kim, et al. Psychological distress is associated with inadequate dietary intake in Vietnamese married immigrant women in Korea. *J Amer Diet Assoc*. 2010; 110(5):779–785.

International Life Sciences Institute Workshop. Mental Energy: Defining the Science. From the International Life Sciences Institute Workshop. Washington DC.: *Special Supplement to Nutrition Reviews* 2006; 64(7)(Part II):(S1–S16).

Kaptchuk, TJ, E Friedlander, JM Kelley, et al. Placebos without depression: A randomized controlled trial in irritable bowel syndrome. *PLoS One* 2010; 5(12) e15591.

Koloski, NS, NJ Talley and PM Boyce. Does psychological distress modulate functional gastrointestinal symptoms and health care seeking? A prospective, community cohort study. *Am J Gastroenterol* 2003; 98(4):789–797.

Kral, JG, LV Sjostrom, and MBE Sullivan. Assessment of quality of life before and after surgery for severe obesity. *Am J Clin Nutr* 1992; 66:611S–614S.

Lieberman, HR. Mental energy: Assessing the cognition dimension. *Nutr Rev* 2006; 64(7–Part 2):S10–3.

Lieberman, HR. Cognitive methods for assessing mental energy. *Nutr Neurosci.* 2007; 10(5–6):229–42.

O'Connor, PJ. Mental energy: Assessing the mood dimension. *Nutr Rev* 2006; 64(7 Part 2):S7–9.

Pohl, J, G Frenzel, W Kerner, and G Fehm-Wolfsdorf. Acute stress modulates symptom awareness and hormonal counter-regulation during insulin-induced hypoglycemia in healthy individuals. *International J Behav Med* 1998; 5(2):89–105.

Puri, BK, SJ Counsell, BM Ross, et al. Evidence from *in vivo* 31-phosphorus magnetic resonance spectroscopy phosphodiesters that exhaled ethane is a biomarker of cerebral *n*-3 polyunsaturated fatty acid peroxidation in humans. *BMC Psychiatry* 2008, 8 (Suppl 1):S2 http://www.biomedcentral.com/1471–244X/8/S1/S2

Sanders, KM, AL Stuart, EJ Williamson, et al. Annual high-dose vitamin D3 and mental well-being: Randomised controlled trial. *Br J Psychiatry.* 2011; 198:357–364.

Schlettwein-Gsell, D. Nutrition and the quality of life: A measure for the outcome of nutritional intervention? *Am J Clin Nutr* 1992; 55:1263S–1266S.

Selye, H. *The Stress of Life*, Revised edition.1984; p.55. McGraw Hill, New York.

Smriga, M, S Ghosh, Y Mouneimne, et al. Lysine fortification reduces anxiety and lessens stress in family members in economically weak communities in northwest Syria. *Proc Natl Acad Sci USA* 2004; 101(22):8285–8288.

Soderstrom, LH, SP Johnson, VA Diaz, and AG Mainous III. Association between vitamin D and diabetic neuropathy in a nationally representative sample: Results from 2001–2004 NHANES. *Diabetes Med* 2011 epub.

Tsaluchidu, S, M Cocchi, L Tonello, and BK. Puri. Fatty Acids and oxidative stress in psychiatric disorders. *BMC Psychiatry* 2008; 8 (Suppl 1):55.

Wilhelmsen, I. The role of psychosocial factors in gastrointestinal disorders. *Gut* 2000; (suppl IV)47:iv73–iv75.

World Health Organization. *Constitution of the World Health Organization. Basic Documents*; Forty-fifth edition Geneva: World Health Organization, Supplement, October 2006.

chapter eleven

Additional links between mental status and nutritional status

Introduction

Some interesting topics do not fit neatly into previous categories, but do describe links between nutrients/nutritional status and mental status. These include the accumulation of copper in Wilson's disease, mercury toxicity and fish consumption, selenium and seizures, use of nutrients in treating migraine headaches, the presence of multiple nutrient deficiencies described in patients with agoraphobia, and others. In addition, some nontraditional perspectives such as spirituality and eating habits, and the effects of biofeedback and state of mind on the function of the body are included.

Nutrients

Vitamins

Agoraphobia and vitamin status—level II evidence

The interrelationship of various nutrients is important in the production of pathological changes. Agoraphobia (AGP), defective glucose metabolism, and beriberi are different disorders. However, they all demonstrate symptomatology of disturbances in the glucose-oxidative pathway as reported by Laraine Abbey, a nurse practitioner in private practice in East Windsor, New Jersey (Abbey 1982).

Beriberi, the disease associated with thiamin deficiency, has been described as a typical example of autonomic dysfunction. Symptoms seen in beriberi patients include palpitation and dyspnea with exertion and mental excitement, which are also commonly described as part of the AGP syndrome.

> Autonomic systems are self-regulating—not in our conscious control.

Patients diagnosed with AGP may also report symptoms such as tachycardia, elevated blood pressure, nausea and vomiting, altered energy metabolism, dizziness at heights (vertigo), ataxia, sensory and motor nerve aberrations, respiratory symptoms (dyspnea), and digestive symptoms including sensation of fullness, heartburn, and constipation. Increase in lactic acid with exertion is seen in beriberi and in patients with anxiety neurosis.

In 1989, Abbey published a report on AGP and nutritional data found in a set of her patients. She described AGP as a debilitating phobia, a complex psycho-physiological disorder manifesting as severe anxiety or panic reactions, accompanied by dread of being away from a "safe" place. Abbey suggested it might be called a somato-psychic disorder; that is, a nutritional-biochemical imbalance that creates an emotional effect.

Abbey described these patients' diets as among the worst she had seen. Three-fourths of calories came from processed foods, usually "junk" carbohydrates low in micronutrients. She noted that many symptoms suffered by agoraphobics were identical to those commonly presented in hypoglycemia. Recovery in this context is defined as normal mobility and elimination of panic reactions.

Abbey reported on results from biochemical laboratory tests on 12 patients who had previously failed to improve with (1) standard treatment provided by psychiatrists, (2) dietary improvement, and (3) intake of a moderate-level multivitamin/mineral supplement (Table 11.1). In another group of 136 highly symptomatic patients for whom functional vitamin testing was performed, only 5 failed to show abnormalities in any of the nutrients assayed.

Abbey further included the explanation that blood infusions of lactate ion produced symptoms identical to those experienced in spontaneous anxiety attacks or breathlessness, particularly in crowded or stuffy atmospheres. As blood lactate rises, pyruvate is high (L/P ratio) and Adenosine

Table 11.1 Abnormal Nutritional Status in Agoraphobic Patients ($N = 12$)

Vitamin	Laboratory Test	Enzyme Activity Tested	No. of Clients with Abnormal Tests[a]
Thiamin	TPP% uptake	Erythrocyte transketolase activity	7
Riboflavin	Red cell	Glutamic reductase	1
Pyridoxine	Red cell	Glutamic pyruvic transaminase	6
Niacin		N1-methylnicotinamide	3
Folacin	Serum	FIGLU (formiminoglutamic acid test)	2
Vitamin B_{12}	Serum	MMA (methylmalonic acid)	3

Source: Abbey 1989.

[a] 10 clients had more than one enzymopathy; 2 clients had more than two enzymopathies; 1 client had more than three enzymopathies.

triphosphate (ATP) is decreased in concentration, illustrating a disturbance in transformation of chemical energy to kinetic energy in anxiety-prone patients. Disturbed thiamine biochemistry could theoretically be one mechanism to produce high L/P ratio.

Abbey advises that chronic symptoms even without objective signs must be taken seriously as a sign of biochemical disturbance. The individual may be beyond the "adaptation" stage of nutritional injury and moving into the stages of nutritional injury demonstrated by alterations in biochemistry and behavioral or emotional change, but not demonstrating physical clinical signs.

Pharmacologic doses (generally larger than nutriologic doses) of vitamins, given for individually determined abnormalities, have eliminated or markedly reduced symptoms. Under certain circumstances, massive levels of a vitamin increase saturation of the binding site for an enzyme. Increasing the percentage of functioning coenzymes will override a partial enzyme block and stimulate or allow the enzyme to function at close to normal rate. Nutritionists have described the stages during which clinical or biochemical lesions may precede histologic (physically observable) changes (Arroyave 1979). (See also Stages of Nutritional Injury, Appendix A.5.)

Niacin for treatment of migraines—level III evidence

Prousky and Seely (2005) noted that the National Headache Foundation reported that 45 million Americans suffer from chronic, recurring headaches, and 28 million of these suffer from migraine headaches annually. These authors' review of literature found nine studies that met their meta-analysis inclusion criteria. They concluded that the quality of the evidence is only hypothesis generating, having a low level of external generalizability. Possible mechanisms of positive results included the vasodilation effect of niacin, which may relieve venous pressure. A deficit of mitochondrial energy metabolism may play a role in the pathogenesis of migraines. The possible prophylactic benefits of daily oral niacin may be related to the improvement of mitochondrial energy metabolism and phosphorylation. They mention that riboflavin and coenzyme Q_{10} may also be worthwhile to investigate due to their roles in the mitochondrial respiratory chain.

Riboflavin and migraine headaches—level II evidence

Hospitalized psychiatric patients often have a history of migraine headaches. A group of researchers reported using pharmacological doses of riboflavin (25 to 400 mg) alone or in conjunction with magnesium and coenzyme Q_{10} as treatment for migraine headaches. Treatment groups received 25 or 400 mg riboflavin (the high-riboflavin group also received magnesium and feverfew). Both groups experienced a reduction in the number of migraines, migraine days, and migraine index (Maizels 2004).

Symptoms of vitamin B_{12} deficiency—level II evidence

The mental signs of vitamin B_{12} deficiency include irritability, apathy, somnolence, suspiciousness, and emotional instability. Symptoms also include memory problems, slowed thinking, mood swings, and trouble concentrating. Zucker and colleagues (1981) reported symptoms of B_{12} deficiency in psychiatric patients as organic brain syndrome, paranoia, violence, and depression.

"An estimated 10–20% of people in their late 60s or older have some degree of B_{12} deficiency," wrote Ralph Carmel, professor of medicine at Weil Medical College of Cornell University in New York and the director of research at New York Methodist Hospital (Mt. Sinai School of Medicine 2002). Up to 30% of older adults lack the stomach acid and other components necessary to absorb adequate amounts of this nutrient from food. In fact, 3 to 12% of older adults have a degree of B_{12} deficiency that is potentially harmful and might require treatment.

Neuropsychiatric disorders in the absence of anemia have been found in some patients suffering from a vitamin B_{12} deficiency. The cause of neuropathy appears to be directly related to the adequacy of methionine (an essential amino acid) in the body (Hunt 1990). This symptom of vitamin B_{12} deficiency can be treated only through an increased supply of methionine or an accelerated production of methionine from homocysteine, a reaction that requires vitamin B_{12}.

An inadequate amount of methionine caused by a deficiency of vitamin B_{12} decreases the availability of S-adenosyl methionine (SAM). SAM is required for important methylation reactions, such as choline formation, that are essential to the maintenance of myelin and prevention of neuropathies.

Vegetarianism can cause vitamin B_{12} deficiency. In a study in India of 17 affluent patients treated for vitamin B_{12} deficiency, all had impaired muscle function and decreased nerve conduction. All the patients had impaired joint position and vibration sensation in the lower limbs, and four had these symptoms in the upper limbs as well. Lower limbs were spastic in 13 patients, and upper limbs were spastic in 2 patients. At a 6-month follow-up, 2 patients had improved completely, 7 were partially improved, and 3 had poor recovery. The evoked potential studies and MRI changes in vitamin B_{12} deficiency syndrome are consistent with focal demyelination of white matter in the spinal cord and optic nerve. Of the 17, 3 were females and 12 were lacto-vegetarian (a vegetarian who includes dairy products in his or her diet) (Misra 2003).

Inositol: Review of effects—level III evidence

A review of controlled trials of inositol supplements concluded inositol has therapeutic effects in the spectrum of illness responsive to serotonin selective re-uptake inhibitors, including depression, panic, and obsessive-compulsive disorder (OCD), and was not shown to be beneficial for schizophrenia, Alzheimer's, ADDH, autism, or electroconvulsive therapy (ECT)-induced cognitive impairment. A trend was noted related to aggravation of ADHD syndrome using myo-inositol as compared to placebo (Levine 1997). (See also Chapter 7.)

Minerals

Copper and Wilson's disease—level II evidence

Brewer and colleagues (2006) conducted a randomized study of 48 patients with neurologic presentation of Wilson's disease (Table 11.2). Participants in arm one of the study received 20 mg of tetrathiomolybdate (TM) three times a day with meals and 20 mg of TM three times a day between meals. In arm two of the study, they received 500 mg two times a day between meals for 8 weeks while hospitalized. All patients were given 50 mg of zinc twice a day and were discharged taking 50 mg of zinc three times a day. Neurologic deterioration occurred in six (26%) patients taking Trientine; one (4%) patient taking TM had deterioration. High baseline speech score was found predictive of neurologic deterioration during treatment. Potential side effects include anemia, leucopenia, or elevated transaminase, prompting the authors to recommend weekly follow-up especially after week three. If side effects occur, they further recommend temporarily stopping the drug and resuming at half dose after a few days (Brewer 2006; Aurangzeb 2009). (See Table 11.2.)

Mercury toxicity

Mercury is found in several forms: (1) elemental mercury (found in mercury vapor, thermometers, etc.), (2) inorganic mercury (found in mercury salts, topical medicine, manufacturing processes, and in some foods), and (3) organic mercury (commonly the methyl form). Organic mercury is the most potentially dangerous form of mercury and is found in paints, preservatives, fungicides, seeds, foods, medicines, cosmetics, and wood preservatives.

Mercury toxicity usually occurs because of gastrointestinal exposure. Absorbed mercury vapor is lipid-soluble and readily crosses the blood–brain barrier and the placenta as well as enters breast milk of lactating women. Mercury compounds penetrate the membrane of red blood cells and bind to hemoglobin. It may concentrate in the kidney or central nervous system. Organic mercury is largely excreted in the urine, with a half-life of about 70 days.

Mercury poisoning from vapors (such as historically occurred in the manufacture of felt hats, leading to the phrase "mad-hatters") or from chronic inorganic poisoning produces mental symptoms such as lassitude, anorexia, timidity, memory loss, insomnia, excitability, and

Table 11.2 Copper: Facts, Signs, and Symptoms, and Treatments for Wilson's Disease

Basic Facts about Copper	Wilson's Disease	
	Signs and Symptoms	Treatments
Plasma Cu: N = 0.91–1.0 µg/ml adult males; N = 1.07–1.23 µg/ml adult females; plasma level is not highly correlated with intake. Serum ceruloplasmin in Wilson's disease: <23 µg/dl. Urine: N = 0.01–0.06 mg/24 hours. In Wilson's disease ≥1.5 mg/day; <50 µg excludes Wilson's disease. Total body Cu = 50–100 mg. Average daily intake: 1–2 mg/day. Intake of 1.5–3 mg is recommended for adults. Usual Western diet provides 24 mg/day. Zinc interferes with Cu absorption and increases fecal excretion of Cu. Foods high in Cu include liver, broccoli, legumes, chocolate, nuts, mushrooms, and shellfish, particularly oysters, crab, and lobster. Well water is variable and should be tested for Cu levels. Phytates (fiber) and ascorbic acid decrease absorption. Intake of >15 mg elemental Cu produces nausea, vomiting, diarrhea, and cramps (direct mucosal toxicity).	Wilson's disease occurs at a rate of 1 per 30,000. Most common age of manifestation is 8–18 years. Signs and symptoms: Kayser-Fleisher (KF) rings in the cornea occur in 90% of individuals; may occur in other chronic conditions as well. Progressive movement disorder is a hallmark of neurologic Wilson's disease. Bluish discoloration occurs at the base of the fingernails. One-third of patients present with psychiatric symptoms such as depression, labile mood, impulsiveness, disinhibition, self-injurious behavior, or psychosis. Free plasma Cu (in contrast to bound plasma Cu) shows significant inverse relationship with Mini-Mental State Examination (MMSE) and attention-related neuropsychological test scores. Positron emission tomography (PET) has revealed a significantly reduced regional cerebral metabolic rate of glucose consumption.	Alcohol should be avoided due to involvement of the liver in Wilson's disease. Zinc monotherapy — induces intestinal cell metallothionein, which binds Cu from food and endogenous secretions, preventing its transfer to the blood. Takes 4–6 months to control effects of Cu. D-penicillinamine — chelates copper, increases urinary excretion. Adverse effects: initially mobilizes large stores of Cu from liver, which elevates blood and brain Cu with potentially severe consequences. D-penicillinamine is an anti-metabolite of pyridoxine (vitamin B_6), resulting in the recommendation for a supplement of B_6 of 12.5–25 mg or 75–150 nmol/day. Trientine — chelates copper, increases urinary excretion, better tolerated than D-penicillinamine. Tetrathiomolybdate (TM) — absorbed into the blood, complexes with albumin, making it unavailable for cellular uptake. TM may be used along with an oral supplement of zinc.

delirium. Symptoms may be seen with blood mercury levels above 1 μmol/L (20 μg/dl) and urine mercury levels above 3 μmol/L (600 μg/dl). Multiple physical symptoms such as headache and lack of coordination co-occur (Graef 1994).

Mercury in fish—level II evidence

Burger, Stern, and Gochfeld (2005) examined mercury levels in tuna, flounder, and bluefish, commonly available in New Jersey stores. They found significant species differences, with tuna having the highest levels and flounder the lowest levels. They also examined mercury levels in six other commonly available fish and two shellfish from central New Jersey markets. Both shrimp and scallops had total mercury levels <0.02 ppm wet weight. Large shrimp had significantly lower levels of mercury than small shrimp. For tuna, sea bass, croaker, whiting, scallops, and shrimp, the levels of mercury were higher in New Jersey samples than did those mentioned in an FDA report dated 1990–1992.

The fish most available to consumers were flounder, snapper, bluefish, and tuna. Tuna had the highest mercury value: >2 ppm of mercury. In fresh tuna, 42% of the fillets had mercury levels >0.5 ppm. The best prices for consumers were for whiting, porgy, croaker, and bluefish (all with average mercury levels <0.3 ppm wet weight). Flounder was the fish with the best relationship among availability, cost, and low mercury levels.

The authors included a 2001 FDA consumption advisory regarding methyl mercury that suggested pregnant women and women of childbearing age who may become pregnant should limit their fish consumption, avoiding four types of marine fish (shark, swordfish, king mackerel, and tilefish) and limit their consumption of all other low-mercury fish to 12 oz./week (Burger et al. 2005).

Mercury in consumers of large amounts of fish—level II evidence

The Environmental Protection Agency (EPA) and the National Academy of Sciences recommend keeping whole blood mercury <5.0 μg/L or hair level at <1.0 μg/g. This corresponds to a reference dose (RfD) of 0.1 μ/kg body weight. Approximately 95% of methyl mercury consumed in fish is absorbed. It is taken up by all tissues, but accumulates to greater concentrations in the brain, muscle, and kidney within 1 to 2 days after being consumed. In the brain, methyl mercury is transformed to inorganic mercury. The 70-day half-life of methyl mercury in blood compares to the half-life of inorganic mercury in the brain in years. Fetal mercury concentration is approximately double that of maternal concentration.

Hightower and Moore (2003) reported an assessment of mercury of patients being seen in an internal medicine practice in San Francisco, where 720 patients were evaluated for excess mercury. The investigators conducted an extensive and detailed estimate of dietary fish and mercury intake. If an

Table 11.3 Screening of Patients for Blood Mercury Levels

Number T = 116	% of Total Screened	Mercury Levels
103	89	≥5.0 µg/L
63	54	≥10 µg/L
19	16	≥20 µg/L
4	3	≥50 µg/L

Source: Hightower and Moore, 2003.

Note: National Academy of Sciences recommends keeping whole blood mercury <5.0 µg/L.

individual's intake of fish for 1 month suggested greater than RfD of 0.1 µ/kg body weight, it was suggested he or she be screened for whole blood mercury level. Individuals were also screened if they complained of symptoms of mercury excess: fatigue, headache, decreased memory, decreased concentration, and muscle or joint pain. One hundred sixteen patients were screened for whole blood levels; seven had hair analysis (Table 11.3).

Shark, with 1.3 µg/g, and swordfish, with 0.95 µg/g, had the highest estimated levels of mercury. Reducing fish intake produced a decrease in whole blood levels in participants. All but two patients had obtained levels <5.0 µg/L by 41 weeks.

Authors noted variation could occur due to different kinetics in individuals, the wide range of mercury in fish, the timing of the laboratory test related to the consumption of fish, and inaccuracy of dietary recall. Other possible environmental contaminants and other health conditions could also contribute to symptom overlap. The authors recommended that diet histories that include intake of fish should be part of a comprehensive health screening (Hightower 2003). In the August 2003 issue of *Reader's Digest*, several case studies and other findings of interest to the public were described in an article entitled "How Safe Is Your Food? One fish, two fish, red snapper, swordfish: A menace lurks in your 'healthy food.'" (Jeter, 2003). (See also Chapter 6—Intellect, Cognition and Dementia; Mercury in Older Adults.)

Selenium

Selenium (Se) is an essential trace element, although the level of selenium in food items reflects the soil in which they were grown and thus varies markedly between different parts of the world. While small amounts are essential, excess amounts of selenium are toxic.

There is greater uptake of selenium in the brain than in other tissues. The metabolism of selenium by the brain differs from other organs. Preferential retention of selenium in the brain at times of deficiency suggests that selenium plays an important function. The function of selenium in the brain is related to functions of selenium-dependent enzymes glutathione peroxidase (Se-GPX) and type I thyroxine deiodinase, a selenocysteine-containing enzyme. Selenium functions in an oxidative damage-protective system, in the regulation of peroxide concentration, and in the biology of aging. Se-GPX activity occurs in myelin tissue. Polyunsaturated fatty acids (PUFA) are protected from peroxidation by Se-GPX (Beard 1996).

Savaskan et al. (2003) suggest that selenium in selenium-containing proteins plays an additional role in anti-apoptosis and has a function beyond that of other antioxidants in protecting against cell death seen in seizures and CNS trauma. An Se-deficient diet in rats first depleted the liver while the brain was depleted 10%. This level of brain depletion resulted in a remarkably higher seizure rate, loss of neurons, and structural damage when compared to animals fed an Se-adequate diet (Savaskan et al. 2003).

Although the mechanism is not clear and needs may be higher than the needs for normal levels of glutathione peroxidase, mood, attention, arousal, and memory may be influenced by selenium. Since selenium is involved in protecting against oxidation reactions, it could possibly play a role in aging, Parkinson's disease, and Alzheimer's disease (Benton 2002).

Selenium levels and enzyme activity in epileptic and healthy children—level II evidence. Serum Se levels and red cell glutathione peroxidase activity in 53 children with epilepsy and 57 controls demonstrated a significant difference in the two groups (Table 11.4) (Ashrafi 2007).

Selenium and seizures: a case study—level IV evidence. In a report of two children with intractable seizures and low selenium status, the older child was given selenium supplementation (3 to 5 µg/kg), which reduced seizures and improved EEG after two weeks. The younger child, who had seizures from day four of life, died at age 10 months (Ramaekers 1994).

Selenium and quality of life—level II evidence. In a randomized, controlled study of 501 volunteers aged 60 to 74, participants were provided three levels of selenium supplementation (100, 200, or 300 µg/day) for 6 months; mood states were assessed at baseline and at 6 months. The authors found no evidence of benefit to mood or quality of life. They felt the findings were evidence that the brain is a privileged site for selenium retention. A U-shaped curve for safety and effectiveness may be just as applicable for mental health as it is for physical health: supplementation

Table 11.4 Selenium and Epilepsy

	Serum Se	Enzyme Activity
Controls	86.0 mcg/L	801.0 nmol/min/ml
Children with epilepsy	72.9 mcg/L	440.57 nmol/min/ml

Source: Ashrafi et al. 2007.

Magnesium

Magnesium and migraine headache—level I evidence. A group of researchers led by Demirkaya described a randomized, single-blind, placebo-controlled trial of 1 g of intravenous magnesium sulfate for treatment of moderate or severe migraine headaches. Patients were assessed immediately after treatment, then 30 minutes and 2 hours later. Pain and accompanying symptoms either decreased or disappeared (pain decreased 86.6%; accompanying symptoms decreased 100%). Three patients in the placebo group reported a decrease in pain severity after the saline placebo, but subsequently all reported an end of the migraine headache after being given a dose of magnesium sulfate (Demirkaya et al. 2001).

Since thiamin-dependent enzymes require magnesium for activation, it is possible that the influence of a magnesium deficiency is less significant when a deficiency in thiamin is also present. If both were deficient, restoring thiamin levels would cause the magnesium deficiency to become a limiting factor for thiamin-dependent enzymes (Hunt 1990).

Manganese

"Manganese Madness" was the first name given to manganism, or manganese (Mn) toxicity. Compulsive behavior, emotional lability, and hallucinations were the psychiatric symptoms occurring with common Parkinson's-like syndrome.

Food intake of Mn is estimated to be 2 to 9 mg/day. Particles of Mn_3O_4 are released during combustion of an additive to gasoline (Barceloux 1991). Manganese in drinking water has been shown to decrease intellectual function in children (Sharma 2006). Low absorption rates seem protective against Mn toxicity. Patients most likely to develop Mn toxicity include those with chronic liver disease, cholestasis (diminished bile flow), individuals on dialysis, those with other malabsorption syndromes, or individuals on long-term total parenteral nutrition (TPN). Cognitive disturbances include mental confusion, reduced concentration, anxiety, and hallucinations (Neville 2002).

Other

Choline, cognition, and dementia

Choline is an amine, is essential for human growth and metabolism, is produced endogenously, is a source of methyl groups, and is a precursor of acetylcholine and phosphatidylcholine. It is especially important in the third trimester of pregnancy and thought to be important for development of the central nervous system and future cognition of the fetus. It is often not included in vitamin-mineral supplements, and so must be obtained from food if not present in sufficient amounts from endogenous production (Caudill 2010). Requirements for choline may vary considerably, depending on an individual's genetics (Zeisel 2006). The involvement of choline and lecithin (a common dietary source) has been investigated in relation to dementia but resulted in little confirmatory evidence (Hollenbeck 2010).

> Amine refers to an organic compound in which nitrogen has replaced one or more hydrogen atoms. Biogenic amines are synthesized by plants and animals and include neurotransmitters, hormones, vitamins, and phospholipids.

Obsessive compulsive disorder and micronutrients following cognitive behavioral therapy—level IV evidence

Rucklidge (2009) reported on an 18-year-old boy who had a modest response to 1 year of cognitive behavioral therapy. In the following year, he reverted to severe anxiety and major depression. An ABAB design (8 weeks each phase) of a nutritional formula resulted in stabilization of mood, reduced anxiety, and remission of obsessions.

Low fat vs. low carbohydrate diet: Effect on mood and cognitive function—level II evidence

A 1-year study of 106 participants divided into three groups—one group on an energy-restricted isocaloric diet plan, one group on a very low carbohydrate, high-fat diet plan, and one group on a high carbohydrate, low-fat diet plan. Psychological mood and well-being were also assessed. There was no significant difference in weight loss between the groups. There was a greater improvement in the psychological mood states (anger-hostility, confusion-bewilderment, and depression-dejection) on the low-fat diet plan (Brinkworth 2009).

Unexplained medical symptoms not necessarily psychiatric problems—level III evidence

Unexplained symptoms are often referred to as "functional" (medical syndromes) or "somatoform disorders" (psychiatric syndromes). Escobar (2002) writes, "Lack of a medical explanation by itself does not necessarily qualify a symptom as psychiatric."

> Somatization refers to the manifestation of psychological distress as unexplained physical symptoms.

There is high overlap of symptoms among medically unexplained and psychiatric conditions such as chronic fatigue syndrome, low back pain, irritable bowel syndrome, chronic tension headache, fibromyalgia, temporomandibular joint disorder, major depression, panic attacks, and posttraumatic stress disorder. A study of interrelationships in 3982 twins found comorbidity of these nine conditions far exceeded expectations. The researchers stated that results support theories suggesting that medically unexplained conditions share a common etiology (Schur 2007).

Would further thought and research produce a theory in the domain of psychiatric/psychological conditions similar to Selye's general adaptation syndrome in the domain of physical conditions?

Gluten sensitivity and psychiatric presentation:
A case study—level IV evidence

A 38-year-old male in treatment was described as presenting with diverse worsening behavioral, cognitive, and affective symptoms accompanied by neuromotor impairment. Weight loss and hypoferremic anemia also developed. The diagnostic process included gluten sensitivity tests, which were positive. An appropriate diet led to remission of symptoms and lesions that had been detected by single photon emission computed tomography (SPECT) (Poloni 2009).

Biochemical and metabolic changes (present at stage III and IV of the stages of nutritional injury model, before clinical signs appear) may be accompanied by symptoms related to nutritional status, explainable only with a comprehensive nutritional assessment.

Another dimension

In addition to the biological, neurological, and genetic models linking mental health and the physical body to nutrition, many writers, teachers, and scientists have other perceptions concerning the relationship between the physical, psychological, and spiritual aspects of a person. The body can be thought of as the vehicle for the expression of the deeply personal essence, the spiritual self. Influencing factors may include beliefs, consciousness, gratitude, intuition, love, mindfulness, religion, the soul, and personal values. Perhaps in one sense, these factors may be considered causes of well-being, mental health, and behavior.

Spiritual traditions and diet

A common thread of worldwide traditions is to encourage treating food with reverence and cooking it with love. Examples of food as an integral part of societies' body–soul connection include dietary laws of Judaism that reflect respect for creation and Christianity's ritual of Holy Communion, which nourishes a multidimensional hunger through the bread and wine and what it represents. Islam teaches that the consumption of wholesome food is a religious obligation and teaches advocates to approach food with thankfulness and awareness of God's bounty expressed in food. Buddhism maintains that enlightenment may be experienced by bringing a mindful, meditative awareness to all aspects of our lives, including the selection, preparation, serving, and eating of food. Sawyer-Morse (2004) writes that mindfulness requires only the willingness to shift to being fully aware of the moment, rather than being involved in an unconscious habit. This involves changing one's state of mind, maintaining present-moment awareness throughout the preparation and eating of an entire meal.

Deborah Keston, author of *Feeding the Body, Nourishing the Soul*, discussed the sense of connection between food and consciousness. She hypothesized that severing the link between food and spirit is at the root of eating disorders. Keston believes that the focus on the biological aspect of nutritional science (vitamins, calories, fat, etc.) is shifting to include the psychological and emotional aspects of food and eating. She suggested that spiritual nourishment is basic to a new nutritional paradigm (Keston 1997).

Biofeedback

In *Beyond Biofeedback*, Elmer Green and Alyce Green express the following principles:

- Our bodies are equipped with regulating mechanisms and systems that work automatically—breathing and circulation, for example.
- Such systems can be influenced by various thoughts and experiences.
- It is important to realize that attempting to make a physiological change through focus of attention will not be accomplished by force or an act of will.

A physiological change is accomplished by imagining and visualizing the intended change while in a relaxed state known as passive volition. The Greens explain that it is a psycho-physiological principle that every change in the physiological state is accomplished by an appropriate change in the mental-emotional state, conscious or unconscious, and conversely, every change in the mental-emotional state, conscious or unconscious, is accompanied by an appropriate change in the physiological state.

> Free will may be defined as the sustaining of a thought because I choose to, when I might have other thoughts.
> —Author unknown

When these psycho-physiological operations are coupled with volition, psychosomatic self-regulation occurs. According to the Greens, the most unexpected thing that many psychiatrists, medical doctors, and clinicians have observed in patients using biofeedback for self-regulation of physiological problems is that personality changes accompanied the physiological changes.

When trying to modify a habit, the principal of psychological homeostasis comes into play. It seems everything in one's nature conspires to prevent change. Eating habits are an example of habits that are notoriously difficult to change. Change is opposed by the mind itself. From the yogic point of view, it is a physiological state of homeostasis as well, and that is why fasting diets, physical exercise, and emotional and mental control exercises are useful or necessary in bringing about personality change and development. Once we become conscious of the workings of homeostasis, we find it easier to change. We can choose a thought over another thought. It is possible to apply these principles to changing eating habits and how one thinks of oneself in relationship to food. Making one change may cause a ripple effect in one's life (Green and Green 1977).

Biology of Belief

Bruce H. Lipton, cell biologist and research scientist, described the cell's response to its environment and the activities necessary to life by saying these responses are reflected by the whole body in relationship to its environment. He maintains that the laws of quantum physics, not Newtonian laws, control a molecule's life-generating movement and that biomedicine is still working under the principles of Newtonian science, which decreases biomedicine's ability to predict and control disease.

Cells can live for some time with no nucleus, no DNA, and no genes, but they are unable to repair or replicate themselves. A cell's membrane allows that cell to maintain the survival function of responding to signals from the environment. Cells can be thought of as responding to basic "perceptions" such as ions of potassium, calcium oxygen, glucose, toxins, light, etc., which are present in the environment. He described two types of response by cells: *either* protection *or* growth (if the cell contains DNA) but not both at the same time with equal efficiency. Nutrients are included in the environmental molecules to which cells respond by letting them in or keeping them out. Growth (which includes repair of damage due to normal wear and tear) requires an open exchange between the organism/cell and its environment. This open exchange includes an adequate supply of necessary nutrients in the right place, in the right concentration, and at the right time. Protection requires closing the organism off from a perceived threat. Situations of stress result in a shift to the "protect/ defend" mode, constricting the blood vessels that supply the digestive tract (the input source for nutrients) as well as affecting the immune system. One theory gaining support is that the inhibition of neuronal growth by stress hormones is the source of depression. The hippocampus and prefrontal cortex become physically shrunken in chronically depressed individuals. (See also Chapter 9.)

Lipton explained that energy-signaling forces are 100 times more efficient in relaying environmental information than physical signals such as hormones and neurotransmitters. Survival depends on speed and efficiency of energy transfer. Chemical coupling to transfer information is accompanied by a loss of energy (as heat) in making and breaking chemical bonds, limiting the amount of information that can be carried. The speed of a diffusible chemical is less than 1 cm/s. The speed of electromagnetic energy is 186,000 mi/s. Energy as vibrational frequencies can alter the physical and chemical properties of an atom as surely as physical signals such as histamine and estrogen. Vibrational energy is influenced by thought, emotions, and worry. Signals from the mind can override local signals regarding the need for the "growth" or "defend" mode.

State of mind has a far-reaching effect on cellular function, the "growth" or "defend" state, and therefore affects the inflow of nutrients through the gastrointestinal tract at the level of organs and through the cell membrane at the cellular level. This model can illustrate how stress and depression can influence the physical state to complement the models describing how the physical state influences mental status (Lipton 2005).

Eating to promote health incorporates the totality of one's being: physical, intellectual, emotional, and spiritual. Food and meals may symbolize greater parts of life, reflecting and expressing an inner state. This state may be chosen purposefully to provide joy, gratitude, and relationships with others connected to the food or circumstance of each meal. If one loses these aspects of eating by not allowing time or thought for them, it can leave a need for nourishment that cannot be met through other parts of life. By trying to obtain nourishment from a substitute that is not capable of this type of nourishment, we are left wanting more. Using more foods that will not fill a need can easily result in obesity.

Explanations such as these seem to be ways of saying that if the inner state is balanced first, aspects of food, eating, and meals are not as likely to be unbalanced. If there is no discomfort, guilt, neediness, stress, or feelings of vague emptiness, food is not needed for relief, and it can fill its rightful role in nourishing our body and brain, making way for our personal essence to be expressed.

Conclusions

1. A variety of additional relationships between nutrients/nutritional status (copper, mercury, selenium, magnesium, manganese, vitamin B_{12}, niacin, riboflavin, folate, and choline) and mental status (agoraphobia, OCD, migraine headaches) are discussed in scientific literature. Some are used clinically; others need more investigation and understanding.
2. Spiritual practices, the principles of biofeedback, and the body's response to an individual's belief system may affect food-related behaviors and one's physiology down to the cellular level.

References

Abbey, LC. Agoraphobia. *Orthomol Psychiatry* 1982; 11(4):243–259.

Abbey, LC. Functional nutrient deficiency in chronically multi-symptomatic people: A pilot study. *J Orthomol Medicine* 1989; 4(2):75–88.

Alpers, DH., WF. Stenson, and DM Bier. *Manual of Nutritional Therapeutics*. 3rd ed. 252–255. Little Brown and Co. Boston, 1995.

Arroyave, G. Sequence of events in the development of clinically evident nutritional disease. Figure 2 in VR Young, and NS Scrimshaw. Genetic and Biologic Variability in Human Nutrient Requirement. *Am J Clin Nutr* 1979; 32:486–500.

Ashrafi, MR, S Sharma, M Nouri, et al. A probable causative factor for an old problem: Selenium and glutathione peroxidase appear to play important roles in epilepsy pathogenesis. *Epilepsia* 2007; 48(9):1750–1755.

Aurangzeb, S. Young girl with clumsiness, dystonia and speech difficulty. http://cme.medscape.com/viewarticle/708214. accessed Sept 2, 2009.

Barceloux, DG. Manganese. *J Toxicol Clin Toxicol* 1999; 37(2):293–307.

Beard, J. Nutrient Status and Central Nervous System Function, *Present Knowledge in Nutrition* 7th ed. 612–622. International Life Sciences Institute. Washington DC. 1996.

Benton, D. Selenium intake, mood and other aspects of psychological functioning. *Nutr Neuroscience* 2002; 5(6):363–374.

Brewer, GJ., F Askarin, MT Lorincz, et al. Treatment of Wilson disease with ammonium tetrathiomolybdate. *Arch Neurol* 2006; 63:521–527.

Brinkworth, GD, JD Buckley, M Noakes, et al. Long-term effects of a very low-carbohydrate diet and a low-fat diet on mood and cognitive function. *Arch Int Med* 2009; 169(20):1873–1880.

Burger, J, AH Stern, and M Gochfeld. Mercury in commercial fish: Optimizing individual choices to reduce risk. *Environ Hlth Persp* 2005; 113(3):266–271.

Caudill, MA. Pre-and Post-natal health: Evidence of increased choline needs. *J Amer Diet Assoc.*; 2010; 110(8):1198–1206.

Demirkaya, S, O Vural, B Dora, and MA Topcuoglu. Efficacy of intravenous magnesium sulfate in the treatment of acute migraine attacks. *Headache* 2001; 41(2):171–177.

Escobar, JI, C Hoyos-Nervi, and M Gara. Medically unexplained physical symptoms in medical practice: A psychiatric perspective. *Env Hlth Persp* 2002; 110 (S4):631–636.

Graef, JW. Heavy Metal Poisoning *Harrison's Principles of Internal Medicine*. ed. KJ Isselbacher, E Braunwald, JD Wilson, et al. 2461–2466. McGraw-Hill, Inc. San Francisco. 1994.

Green, E and A Green. *Beyond Biofeedback*. 33–180. Dell Publishing Co, Inc. New York. 1977.

Hightower, JM and D Moore. Mercury levels in high-end consumers of fish. *Environ Hlth Persp* 2003; 11(4):604–608.

Hollenbeck, CB. The importance of being choline. *J Amer Diet Assoc*. 2010; 110 (8):1162–1165.

Hunt, SM and JL Groff. *Advanced Nutrition and Human Metabolism*. 212–280. West Publishing Company, Los Angeles. 1990.

Jeter, A. How Safe Is Your Food? One fish, two fish, red snapper, swordfish: A menace lurks in your "healthy" food. *Reader's Digest*. August 2003.

Kesten, D. *Feeding the body, nourishing the soul: Essentials of eating for physical, emotional and spiritual well-being*. 149–217. Conart Press, Berkeley 1997.

Levine J. Controlled trials of inositol in psychiatry. *Eur Neuropsychopharmacol* 1997; 2:147–155.

Lipton, BH. *The Biology of Belief*. p. 110–112, 128–129, 145–147, Mountain of Love Productions/Elite Books, Santa Rosa, California. 2005.

Maizels, M, A Blumenfeld, and R Burchette. A combination of riboflavin, magnesium, and feverfew for migraine prophylaxis: A randomized trial. *Headache* 2004; 44(9):885–890.

Misra, UK., J Kalita, and A Das. Vitamin B12 deficiency neurological syndromes: A clinical, MRI and electrodiagnostic study. *Electromyograph Clin Neurophysiology* 2003; 43(1):57–64.

Mt. Sinai School of Medicine. Running Low on B-12? *Focus on Healthy Aging*. New York, 2002; 5(11):3.

Neville, MJ and V Daley. Developing a risk profile for patients at risk for manganese (Mn) toxicity. Coram Healthcare, Inc. Denver, Colorado. 2002.

Poloni, N, S Vender, E Bolla, et al. Gluten encephalopathy with psychiatric onset: Case report. *Clin Pract Epidem in Ment Hlth* 2009; 5:16.

Prousky, J and D Seeley. The treatment of migraines and tension-type headaches with intravenous and oral niacin (nicotininc acid): Systematic review of the literature. *Nutr J* 2005; 4:3.

Ramaekers, VT, Calomme M, D Vanden Berghe and W Makropoulos. Selenium deficiency triggering intractable seizures. *Neuropediatrics*. 1994; 25(4):217–223.

Rayman, M, A Thompson, M Warren-Perry, et al. Impact of selenium on mood and quality of life: A randomized, controlled trial. *Biol Psychiatry* 2006; 59(2):147–154.

Rayman, MP. Selenium and human health. *Lancet*. 2012 doi: 10.1016/S0140-6736(11)61452-9.

Rucklidge, JJ. Successful treatment of OCD with a micronutrient formula following partial response to Cognitive Behavioral Therapy: A case study. *J Anxiety Disord* 2009; 23(6):836–840.

Sawyer-Morse, MK. Food, spirituality, and mindful eating. *Today's Dietitian* 2004; 6(3):29–31.

Savaskan, NE, AU Brauer, M Kuhbacker, et al. Selenium deficiency increases susceptibility to glutamate-induced excitotoxicity. *FASEB J* 2003; 17(1):112–114.

Schur, EA, N Afari, H Furberg, et al. Feeling bad in more ways than one: Comorbidity patterns of medically unexplained and psychiatric conditions. *J Gen Intern Med* 2007; 22(6):818–821.

Sharma, DC. Manganese in drinking water. *Environ Health Perspec* 2006; 114(1):A50.
Turnland, J. Copper. *Modern Nutrition in Health & Disease*, 297. ed. ME Shils, M Shike, AC Ross, et al. Lippincott, Williams and Wilkins, Philadelphia, 2006.
Zeisel, SH. Choline: Critical role during fetal development and dietary requirements in adults. *Ann Rev Nutr.* 2006; 26:229–250.
Zucker, DK., RL Livingston, R Nakra, and P.J. Clayton. B12 deficiency and psychiatric disorders: Case report and literature review. *Biol Psychiatry* 1981; 16(2):197–205.

chapter twelve

Conclusions and recommendations

Introduction

Historical and current evidence illustrates that nutrients and nutritional status potentially influence variations in, and degrees of, mental status, mental health, and mental illness. Relevant aspects of mental state and methodologies for measuring them are reviewed by Westenhoefer (2004).

Measurable Scientific Concepts of Mental State (Westenhoefer 2004)		
Mood	Arousal	Activation
Vigilance	Attention	Sleep
Motivation	Effort	Perception
Memory	Intelligence	

Relationships may be causal or correlational, bidirectional or multidirectional. Nutritional influences may be dependent on overt deficiencies, less-than-optimal levels, excess or toxic levels, or ratios of nutritional factors. Evidence is largely observational. Mental status and nutritional status are both complex; combining them magnifies the difficulty. Causal relationships may never be established unequivocally because depriving or providing an excess of individual of nutrients or mental health care is considered unethical. Clinical care and personal use of nutrients may not be driven by gold-standard evidence while the scientific process proceeds. (See Appendix B, Table B.3 "Level I Evidence Reports.") The application of available knowledge in clinical care and daily life is of great interest to mental health caregivers, families, and individuals who live with altered mental status. Conflicting or incomplete evidence is a call for additional, creative research; it may be necessary to develop new methods. (See Appendix B, Table B.1 "Mental Health Concerns: Nutritional Concerns Matrix" for a summary of nutrients related to mental health topics found in this book.)

Although not in time for publication of the *Diagnostic and Statistical Manual of Mental Disorders*, fifth edition (DSM-V), the probability of psychiatric diagnosis augmented and refined by use of biochemical, metabolomic, and proteomic analysis of plasma or cerebrospinal fluid may be standard within five years, according to Jeffrey

A. Leiberman, MD, at Columbia University (http://www.medscape.com/viewarticle/750288?src=ptalk). Nutritional parameters could well be included in forming these diagnostic signatures.

Nutrition education for mental health professionals

David Horrobin, MD, a neuroendocrinologist from Scotland, wrote,

> Nutritional biochemistry is soon going to be essential for anyone working with mentally ill patients ... Without an adequate intake of all the required essential nutrients, the brain simply cannot function normally. Trying to apply any treatment modality, whether psychological, pharmacological or social, to a brain that cannot function normally because of lack of an essential nutrient is like trying to run a 220-volt electrical appliance on a 120-volt system. (Horrobin 2002)

A survey of 100 mental health counselors found that 90% felt diet influenced their clients and played a role in major mental illnesses and one-third felt clients could benefit from nutrition counseling. However, 5 of 6 nutritional questions on the survey were answered incorrectly by 75% of participants, even though 50% felt confident in their ability to talk to their clients about nutrition (Lacey 2001). The majority felt more training in nutrition was necessary. Inclusion on research or treatment teams of a registered dietitian with training in the specialty area of behavioral health, or other cross-trained professionals, could help meet the need for including nutritional aspects in assessment, care, and research.

Shekar Saxena, program manager in the Department of Mental Health and Substance Abuse at World Health Organization, and colleagues write, "Mental health professionals have an important role to play in improving the evidence on prevention and promotion in mental health, in engaging relevant stakeholders for developing programmes, and as professional care providers in their practice" (Saxena et al. 2006).

Nutritional assessment

Nutritional assessment and status are not routinely described as parameters of equality of groups or individuals in research of mental health conditions, assuming no difference could be a source of confounding or bias. Inferences concerning nutrients and nutritional concerns may be afterthoughts rather than included in the original purposes and methods of research. Addressing these issues would lift the quality of research in nutrition and mental health. Individuality as described by genetic descriptions and imaging techniques will give scientists of the future additional exciting means of predicting, diagnosing, and treating conditions related to nutrition and mental health.

Consistent records of assessment of nutritional status and mental status may uncover links between subclinical and nonspecific symptoms that connect mental status with nutritional status. "Partial and acute vitamin deprivation suggest that the earliest impairments occur in measures of mood rather than mental performance" (Haller 2005). Assessment for subclinical nutrient deficiencies may contribute to prevention or early treatment. "We don't know the biochemical basis ... or always anticipate mechanisms of action [of nutrients], so we may be limiting ourselves if we only look first to the biochemical function and then to match a behavioral outcome" (Penland 1997). Researchers and clinicians need to examine a variety of cognitive processes and response characteristics when assessing the impact of nutritional intervention or suboptimal nutrition on cognitive function.

Assessment for nutritional status

Methods for assessing diet and nutrition are not precise, but can be used as indicators for further exploration. Since food and eating are part of every human life, they are appropriately included in treatment recommendations. Levels of assessment are described next. (See Appendix A for Assessment of Nutritional Status form; for downloadable forms, please visit www.RuthLeyseWallace.com)

Diet assessment

Assessment of diet could be a triage point for preliminary establishment of uniformity of a research population or possible adequacy of nutritional intake. Some food intake studies have linked biochemical levels of some nutrients to reported food intake (Weinstein 2004). This is a long-term concern of nutritionists, with methods improving over time. People frequently cannot remember what they ate, and often make inaccurate estimates of amounts and frequency of foods and beverages consumed. According to the Report of the Dietary Guidelines for Americans Committee (DGAC) 2010 section D2-22, if intake data are available for at least two days, statistical methods can be used to estimate usual intake. It is wise to remember that values for the Dietary Recommended Intakes (DRI) are based on meeting the needs of healthy people, not those with health conditions or genetic alterations.

Nutrition-focused physical examination

A nutrition-focused physical examination and observation of the physical effects of nutrient excess or inadequacy can demonstrate effects of altered nutritional status. It takes a few weeks for depletion of water-soluble vitamins, and months or years for fat-soluble vitamins for signs and symptoms to manifest. Psychological and mental changes will frequently occur before physical change is observed. A nutrition-focused physical examination could serve as a prompt for confirmation by laboratory assessment and as a secondary level of assessment for assumption of equality between individuals for research purposes. (See also Appendix A, Table A.2 "Clinical Signs Potentially Related to Nutritional Deficiencies.")

Laboratory assessment
Biochemical laboratory assessment may include the following:

- Serum or plasma levels of a nutrient or nutrient metabolite.
- Blood component assessment (erythrocytes, leukocytes, etc.).
- Functional assessment (nutrient-dependent enzyme activity, etc.).
- The ratio of one nutrient to another (minerals, essential fatty acids). Ratios should be used with caution, as identical ratios can be created at widely different levels (e.g., 200:2 yields the same ratio as 1000:10).
- Different forms of a nutrient (ionized, α vs. γ form, etc.).
- Biological/biochemical markers as indicators of activity or function. Metabolic markers are more reliable than participant's verbal or written reports to validate compliance.

When compromised nutritional status is contributing to an individual's mental health concerns, laboratory measures can lead to diagnosis that is more accurate. Laboratory assessment of nutritional status would be a conclusive step establishing uniformity of nutritional status and an assumption of no difference in comparing populations under investigation. Research needs to include biochemical nutritional assessment to differentiate individuals who are deficient or not deficient to compare with those who do or do not change behavior, mood, symptoms, or other characteristics. Unfortunately, more knowledge is needed regarding which measures are valid indicators of health. For example, current blood levels may reflect most recent intake rather than stored, sequestered, or available levels of a nutrient. Hair analysis may act as an indicator of long-term past intake of toxic metals, but not current intake. (See also Appendix B, Table B.1 "Mental Health Concerns: Nutritional Concerns Matrix" and Table B.2 "Health Conditions or Health Histories that Warrant Evaluation for Nutritional Status.")

Suggestions for research in nutrition and mental health

Even as current knowledge is integrated into life, scientists continue to seek in-depth understanding and answers to questions of what, how, and why regarding the connections between mental status and nutrients. Research issues could be broadly summarized as future inclusion of nutritional status descriptions in psychiatric literature and inclusion of mental status in nutrition literature. Specific suggestions follow:

1. Including and reporting consistent records of assessment of nutritional status and mental status may uncover links between subclinical and nonspecific symptoms that further connect mental status with nutritional status. Assessment for subclinical nutrient deficiencies may contribute to prevention or early treatment.
2. Research studies need to fully report the use, dosage, and compliance regarding any nutritional supplements included or excluded in the protocol. Supplements may be multinutrient or single nutrient and may be at high, moderate, or low levels. Compliance validated by metabolic markers is more reliable than participants' reports.
3. Available evidence sometimes comes from studies that were originally designed and conducted to investigate other conditions. Defining primary nutritional and mental health outcomes while accounting for other medical conditions will increase usefulness of results and facilitate comparison with other research (e.g., target the PsychoNutriologic Person). Defining conditions of research needs to include the *form* of nutrient used; for example, whether the alpha or gamma form of tocopherol is used as an intervention.
4. Double-blind conditions are difficult or impossible to manage with complex natural foods or provision of individual psychiatric or nursing care. Pessler and colleagues (2011) describe a method for combining partially blinded and partially unblinded portions of a study of ADHD, diet, individualized food challenges, and behavior. As an alternate to controlled clinical trials, the Consecutive Patient Self-Report Questionnaire Database and the limitations of this method, compared to the limitations of controlled clinical trials, are described by Pincus (1997). Comments on this method by Dachert, Hebert, and Proudfit (1997) are found in a subsequent volume. This method provides a type of evidence from clinical practice gathered over the long term and could possibly complement other levels of evidence in the fields of nutrition and psychiatry/psychology.
5. Literature reviewers often commented on the need for an adequate number of research participants to achieve power to detect statistically significant change.
6. Tracking an individual's consistent personal metabolite levels while experiencing good health would be a means to determine if or when nutrients are low, high, or unbalanced during states of poor health (Williams 1998).

7. Randomization of groups instead of individuals and analysis of long-term dietary patterns instead of single nutrients could provide an additional perspective for investigating nutrition and psychiatry (Freeman 2010).
8. Perhaps not far into the future, assessment for alterations in DNA related to increased nutrient needs or changes in metabolism involving a nutrient and the consequences for mental status will become meaningful, affordable, and common enough to play a role in clinical care and research. Methods such as single-subject research (within subject; N = 1, possibly combined with ABAB treatment) may be relevant (Horner 2005) and applicable to clinical care (Reynolds 1983). See http://www.practicalpress.net/updatenov05/SingleSubject.html by John B. Wasson and a number of texts on this method that are available.
9. A European commission coordinated by the International Life Sciences Institute (ILSI Europe) has agreed upon criteria for evaluating research and health claims of functional foods. The Process for Assessment of Scientific Support for Claims on Foods (PASSCLAIM) can be found on the Website of the Office of Dietary Supplements (Rechkemmer 2005).

Selected topics of interest

1. Redefining success following bariatric surgery to include problematic eating behaviors and psychological well-being as well as loss of excess weight seems advisable (Kruseman 2010).
2. Follow-up with psychological and nutritional care longer than 2 to 3 years with patients who had gastric bypass surgery seems warranted in light of possible reappearance of psychiatric conditions and reported noncompliance with nutritional recommendations (Waters 1992; Greenburg 2003; Tindle 2010).
3. While acknowledging that eating disorders are psychiatric diagnoses, additional exploration and publication of nutritional consequences found in patients with eating disorders would be informative, including the BMI values associated with hospitalization and death. Studies report mean BMI at admission to hospital for intense treatment as 12.5 ± 0.9 g/m^2 (Gentile 2008), 12.09, and 13.2 kg/m^2 (Thiels 2008). In a long-term study, patients with a BMI over 11.5 had an average standardized mortality ratio (SMR) of about 7 and for those with BMI lower than 11.5 had SMR above 30 (Rosling 2011). The critical level for observing nutrient-based lesions was reported as a BMI of 16 (Heidiger 2000).
4. "Normal variability," "subthreshold" conditions, and lack of tolerance for variability may influence whether an individual is diagnosed

with a psychiatric condition (Frances 2010). Understanding how and when nutritional factors influence normal variation in mental status could help in the process of diagnosing psychiatric problems.
5. Identification of alterations in nutrients (such as fatty acids) as biological "trait" markers, as well as genetically determined mitochondrial dysfunction and oxidative stress, may also be informative (Assies 2010).
6. High overlap of symptoms among medically unexplained and psychiatric conditions such as chronic fatigue syndrome, low back pain, irritable bowel syndrome, chronic tension headache, fibromyalgia, temporomandibular joint disorder, major depression, panic attacks, and posttraumatic stress disorder suggest medically unexplained conditions share a common etiology (Schur 2007). Could that etiology include alterations in nutritional status, possibly related to genetics?
7. Providing psychological care for those with diabetes is being assessed in the UK due to the recognized comorbidity of depression and diabetes. A Dialogue on Diabetes and Depression meeting was held in 2007 and attended by professionals from around the world (Lloyd 2010; Nicholson 2009). The World Federation of Mental Health provides a downloadable educational booklet that may be of interest to the nutrition and mental health communities (see http://www.diabetesanddepression.org/).
8. Basic differences between nutraceuticals and pharmaceuticals result in differences in funding for research. The relatively recent social priority for supporting mental health may make nutrition-related research in mental health more fundable.

Notes regarding specific nutrients

Lipids: EPA, DHA, essential fatty acids, and omega-3 fatty acids

In 2006 the American Psychiatric Association Committee on Research on Psychiatric Treatments, the Council on Research, and the Joint Reference Committee concluded that studies support a protective effect of omega-3 essential fatty acids (EFA) intake (particularly EPA and DHA) for unipolar and bipolar depression. Use of EPA and DHA appear to have negligible risk. It was also reported that the evidence for the protective effect of omega-3 fatty acids for schizophrenia was less conclusive (Freeman 2006).

A healthy dietary allowance for omega-3 long chain fatty acids for current U.S. diets was estimated at 3.5 g/day for a 2000-kcal diet (see also Table 12.1). This allowance for omega-3 fatty acids can likely be reduced to one-tenth of that amount by consuming fewer omega-6 fats (Table 12.1) (Hibbeln 2006).

Fish with the highest levels of EPA and DHA are herring, salmon, mackerel, and bluefin tuna, with approximately 1500 mg of EPA and DHA

Table 12.1 Recommendations for Intake of Essential Fatty Acids[a]

Hibbeln 2005	Estimated U.S. intake EPA and DHA: 180 mg/day 180–500 mg/day inadequate 750 mg/day possibly adequate 1000 mg/day clearly adequate for lowering risk of psychiatric disorders
Food and Nutrition Board IOM 2002	DRI adequate intake: Linoleic acid (omega-6): 5–10% of an individual's calories 14–17 g/day for men 11–12 g/day for women Linolenic acid (omega-3): 0.6–1.2% of an individual's calories 1.6 g/day for men 1.1 g/day for women
The National Institutes of Health	650 mg per day of omega-3 fatty acids
The American Heart Association	1 g (1000 mg) of omega-3 in enriched foods daily 5–10% of daily calories from omega-6 fatty acids, primarily linoleic acid

[a] See controversy regarding research leading to recommendations for increase of omega-6 fatty acids for the prevention of heart disease (Ramsden 2010; Calder 2010).

in 3 to 4 oz. of the cooked fish. Other types of tuna provide only about one-third this amount (Table 12.2).

Fish oil supplements commonly provide approximately 300 mg of omega-3 fatty acids per day, divided between EPA and DHA. A common ratio of EPA to DHA in standard fish oil is 2:1. Some supplements may provide up to 850 mg per recommended dose. Some foods are fortified with omega-3 fatty acids with several hundred milligrams per serving.

Precautions regarding essential fatty acids

Excess EFAs may lead to decreased coagulation time, interfere with wound healing, raise low-density lipoprotein (LDL) levels, and suppress immune function. Individuals with these health issues need to coordinate any extra consumption of omega-3 fatty acids with their physician and dietitian.

Protein and amino acids

Protein

The need for protein in the diet is based on weight, health, age, and physical activity such as muscle building. An average, healthy adult needs 0.8 g/kg of weight or 0.36 g/lb of weight. Protein and amino acids are

Table 12.2 Food Sources of Essential Fatty Acids

Omega-6 (linoleic) fatty acid	Oils from corn, sunflower, safflower, soybean, cottonseed, poultry fat, nuts, and seeds	
Omega-3 (linolenic) fatty acid ALA (Alpha-Linolenic Acid or α-Linolenic Acid)	Oils from flaxseed, canola, walnut, wheat germ, and soybean; nuts and seeds such as flaxseeds, walnuts, soybean kernels, and butternuts	Some fortified chicken feed results in increased omega-3 in eggs the chickens produce (3 eggs = 1 fish meal) (Lewis 2000) Some cereals are fortified with 300+ mg of omega-3 fatty acids per serving Rate of conversion of ALA to DHA is <1% (Arterburn 2003)
EPA and DHA	Pacific oysters, mackerel, salmon, bluefish, mullet, sablefish, menhaden, anchovy, herring, lake trout, sardines, and tuna	The amount of omega-3 fatty acids found in fish may vary with the genetics of the fish, the environment and food supply when it was growing, and whether it was canned in oil or water
Omega-3 supplements	Use of supplements with concentrations of natural fish oil should be checked to assure that a toxic amount of vitamin A is not consumed	Monitoring for drug interactions and treatment effect is especially important for those taking pharmaceutical doses

found in large amounts in animal foods, and in smaller amounts in grains and some vegetables (Table 12.3). Foods with the highest-quality proteins (those most efficiently utilized by the human body) are egg white and milk. Only a small amount of protein (4% of calories) in a high-carbohydrate meal is sufficient to block any meal-induced increases in the ratio of plasma tryptophan to amino acids (Rogers 1994).

Conversion Equivalents

28 g = 1 oz.
448 g = 1 lb
16 oz. = 1 lb
16 fluid oz. = 2 cups; ~1 lb
1 kg = 2.2 lb
1 lb = 0.45 kg

Table 12.3 Approximate Protein Obtained from a Variety of Food Groups

Food	Quantity	Approximate Grams of Protein
Meat, poultry, fish	5 oz. (1 moderate serving)	35
Milk	8 oz. (1 cup)	8
Beans	1 cup	8
Rice, corn, cereal, or bread	1 cup or 2 slices	4
Vegetables	1 cup	1–4
Nuts	1 T	1–4
Fruit and fats		Negligible amounts of protein

Table 12.4 Recommended Intakes of Essential Amino Acids for Adults

Essential Amino Acid	mg/kg of weight/day	mg/day for a 150-lb Person
Histidine	14	555
Isoleucine	19	1296
Leucine	42	2864
Lysine	38	2592
Methionine + cysteine	19	1296
Phenylalanine + tyrosine	33	2251
Threonine	20	1364
Tryptophan	5	341
Valine	24	1637

Source: Matthews 2006.

Amino acids

Amino acids in foods are rarely excessive or unbalanced (Table 12.4). Amino acids from supplements are often not balanced. A vegetarian diet requires attention to amino acids that may be insufficient or unbalanced in some foods. Methionine and tryptophan found in grains complement the lysine and isoleucine found in legumes.

Tryptophan doses of at least 1 g help mildly insomniac patients fall asleep more quickly and wake less frequently (Fernstrom 1971).

> Mild insomnia: taking ~30 minutes to fall asleep and waking during the night

Carbohydrates

The WHO/FAO Independent Expert Report on Diet and Chronic Disease suggests that carbohydrate should provide the bulk of energy

Table 12.5 Comparison of Measures of Blood Glucose Levels

Hemoglobin A1c	Blood Glucose (mg/dl)	Blood Glucose (mmol/L)
3	30	1.5
5[a]	100[a]	5.5[a]
7	170	9.5
9	240	13.5
11	310	17.5

[a] Considered a healthy fasting blood glucose level.

requirements (between 55 and 75% of daily intake) and that free sugars should remain below 10% of calorie intake.

The Institute of Medicine suggests not more than 25% of energy intake come from low or very low nutrient sources (sugar, sweets, desserts, soft drinks, alcohol, etc.).

Blood glucose levels

Hypoglycemia is often defined as a blood glucose level below 50 mg/dl (Table 12.5). Prediabetes may be diagnosed at levels designated as follows:

Impaired fasting glucose between 100 and 125 mg/dl
Impaired glucose tolerance between 140 and 199 mg/dl

Vitamins

Vitamin B_{12} assessment

Anemia can occur without macrocytosis and both neurologic and cognitive dysfunctions often develop in the absence of hematologic change. Early recognition of cobalamin deficiency is important to prevent progressive, irreversible neurologic damage and cognitive impairment. Solomon (2005) evaluated 95 patients and reported if therapy had been restricted to patients with symptoms, such as both low and intermediate cobalamin levels and increased methylmalonic acid (MMA), 63% of patients who responded to treatment would not have been treated. Of the patients with abnormal values, 25 did not respond to treatment with cobalamin. Solomon concluded that cobalamin levels, high homocysteine, and MMA tests fluctuate over time and neither predict nor preclude the presence of cobalamin-responsive hematologic or neurologic disorders. More work on the diagnosis and treatment of vitamin B_{12} deficiency is needed (Solomon 2005).

The normal blood level for vitamin B_{12} is 200 to 1100 pg/ml. Subclinical deficiency of serum B_{12} levels of 200 to 300 pg/ml can cause neuropsychiatric or neurological abnormalities such as paresthesia, sensory loss, ataxia, and dementia without anemia or macrocytosis. Mental changes occur before blood changes (Maleskey 2005).

Vitamin B_{12} treatment

> The response rate to vitamin supplements supports the notion that metabolic evidence of vitamin deficiency is common in the elderly, even in the presence of normal serum vitamin levels. Metabolite assays permit identification of elderly subjects who may benefit from vitamin supplements. (Naurath 1995)

Acute treatment for pernicious anemia is 1000 µg intramuscular (IM) injection followed by weekly injections for a month, then monthly injections for the remainder of the individual's life. Recovery may be complete if treatment is instituted within a few weeks of onset. If symptoms are present for longer than 1 to 2 months, only partial recovery can be expected. For long-standing cases, an arrest of the progression of the disease is the best that can be expected (Victor 1994).

Low-dose vitamin B_{12} supplementation

A study in which 240 community-dwelling, healthy, older adults consumed a low-dose supplement (2 to 37.5 µg/day) of oral cobalamin showed that the supplement was associated with higher serum levels of cobalamin and improved or normalized cobalamin function, as indicated by lower concentrations of metabolites. Use of low-dose oral cobalamin replacement therapy (as compared to injections) might be sufficient to prevent cobalamin deficiency in a large proportion of this population (Garcia 2002).

Hirsch et al. (2006) found that well-nourished elderly receiving 3.8 µg B_{12} for 6 months experienced a rise in serum B_{12} from ~350 pmol/L to 409 pmol/L, while the control group decreased from ~391 pmol/L to 290 pmol/L.

No toxicity or benefit has been recorded for unneeded high doses of vitamin B_{12}.

Thiamin

Patients who report missing meals or having a disease or condition that interferes with absorption of thiamin have high likelihood of developing Wernicke–Korsakoff syndrome. After a confirmatory laboratory assessment, Roman (2006) recommends treatment with an injection or IV dose of 100 mg of thiamin followed by 100 mg given IM daily for five days and oral doses at maintenance levels plus balanced meals.

For suspected Wernicke's encephalopathy, the European Federation of Neurological Societies (EFNS) recommends 200 mg of thiamin given IV three times a day before any carbohydrate intake (Galvin 2010).

Minerals

Chromium

The Normal plasma level of Chromium is 0.1–2.1 µg/mL; the Normal level for whole blood is 2.0–3.0 µg/100 ml (Agency for Toxic Substances and Disease Registry 2011).

The usual U.S. diet is estimated to provide 15 µg of chromium per 1000 calories. The dietary reference intake is 20 to 30 µg/day; the tolerable upper intake limit (TUIL) has not been determined.

One form of chromium found in supplements is chromium picolinate. Multivitamin supplements often recommend 1.6 mg Cr picolinate/day, which is equal to 200 µg Cr(III)/day. Supplements may also contain Cr(III). Picolinic acid is a minor metabolite of the tryptophan/kynurenine metabolic pathway.

In limited data, reported neuropsychiatric adverse reactions to Cr supplements included seizures, convulsions, agitated psychotic behavior, mood swings, nervousness, and high blood pressure.

Magnesium

Erythrocyte magnesium is decreased in major depression, acute schizophrenia, and bipolar disorder. Treatment with sertraline, amitryptiline, haloperidol, carbamazepine, or sodium valproate improved magnesium status. Improvement of magnesium status enhanced the clinical state (Nechifor 2008).

The activity of vitamin B_6 is Mg dependent. Vitamin B_6 is a cofactor in serotonin and GABA synthesis.

Magnesium inhibits the release of acetylcholine. Increased cholinergic activity and decreased serotonin are biochemical markers of negative symptoms of schizophrenia (Kanofsky 1991).

In patients with eating disorders, magnesium supplementation was found to not normalize serum levels after 3 weeks.

Mercury

Inorganic mercury (Hg) from the environment is methylated in the human body and is implicated as a neurotoxicant, a mutagen, and a teratogen. Fish is a major source of mercury, but drinking water, cereal, vegetables, meat, and the release of mercury in dental amalgam by gum chewing may also be sources. Interaction with milk, bran, selenium, alcohol intake, cysteine, the positive or negative effect of vitamin/mineral supplementation, and other nutritional factors have been found in animal studies to influence the metabolism of Hg and its entry into the brain.

The Food and Agriculture Organization and the World Health Organization (FAO/WHO) have defined a provisional tolerable weekly intake of mercury as 3.3 µg/kg/week (~200 µg/week) for adults and breast-fed infants. Health Canada has defined the provisional tolerable daily intake for women of reproductive age and infants as 0.2 µg/kg/day.

Inclusion of detailed information on dietary intake and exposure to mercury could clarify inconsistencies in epidemiological studies, suggest Chapman and Chan (2000).

Converting Oz, Mg, Mcg of Mercury

1 oz. = 28 g = 1 oz. = 28,000 mg
1 oz. = 28,000,000 µg (mcg)
200 µg/kg/week for a person weighing 70 kg (154 lb) = 14,000 µg/week
14,000 µg/week/7 = 2000 µg/day (tolerable level)
2000 µg/day = 0.0000714 oz./day (a tiny bit of mercury)

Mercury in fish

- High-mercury fish include shark, king mackerel, swordfish, and tilefish.
- Mid-range in mercury content are bluefish, Chilean sea bass, and tuna.
- Lowest mercury fish are whiting, flounder, porgy, cod, and croaker (Burger 2005).

To balance the benefits of essential fatty acids found in fish with the possible harm from mercury, the American Heart Association, WHO, and the USDA Dietary Guidelines recommend that women of childbearing age and others eat two low-mercury fish meals per week (Kris-Etherton 2005).

Selenium

In mammalian bodies, spermatozoon tissue contains the highest levels of selenium. Selenium concentration in gonads increases during maturation in males, while serum levels decrease (Holben 1999).

The adult DRI for selenium is 55 µg/day. The daily safe upper range of Se is 350 to 600 µg. Inorganic forms of selenium (sodium selenite) are toxic. Acute toxicity is seen with serum levels of 1000 µg/L. Chronic selenosis (excess) results in hair and nail loss (Sutter 2008).

Concluding statement

Continued research and application of the science linking nutrition and nutrients to mental health and mental illness is burgeoning around the world. Literature reporting trials and advances is distributed through numerous journals and professional disciplines. This book is but a representative segment of areas being investigated. Communication and dispersal of new knowledge will be a major undertaking. Progress needs to be integrated into the education of nutritionists, dietitians, and mental health professionals. As development of the science and applications proceeds, individuals and families around the nation and the world could benefit from ready access to education illustrating how foods, diet, nutrients, and eating habits can influence their state of mind and state of mental well-being.

References

Assies, J, F Pouwer, A Lok, et al. Plasma and erytrocyte fatty acid patterns in patients with recurrent depression: A matched case-control study. *PLoS One* 2010; 5(5):e10635. Accessed September 15, 2010.

Dacher, E. All Healing is Local. *Advances: J Mind-Body Hlth* 1997; 13(4):59–60.

Frances, A. The DSM5 subthreshold disorders: Not ready for prime time 2010. http://www.psychologytoday.com/blog/dsm5-in-distress/201004/the-dsm5-subthreshold-disorders-not-ready-prime-time. (accessed 10/9/2011)

Freeman, M. Nutrition and Psychiatry Editorial. *ajp.psychiatryonline.org* 2010; 167:3.

Gentile, MG, GM Manna, R Ciceri, and E Rodeschini. Efficacy of in-patient treatment in severely malnourished anorexia nervosa patients. *Eat Weight Disord* 2008; 13(4):191–7.

Greenberg, I, Psychological aspects of bariatric surgery. *Nutr in Clin Pract* 2003; 18:14–130.

Haller, J. Vitamins and Brain Function. *Nutr Neurosci.* ed. HR Lieberman, RB Kanarek, and C Prasad, 211. CRC Press: Taylor & Francis Group Boca Raton, FL. 2005.

Hebert, JR. Randomization and measurement in efficacy studies of behavioral and mind-body interventions. *Advances: J Mind-Body Hlth* 1997; 13(4):60–64.

Hediger, CB, B Rost, and P Itin. Cutaneous manifestations in anorexia nervosa. *Schweizerische Medizinische Wochenschrift* 130(16) (2000):565–575.

Horner, RH, EG Carr, J Halle, et al. The use of single-subject research to identify evidence-based practice in special education. *Exceptional Children* 2005; 71(2):165–179.

Horrobin, DF. Food, Micronutrients, and Psychiatry. Editorial, *Int'l Psychogeriatric* 2002; 14(4):331–334.

Kruseman, M, A Leimgruber, F Zumbach, and A Golay. Dietary, weight, and psychological changes among patients with obesity, 8 years after gastric bypass. *J Am Diet Assoc* 2010; 110:527–534.

Lacey, Janet and Rick Houser. Dietetics and mental health counseling: Time for partnership. *J Amer Diet Assoc* 2001; 101(7):744.

Matthews, DE. Proteins and amino acids. In *Modern Nutrition in Health and Disease*. Ed. ME Shils, M Snike, AC Ross, et al. 54. Lippincott Williams & Wilkins. Philadelphia, 2006.

Penland, JG. 1997. Trace elements, brain function and behavior: Effects of zinc and boron. In *Trace Elements in Man and Animals – 9: Proceedings of the Ninth International Symposium on Trace Elements in Man and Animals*. 213–216. ed PF Fischer, MR L'Abee, KA Cockell and RS Gibson. NRC Research Press, Ottawa, Canada.

Pessler, LM, K Frankena, J Toorman, et al. Effects of a restricted elimination diet on the behavior of children with attention-deficit hyperactivity disorder (INCA) study: A randomized controlled trial. *Lancet* 2011; 377(2):494–503.

Pincus, T. Analyzing long-term outcomes of clinical care without randomized controlled clinical trials: The consecutive patient questionnaire data base. *Advances: J Mind-Body Hlth*. 1997; 13(2):3–32.

Proudfit, WL. All clinical questions cannot be answered by clinical trials. *Advances: J Mind-Body Hlth* 1997; 13(4):64–71.

Reynolds, LV and KP Kearns. *Single-subject experimental designs in communicative disorders*. University Park Press. Baltimore. 1983.

Rosling, AM, P Sparén, C Norring, and A-L von Knorring. Mortality of eating disorders: A follow-up study of treatment in a specialist unit 1974–2000. *Int'l J Eat Dis* 2011; 44:304–310.

Saxena, S, E Jane-Lopis, and C Hosman. Prevention of mental and behavioral disorders: Implications for policy and practice. *World Psychiatry* 2006; 5(1):5–14.

Schur, EA, N Afari, H Furberg, et al. Feeling bad in more ways than one: Comorbidity patterns of medically unexplained and psychiatric conditions. *J Gen Intern Med* 2007; 22(6):818–821.

Thiels, C. Forced treatment of patients with anorexia. *Curr Opin Psychiatry*. 2008; 21(5):495–498.

Tindle, HA, B Omalu, A Courcoulas, et al. Risk of suicide after long-term follow-up from bariatric surgery. *Am J Med* 2010; 123(11):1036–1042.

Waters, GS, WJ Pories, MS Swanson, et al. Long-term studies of mental health after the Greenville gastric bypass operation for morbid obesity. *Amer J Surg* 1991; 161:154–158.

Weinstein, SJ, TM Vogt, SA Gerrior. Healthy Eating Index scores are associated with blood nutrient concentrations in the third National Health and Nutrition Examination survey. *J Amer Diet Assoc* 2004; 104(4)576–584.

Westenhoefer, J, F Bellisle, JE Blundell, et al. PASSCLAIM-mental state and performance. *Eur J Nutr* 2004; 43(Suppl 2):II85-II117.

Williams, R. *Biochemical Individuality*. 2nd ed. Keats Publishing, New Canaan, Conn., 1998. (First published in 1956.)

Appendix A: Tools for nutritional assessment

A.1 Food intake record

Enter the names and amounts of the foods and beverages you eat and drink throughout three days. You may think of three actual days or three typical days. Include details such as "fried" "sweetened" or other descriptions as relevant.

Approximate Times of Eating/Drinking	Day 1	Day 2	Day 3
Early Morning 4–9 AM			
Midmorning 9–11 AM			
Midday 11 AM–2 PM			
Midafternoon 2–5 PM			
Early Evening 5–8 PM			
Late Evening 8–11 PM			

Night 11 PM–4 AM			

A.2 Assessment of nutritional status (ANS)

This is a suggested comprehensive assessment of nutritional status form. Noting these nutritional and psychological descriptors has potential use in patient care for creating a treatment plan and in research for consistently discovering and quantifying the links between nutritional status and mental status.

(Circle any that apply; fill in any known values.)

ANS Aspect 0: Risk Factors (family history, potentially inheritable conditions that may affect nutritional status)

Alcohol	Anemia	Anorexia nervosa	Bipolar disorder
Cancer	Depression	Diabetes	Food allergy
Intestinal disorder	Heart Disease	Hemochromatosis	Kidney disease
Migraine Headaches	Osteoporosis	Thyroid disorder	_____

Gene analysis polymorphism (describe) _____

ANS Aspect 1: Physical Status/Body Composition (circle and/or fill in blanks)

Height: ____   Weight: ____   BMI: ____
Yes   No   BMI below 18.5
Yes   No   BMI above 30
Waist: ____   Hips ____   Waist: Hip Ratio ____
Yes   No   Wt. Gain/Loss of ____ lbs in past ____ months.
  Loss of 10% of weight in 6 months is clinically significant.
% body fat _____   Yes   No   Below 20% – Females/10% – Males
                    Yes   No   Above 35%

ANS Aspect 2: Dietary Habits

a. _____ Eats fewer than three times a day
b. _____ Makes food choices that do not meet the Food Guide Pyramid recommendations
    Yes   No   6–11 servings starches
    Yes   No   2–3 3-oz. servings meat/substitute
    Yes   No   3–5 servings vegetables
    Yes   No   2–3 servings fruit
    Yes   No   2–3 servings dairy foods
    Yes   No   Eats mono-/polyunsaturated fats
    Yes   No   Not over 10% calories from sugar
    Yes   No   Not more than (F) 1 (M) 2 drinks alcohol/day
    Yes   No   Low to moderate use of salt

Appendix A: Tools for nutritional assessment 235

    c.   Yes No   Consumes more than 400 mg caffeine/day
    d. _____ Uses nutrient supplements:
                  Yes No   Less than 100% DRI _____
                  Yes No   About or equal to DRI _____
                  Yes No   More than 500% DRI or greater than UL _____

ANS Aspect 3: Laboratory/Biochemical/Metabolic (Above or below normal (N) range for laboratory/biochemistry tests; enter lab value and N value used for comparison)

<u>Carbohydrate</u>
__ Fasting Blood Glucose (FBS)____ ____ 2-hour post-prandial glucose (2 hr PP) ____
__ Hemoglobin A1c (Hb$_{A1c}$) ____ ____ Galactose – enzymes and/or metabolites____

<u>Lipids</u>
____ Total Cholesterol_____ _____ High Density Lipoprotein __
____ Low Density Lipoproteins____ _____ Triglycerides _____
____ EFA and/or metabolites (EPA, DHA, O-3, O-6) _____

<u>Proteins and Amino Acids</u>
____ Albumin _____ ____ Pre-albumin _____ _____ BUN _____
____ Homocysteine____ ____ Phenylalanine–related enzymes and/or metabolites _____
____ Other _____

<u>Vitamins</u> (Blood, Serum levels, or Vitamin-Dependent Enzyme)
____ B$_1$ (Thiamin) (TKA) _____ ____ B$_2$ (Riboflavin) _____
____ B$_3$ (Niacin) (Nicotinamide) ___ ____ B$_6$ (Pyridoxine) _____
____ Biotin _____ ____ B$_{12}$ (Cobalamin) (MMA) _____
____ Folacin (Folic Acid) (FIGLU)__ ____ A (Retinol) _____
____ C (Ascorbic Acid) _____ ____ D (Choleciferol) (Ergosterol) __
____ E (Tocopherol) _____ ____ K (Phylloquinones) _____

<u>Minerals, Elements, Electrolytes, and Heavy Metals</u>
____ Aluminum _____ ____Calcium, DEXA scan _____
____ Chromium _____ ____Copper _____
____ Iodine, T-3, T-4 _____ ____ Iron, Hct, TIBC, Hemoglobin, MCV _____
____ Lead _____ ____ Magnesium _____
____ Mercury _____ ____Potassium _____
____ Selenium _____ ____ Sodium _____
____ Other _____ ____Other _____

ANS Aspect 4: Clinical Signs and Symptoms (Presence of nutrient-based lesions determined by physical examination (a–e) and/or other symptoms reported by client (f–g))

a. Oral   Tongue   Lips   Gums   Teeth _____
b. Skin _____
c. Nails _____
d. Eyes _____
e. Hair _____
f. Yes No Diarrhea (more than two loose bowel movements/day)
g. Yes No Constipation (fewer than one bowel movement every three days)
h. Yes No Dental pain or discomfort that influences eating

ANS Aspect 5: Nutrient/Drug Interaction (Potential for Nutrient/Drug or Drug/Nutrient interaction) (Check those used, enter drug name if known)

___ Antacids _____ ___ Antianxiety _____
___ Antibiotic _____ ___ Antidepressant _____
___ Antidepressant (Tricyclic) ___ ___ Antidepressant (Monoamine Oxidase Inhibitor)_____
___ Antipsychotic _____ ___ Antiseizure _____
___ Diet pills _____ ___ Diuretics _____
___ Hypoglycemic (oral) _____ ___ Insulin _____
___ Laxative _____ ___ Lipid-lowering _____
___ Lithium _____ ___ Methotrexate _____
___ Tobacco _____ ___ Thyroid _____
___ Other _____ ___ Other _____

Nonspecific Signs or Symptoms Reported by Client: (circle any reported; add any additional symptoms)

Appetite ↓ ↑	Concentration reduced	Energy level reduced/increased
Fatigue	Headaches	Irritability
Memory Problems	Sleep Problems	Tearful
_____	_____	_____

Additional Nutritional Observations, Comments:

Appendix A: Tools for nutritional assessment

Assessment of Nutritional Status related to Stages of Nutritional Injury
Summarize findings of ANS by listing the risks in each stage contributing to determination of an individual's Stage of Nutritional Injury.

ANS 0: Risk of nutritional injury _____
ANS 1 _____
ANS 2 _____
ANS 3 _____
ANS 4 _____
ANS 5 _____
Nonspecific signs and symptoms_____

The Stage of Nutritional Injury (a descriptor of nutritional status) may be assigned to each individual based on any or all of the findings from the assessment and the professional judgment of the practitioner.

Use the descriptions below to determine the Stage of Nutritional Injury of the individual assessed. The highest level present is most often the designated Stage of Nutritional Injury.

<u>Stages of Nutritional Injury</u>
  I. Depletion of nutrient stores, adaptation (ANS Aspects 1 and 4)
 II. Reserves exhausted (potential: Stage I indicators of depletion or excesses lasting for six weeks or longer)
III. Physiologic and metabolic alterations (ANS Aspect 2)
 IV. Nonspecific signs and/or symptoms (potential indicated by reports of fatigue, headaches, loss of appetite, decrease in attention, insomnia, etc.)
  V. Illness or specific signs and/or symptoms (ANS Aspects 3 and 5)
 VI. Damage irreversible or nonresponsive to treatment (potentially including but not limited to loss of absorption sites resulting from bariatric surgery, bone loss, vision loss, loss of nerve function)

Stage of Nutritional Injury: _____ (0–VI)
GAF score _____ (date ____)   GAF score ____ (date____)*
DSM-IV Diagnosis Axis I _____   Axis II ____**

* GAF: Global Assessment of Functioning is Axis V of a multiaxial diagnosis by a psychiatrist; a numerical evaluation using the Global of Functioning Scale expresses an individual's level of psychological, social and occupational functioning at a given point in time. (p. 30–32 of the DSM-IV)

** Axis I includes Clinical Disorders and other Conditions;
Axis II includes Personality Disorders and Mental Retardation (p. 25–27 of the DSM-IV).

A.3 Assessment of nutritional status: guidelines with critical values and notes

Below are values that may be used in evaluation of the information collected on the Assessment of Nutritional Status. Since methods and norms vary between locations and institutions, please note for comparison the norms that you may be using for evaluation. Due to biochemical and genetic individuality, professional judgment is always necessary in clinical evaluations of any kind.

ANS 3.1 ANS Aspect 1: Physical Status/Body Composition

BMI: 20–25—: most healthy
Below 20—Underweight; assess cause
Below 16—High likelihood of nutrient-based cutaneous lesions; nutrition-focused physical examination needed
25–30—Overweight; related to less depression in some populations
Above 35—Criteria for morbid obesity; possible metabolic effects and social stigma
Reports for patients with eating disorders:
BMI 12 kg/m: admission to hospital for intense treatment
BMI over 11.5: Standardized Mortality Ratio (SMR) of ~7
BMI below 11.5 had SMR above 30

Waist: Hip ratio: Females: 0.8 to 0.9 critical value for increased health risk; Males: 0.9 to 1.0

Recent weight gain of 5 to 7% of usual body weight is cause for monitoring lipids, glucose, etc., for changes in metabolism secondary to psychotropic medications

Recent weight loss of 10% of weight in 6 months is clinically significant; assess cause

% body fat: Below 20% – Females } May signal presence of eating
Below 10% – Male } disorder

Weight Gain Reported on Selected Atypical Antipsychotic Medications

Medication Status	Weight Gain
Antipsychotic free	0.21 lb/wk (0.09 kg/wk)
Typical antipsychotic	0.61 lb/wk (0.27 kg/wk)
Atypical antipsychotic	0.89 lb/wk (0.40 kg/wk)
Olanzapine treatment	1.70 lb/wk (0.76 kg/wk)

Clozapine treatment   0.50 lb/wk (0.22 kg/wk)
Risperidone treatment 0.34 lb/wk (0.15 kg/wk)

Simpson, M.M., Goetz, R.R., Devlin, M.J., Goetz, S.A., and Walsh, B.T. Weight gain and antipsychotic medication: differences between antipsychotic-free and treatment periods. *J Clin Psychiatry*. 2001; 62(9):694–700.

ANS 3.2 ANS Aspect 2: Dietary Habits

Eats fewer than three times a day—increased likelihood of inadequate nutrient intake

Makes daily food choices that do not meet the Food Guide Pyramid recommendations

    6–11 servings starches — needed for adequate energy, fiber, glucose intake

    6–8 oz. meat/substitute — needed for adequate protein, iron, zinc; 6 oz fish/week for omega-3 fatty acids

    3–5 servings vegetables } needed for fiber, vitamin A, C, folate
    2–3 servings fruit } fruits and some vegetables have significant carbohydrate

    2–3 servings dairy foods — difficult to consume adequate calcium without these or using a supplement, also supplies riboflavin, protein, carbohydrate (lactose)

    Mono- or polyunsaturated fats — balance of omega-3 and omega-6 fatty acids for CNS, neurotransmitters

    % calories from sugar } sugar and alcohol calories require
    Alcohol- not more than (F) } nutrients for metabolism but
    1 (M) 2 ounces/day } supply none; excess use may lead to high-calorie malnutrition

    Low to moderate use of salt — Moderate may be defined as ~2000 mg/day

    Caffeine intake/day — Moderate considered 300 to 400 mg/day; extremely high levels may appear as anxiety, lack of sleep; extremely high levels may result in paranoia, psychosis

        5 to 6 oz. regular coffee = ~100 mg caffeine
        6 oz. tea = ~40 mg
        12 oz. cola = ~35 to 50 mg
        1 can Red Bull® = ~80 mg
        hot chocolate = ~5 mg

Uses nutrient supplements:
  Less than 100% DRI ⎱ Consider individual diet, health,
  About or equal to DRI ⎰ and genetic factors
  More than 500% DRI or greater than UL – Concern regarding toxicity or side effects

ANS 3.3 ANS Laboratory/Biochemical/Metabolic

Whole blood, plasma, erythrocytes, leukocytes, urine, hair, saliva, gas-liquid chromatography, MRI, and other technological methods have all been used to assess health and nutrition in some way. Natural regulation toward homeostasis, influences on absorption, health of the liver and kidney, presence of disease or conditions such as pregnancy or stress, and genetics all may influence the results of a laboratory finding. Recent food intake influences some values; current values do not always reflect body stores and function.

Research reports should include which tests and standards (normal) values used, along with findings. Research regarding methodology for meaningful assessment is ongoing. Values below should be used with the above caveats in mind.

Selected standards for assessment of nutritional status are in Table A.1.

ANS 3.4 Aspect 4: Clinical Signs and Symptoms

  a. Oral   Tongue   Lips   Gums   Teeth _____
  b. Skin _____
  c. Nails _____
  d. Eyes _____
  e. Hair _____
  f. Yes No Diarrhea (more than two loose bowel movements/day) ⎫
  g. Yes No Constipation (fewer than one bowel movement every three days) ⎬ needs further assessment
  h. Yes No Dental pain or discomfort that influences eating ⎭

Presence of cutaneous lesions observed during physical examination (a–e) or other symptoms reported by client (f–g) may be related to nutrients or diet. Abnormal appearance that occurs with a history of poor diet or health conditions that influence nutritional status needs further assessment.

A laboratory test followed by a trial of the appropriate supplement and confirmation of resolution by follow-up laboratory test is the

Appendix A: Tools for nutritional assessment 241

Table A.1 Selected Standards for Assessment of Nutritional Status

Observed	Laboratory Assessment Test	Expected/Normal Value (Blood Levels Unless Noted Otherwise)
	Carbohydrate	
	Fasting glucose	<100 mg/dL; <6.1 mmol/dl
	2-h post prandial glucose	<140 mg/dL; <7.8 mmol/dl
	Prediabetes	100–125 mg/dl
	Hemoglobin A1c (Hb_{A1c})	4–5.9%
	Galactose enzymes or metabolites	18.5–28.5 U/g Hb (units per gram of hemoglobin)**
	Hypoglycemia	<50 mg/dl
	Impaired fasting glucose between impaired glucose tolerance	100–125 mg/dl 140–199 mg/dl
	Lipids	
	Total cholesterol	<200 mg/dL; <5.2 mmol/L
	High density lipoproteins (HDL)	40–59 mg/dL
	Low density lipoproteins (LDL)	<100 mg/dL; <2.59 mmol/L
	Triglycerides	Adults: Male: 40–160 mg/dL; 0.45–1.81 mmol/L Female: 35–135 mg/dL; 0.40–1.52 mmol/L
	Essential fatty acids (EFA)^^	
	EPA	0.51% (±0.43) % total lipids
	DHA	1.65% (±0.67) % total lipids
	DHA red blood cells	~4% of total lipids (1.9–7.9%)++
	DHA plasma	~3.5% total lipids (1.5–7.5%)++
	Arachidonic acid	8.84% (±1.66) % total lipids

(*Continued*)

Table A.1 (Continued) Selected Standards for Assessment of Nutritional Status

Observed	Laboratory Assessment Test	Expected/Normal Value (Blood Levels Unless Noted Otherwise)
	AA: DHA ratio	6.03 (±2.23)
	AA: EPA ratio	23.11 (±11.81)
Proteins		
	Albumin	Adults 3.5–5 g/dL; 35–50 g/L; not a reliable indicator of protein nutritional status*
	Pre-albumin; thyroxine-binding prealbumin; transthyretin	Pre-albumin: >170 mg/L⁺
	Blood urea nitrogen (BUN)	Adults: 10–20 mg/dL; 3.6–7.1 mmol/L
	Homocysteine (Hcy)	4–14 μmol/L
	Phenylalanine, enzymes, metabolites	Normal blood level for phe is ~0.8 to 1 mg/dl. The maximum normal level has also been defined as 0.125 mM/L. ^ Classical PKU as blood phenylalanine may be defined as >20 mg/dl. Others use criteria of 4–15 mg/dl.
	Other	
Vitamins		
	B_1 Thiamin (TKA)	Body stores – erythrocyte transketolase enzyme activity increase: N = 0–15%## Adults urinary excretion thiamin: <65 mg/g creatinine = deficient intake*
	B_2 Riboflavin	Erythrocyte glutathione reductase enzyme activity coefficient > 1.4 = great deficiency Adults urinary excretion: 70–199 μg/g creatinine*
	B_3 Niacin, nicotinamide	Adults: Urine – excretion of N-methlnicotinamide 1.6–4.29 mg/g creatinine##

Appendix A: Tools for nutritional assessment

B_6 Pyridoxine		Erythrocyte transaminase index E-AST <1.5 E-AST 1.9–2.2 marginal status; E-AST >2.2 = deficiency
Biotin		Biotinidase-screening newborns
B_{12} Cobalamin, Methylmalonic Acid (MMA)		160–950 pg/mL 118–701 pmol/L <3.6 µmol/mmol creatinine
Folate, folacin, folic acid, FIGLU, tetrahydrofolate reductase (THFR)		Serum 5–25 ng.mL; 11–7 nmol/L RBC 360–1400 nmol/L
A Retinol; retinol binding protein (RBP)		Urine: 163 µg/24 h Serum vit A >20 µg/dL*
C Ascorbic acid		Plasma ascorbate <0.20 mg/dL* Leukocyte ascorbate <7 mg/L*
D_3 Cholediferol, ergosterol		25–80 ng/mL; <20 = def Toxicity: >150 ng/ml; > 375 nmol/L*
E Tocopherol		Serum adults: 0.47–2.03 mg/dL*
K Phylloquinones		11–12.5 sec prothrombin time
Minerals/Elements/Electrolytes/Heavy Metals Blood or urine levels of many minerals are not good indicators of body tissue stores.		
Aluminum		0–6 ng/mL
Calcium, DEXA scan		Ionized Ca adults: 9–10.5 mg/dl; 2.25–2.75 mmol/L
Chromium		Hair: 440 ppm* Urine: 1–20 nmol/L*

(Continued)

Table A.1 (*Continued*) Selected Standards for Assessment of Nutritional Status

Observed	Laboratory Assessment Test	Expected/Normal Value (Blood Levels Unless Noted Otherwise)
	Copper	Plasma — Adult males: 0.91–1.0 µg/ml Females 1.07–1.2 µg/ml On oral contraceptives: 2.16–3.0 µg/ml
	Iodine, T-3, T-4, TSH	T-4 adult: ~5–12 µg/dL; ~60–154 nnmol/L TSH: 2–10 µU/mL;
	Iron (total)	Adult male: 80–180 µg/dL; 14–32 µmol/L Female: 60–160 µg/dL; 11–29 µmol/L
	Hematocrit (Hct)	Male: 42–52%; 0.42–0.52 volume fraction Female: 37–47% or 0.37–0.47 " "
	Total iron binding capacity (TIBC)	250–460 µg/dL; 45–82 µmol/L
	Lead	<10 mcg/dL
	Magnesium (hypokalemia may be a better indicator of low Mg than serum Mg^a)	Adult: 1.3–2.1 mEq/L 0.65–1.05 mmol/L
	Manganese	Red cells 24 ±8 µg/L; serum 1.48 µg/L*
	Mercury	Inorganic exposure: normal, <20 µg of mercury per liter of urine# Whole blood mercury level <5.0 µg/L% Hair level <1.0 µg/g%

Appendix A: Tools for nutritional assessment

Potassium	Adult: 3.5–5.0 mEq/L 3.5–5.0 mmol/L
Selenium	Blood: 0.1–0.34 µg/ml* Red cells: 0.23–0.36 µg/ml*
Sodium	Adults: 136–145 mEq/L
Zinc	May not be reliable indicators of nutritional status for zinc* Plasma: 115 ± 12 µg/dl* Marginal status: 10.7–3 µmol/L; 0.70–0.85 µg/ml# Neutrophils: 108 ± 11 µg/10^{10*} Response of alkaline phosphatase to zinc supplementation*
Other	

Source: Pagana, K.D. and T.J. Pagana. *Mosby's Diagnostic and Laboratory Test Reference*, 10th ed. 2011. Elsevier. St Louis, MO.

* Alpers, D.H., W.F. Stenson, and D.M. Bier. *Manual of Nutritional Therapeutics*. 1995. Little Brown & Co. New York.
# Verrier, D. and M.I. Greenberg. Care of patients who are worried about mercury poisoning from dental fillings. *J Amer Board Fam Med*. 2010; 23(6):797–798. www.medscape/com/viewarticle/733712
% Hightower, J.M. and D. Moore. Mercury levels in high-end consumers of fish. *Environmental Health Perspectives* 2003; 111(4): 604–608.
^ Scriver, C. R. Phenylketonuria: Paradigm for a Treatable Genetic Disease...? NIH Planning Committee on Consensus Development Conference on Phenylketonuria (PKU): Screening and Management, last update August 28, 2006. http://www.nichd.nih.gov/publications/pubs/pku/sub7.cfm.
+ Ingenbleek Y., Ed. First International Congress on prealbumin in health and disease. *Clin Chem Lab Med* 2002; 40:1189–1369.
** http://www.nlm.nih.gov/medlineplus/ency/article/003645.htm
^^ Conklin, S.M., S.J. Manuck, J.K. Yao, et al. Serum omega-3 fatty acids are associated with depressive symptoms and neuroticism. *Psychosomatic Med* 2007; 69:932–934.
# Sauberlich, H.E. *Laboratory Tests for the Assessment of Nutritional Status*. 2nd ed. 1999. CRC Press, Boca Raton, FL.
++ Arterburn, L.M., E.B. Hall, and H. Oken. Distribution, interconversion, and dose response of omega-3 fatty acids in humans. *Amer J Clin Nutr* 2006; 83(suppl):1467S–1476S.
a Shlamovitz, G.Z. http://www.medscape.org/viewarticle/704606

Table A.2 Clinical Signs Potentially Related to Nutritional Deficiencies

Area Examined	Clinical Observation	Associated Nutrient	Selected Other Causes
Eye	Angular blepharitis	Riboflavin, niacin, B_6	
	Bitot's spots	A	
	Brow, outer 1/3 missing		Hypothyoidism
	Corneal arcus	Dyslipidemia	Aging
	Corneal vascularization		
	Kayser-Fleischer ring	Copper accumulation	Hereditary-altered metabolism
	Keratomalacia	A	Alcoholism
	Night blindness	A	
	Ophthalmoplegia	Thiamin, phosphorous	Brain lesion
	Pallor of everted lower lids	Iron, folic acid	Nonnutritional anemia
	Photophobia, burning, itching	Riboflavin	
	Pterygium		Non-nutritional
	Stare	Thiamin	Alcoholism
	Xerosis	A	Aging, allergies
Mouth, lips, mucous membranes	Angular stomatitis	Riboflavin, niacin, B_6, folate	Poorly fitting dentures, herpes, syphilis
	Cheilosis, vertical fissuring	Riboflavin, niacin	AIDS, environmental exposure
	Dryness	Water	Medications
	General inflammation	C, iron, B-complex	
	Pallor	Iron	
	Undifferentiated mucocutaneous border	Riboflavin	
	Red, swollen, interdental gingival hypertrophy	C, folate, B_{12}	Medications (Dilantin)
	Inflammation, generalized stomatitis		Oral hygiene, dry mouth
	Caries	Fluoride, phosphorous	
	Pitting, mottling		Excess floride

Table A.2 (Continued) Clinical Signs Potentially Related to Nutritional Deficiencies

Area Examined	Clinical Observation	Associated Nutrient	Selected Other Causes
Tongue color	Beefy red	Niacin, folate, roboflavin, iron, B_{12}	Diabetes
	Magenta, purplish red	Riboflavin	Crohn's disease, infection
	Scarlet	Niacin, folate, possibly B_{12}, B complex	
	Dysgeusia	Zinc	Trauma, syphilis, dry mouth
	Hypogeusia	Zinc, A	Poor fitting dentures, hypothyroidism
Tongue texture	Aphthous (ulcer) stomatitis	Folate, B_{12} Niacin	
	Fissuring, edema Geographic tongue, pallor, patchy atrophy	Biotin	
	Glossitis	Niacin, riboflavin, B_{12}, folate	
	Leukoplakia	A, niacin, folate, B_{12}	
	Lobulated with atrophy	Folate	
			Cancer therapy, dehydration, diabetes, influenza, polypharmacy
	Papillary atrophy	General under-nutrition and deficiencies	
	Pebbly, granular, cobblestone dorsum	Riboflavin, possible biotin	
	Cellophane-like	Protein, energy, essential fatty acids	

(Continued)

Table A.2 (Continued) Clinical Signs Potentially Related to Nutritional Deficiencies

Area Examined	Clinical Observation	Associated Nutrient	Selected Other Causes
Skin	Ecchymosis, subcutaneous with minor trauma	K, C, protein, energy	
	Decubitus ulcers, delayed wound healing	C, zinc, protein, possibly linoleic acid	Malignancy, steroid use, immobility, diabetes, AIDS
	Delayed wound healing	Essential fatty acids, zinc, niacin, riboflavin	Addison's disease, burns, hyper-sensitivity reactions, connective tissue disease
	Eczematous dermatitis—scrotum, vulva	Riboflavin	
	Pellagrous dermatitis	Niacin, tryptophan	
	Casal's necklace	Niacin	
	Flaky-paint dermatosis	Protein	
	Dry, scaling	A, essential fatty acids, zinc	Hypothyroidism, psoriasis, environmental factors, hygiene
	Edema, pitting	Protein, energy	Liver disease
	Follicular hyperkeratosis	A, essential fatty acids	
	Hyperpigmentation	Protein-energy, folate, B_{12}	
	Nasolabial seborrhea	Riboflavin, niacin, B_6, EFA	
	Petechiae	C, K, possibly A	
	Poor wound healing		
	Xanthoma	Lipids	Diet or inherited disorder

Appendix A: Tools for nutritional assessment

Table A.2 (Continued) Clinical Signs Potentially Related to Nutritional Deficiencies

Area Examined	Clinical Observation	Associated Nutrient	Selected Other Causes
Fingernails	Koilonychia	Iron	
	Pale	Iron, folate, B_{12}	COPD, heart disease, non-nutritional anemia
	Splinter-type hemorrhages under nails	C	
	White-spotting	Zinc, possibly selenium	
Hair	Corkscrew hair	Copper, C	
	Dull, thin, sparse	Protein, iron, zinc, essential fatty acids	Chemicals, chemotherapy, hypothyroidism, hereditary

most accurate method for determining whether a lesion is nutritionally caused.

Color, texture, shape, timing, and departures from common appearance/aberrations of appearance require familiarity with usual healthy human features. Awareness of other causes for a change in appearance is essential for ruling out nutritional causes for change (Table A.2).

ANS 3.5 Aspect 5. Potential for Nutrient/Drug or Drug/Nutrient Interactions

Drugs can change the appetite, metabolism, requirement for, action, and excretion of nutrient and vice versa. Effects may be to increase or decrease in either direction. A few examples are given in Table A.3. Knowing the effects of a specific drug is advised.

Many drugs recommend avoiding consumption of alcohol. Alcohol intake changes nutrient intake, nutrient metabolism, needs, and excretion.

Many drugs are carried throughout the body bound to albumin. Poor protein status may affect the drug's effectiveness. Adequate but not excess protein intake is advised.

Table A.3 Selected Drug: Nutrient/Food Interactions

ANS 5	Potential for Nutrient/Drug or Drug/Nutrient Interaction	
	Antacids	May lower absorption of folate, iron, phosphorous; may raise aluminum, magnesium; take separately from citrus fruit, juice, or calcium citrate
	Antianxiety	Limit caffeine to <400 mg/day; avoid stimulant or sedative herbs; avoid grapefruit juice
	Antibiotic	PenV-K may raise potassium, sodium levels, give false positive glucose test, may produce black hairy tongue, oral candidiasis
	Antidepressant — Tricyclic	Incompatible with carbonated beverages, grape juice; limit caffeine, may increase need for riboflavin; black tongue possible; appetite up for sweets
	Antidepressant — Monoamine Oxidase Inhibitors	Limit licorice, avoid tryptophan supplements, limit foods with tyramine
	Antipsychotic	Appetite changes, elevated cholesterol, glucose up or down, weight increase; take Mg supplement separately, may increase riboflavin need
	Antiseizure/ anticonvulsants	May need Ca, D, B_1, carnitine supplement; folate supplement frequently Rx; folic acid is an antagonist of phenytoin (dilantin), phenobarbitol, methotrexate, and other medications; appetite change
	Diet pills	If product mechanism is to decrease fat absorption, may interfere with fat-soluble vitamins A, D, E, K; some decrease appetite temporarily and are potentially addictive
	Diuretics	Monitor potassium, magnesium levels, avoid natural licorice; caution with supplements of D and calcium; may interfere with B_6 metabolism
	Hypoglycemics (oral)	Metformin may lower B_{12}, folate, raise homocysteine while lowering lipids
	Insulin	May lower potassium, magnesium phosphorous, may raise T-4

Appendix A: Tools for nutritional assessment

Table A.3 **(Continued)** Selected Drug: Nutrient/Food Interactions

ANS 5	Potential for Nutrient/Drug or Drug/Nutrient Interaction	
	Laxative	May lower potassium, calcium; monitor electrolytes with excess use
	Lipid-lowering	With statins avoid grapefruit and its juice; avoid high-dose niacin, red rice yeast
	Lithium	Requires consistent fluid and sodium intake
	Methotrexate	Lowers absorption of folate; gingivitis
	Selective serotonin reuptake inhibitor (SSRI)	Avoid tryptophan supplements; caution with grapefruit juice
	Tobacco	Increases need for C
	Thyroid	Absorption lowered with iron, calcium, magnesium; supplement with soy milk, soy foods, walnuts

Source: Pronsky, Z.M. *Food Medication Interactions*, 13th edition. 2004. Birchrunville, PA.

See also Boullata, J.B., and L.M. Hudson. Drug-nutrient interactions: A broad view with implications for practice. *J Acad Nutr Diet*. 2012; 112(4):506–517.

A.4 Recommendations for macronutrient intake

Essential fatty acids

Laboratory tests for lipids should be within normal limits (WNL); diet should contain 6 to 10 oz. fish per week (Table A.4). General recommendations for fat intake are ~20 to 30% of calories (e.g., 25% of a 2000-calorie intake is 500 calories in fat: 500/9 calories/g = 55 g fat per day). Fats from all sources should be considered (visible, invisible, as ingredients, use in cooking, etc). Omega-6 fats need to be balanced with omega-3 fats (EPA, DHA). Fats high in monounsaturated fatty acids (olive oil, safflower oil, avocados, sunflower seeds, and macadamia and filbert nuts) are preferable to saturated animal fats.

Protein and amino acids

For a rough estimate of protein needs, multiply the body weight (in lb) by 0.4. That number in grams is the recommended protein intake. For example, 150 lb × .4 = 60; ~ 60 g protein/day for a physically healthy adult. Protein from all sources should be considered: meat, fish, poultry, dairy protein, legumes, grains, and nuts.

Carbohydrate

A minimum of 100 to 130 g of carbohydrate insert is needed per day. Carbohydrate may easily be obtained by eating the recommended grains, dairy, fruits, and certain vegetables. This intake will supply the glucose

Table A.4 Selected Food Sources of EPA (20:5n-3) and DHA (22:6n-3)

Food	Serving (oz.)	EPA (g)	DHA (g)	Amount (oz.) Providing 1 g EPA + DHA
Herring, Pacific	3	1.06	0.75	1.5
Salmon, Chinook	3	0.86	0.62	2
Sardines, Pacific	3	0.45	0.74	2.5
Salmon, Atlantic	3	0.28	0.95	2.5
Oysters, Pacific	3	0.75	0.43	2.5
Salmon, sockeye	3	0.45	0.60	3
Trout, rainbow	3	0.40	0.44	3.5
Tuna, canned, white	3	0.20	0.54	4
Crab, Dungeness	3	0.24	0.10	9
Tuna, canned, light	3	0.04	0.19	12

Higdon, J. (2005), update Drake, V.J. (2009). Micronutrient Information Center. Essential Fatty Acids. Linus Pauling Institute Oregon State University. http://lpi.oregonstate.edu/infocenter/othernuts/omega3fa/ (Copyright 2003-2011).

for the brain and central nervous system. More carbohydrate is needed to supply the glucose for vigorous activity. Some athletes need over 500 g of carbohydrate per day (Table A.5).

Often hypoglycemia is preventable with eating patterns of small frequent meals and snacks. Eating a combination of carbohydrate and

Table A.5 Brief Assessment of Dietary History

Food Group	Major Nutrients Contributed by This Group May be Deficient If Omitted; Evaluate Other Sources
Dairy (milk, yogurt, cheese, ice cream)	Calcium, riboflavin, protein, carbohydrate (except in cheese), potassium (in milk and yogurt), and vitamins D and A (in milk if fortified)
Beef, pork, poultry, fish, eggs	Protein, fat, iron, niacin, zinc, fish, omega-3 fatty acids, D, eggs (choline)
Carbohydrates	Thiamin, riboflavin, niacin, folic acid, iron if enriched carbohydrates omitted; also zinc and magnesium if *whole* grains omitted
Vegetables	Vitamins C and A, folic acid, carbohydrate (in some vegetables), potassium, magnesium (in green vegetables)
Fruits	Carbohydrate, vitamin C, vitamin A, potassium
Fats	Omega-3 and omega-6 fatty acids, vitamin E Selected nuts (Mg, Phos, E, Se, PUFA)

Note:

* Using a completed Food Intake Record (A.1) + ANS, Section 3.2 check for low or missing food groups for initial assessment of nutritional adequacy. The body pool of ascorbic acid can be depleted in one month on a vitamin C deficient diet.

** Psychological changes occur before physical impairment. Personality changes occurred at whole blood levels of 1.21–1.17 mg/100 ml. Physical impairment occurred at 0.67–0.14 mg/100 ml. Clinical signs of scurvy were observed at these levels.

*** Dermatological changes due to essential fatty acids occur in three weeks of fat-free TPN. Biochemical indications of deficiency are observed long before scaly dermatitis. three weeks on a B_6 deficient diet decreased blood levels and enzyme activity. Seven weeks of folic acid deficiency can result in megaloblastic anemia. Depletion times: duration of vitamin deficiency required before body stores are depleted.

Vitamin A	1–2 years
α-Tocopherol	6–12 months
Pyridoxine	2–6 weeks
Riboflavin	2–6 weeks
Ascorbic acid	2–6 weeks
Thiamin	1–2 weeks

* Sauberlich, H. E. *Laboratory Tests for the Assessment of Nutritional Status*. CRC Press, Boca Raton, FL, 1999, p. 20.

** Kinsman, R.A. and J. Hood. Some behavioral effects of ascorbic acid deficiency. *Amer J Clin Nutr* 1971; 24:455–464.

*** Haller, J. Biokinetic parameters of vitamins A, B_1, B_2, B_6, E, K and carotene in humans. *Nutritional Neuroscience*, H. R. Lieberman, R. B. Kanarek, and C. Prasad, Eds. CRC Taylor and Francis, New York, 2005, p. 229.

protein during each meal may also delay digestion and absorption time. "Frequent" is defined as eating small amounts six times a day, spaced 2.5 to 3.5 hours apart over waking hours. Consuming concentrated sweets only in small quantities, at the end of a meal providing a mixture of protein, fats, and complex carbohydrates is recommended and not more than 10 to 20% of total calories.

Appendix A: Tools for nutritional assessment

A.5 Stages of nutritional injury

Stage 0: Healthy or Possible Risk of Nutritional Injury (Genetics:Genotype Host:Phenotype Environment:Agents)

Stage I: Diminishing Reserves/Building Excess

Stage II: Reserves Exhausted

Stage III: Physiologic and Metabolic Alterations

Stage IV: Nonspecific Signs and Symptoms

Stage V: Illness

Stage VI: Permanent Change

	Stage 0	Stage 1 and 2	Stage 3	Stage 4	Stage 5	Stage 6
Examples of conditions:	Family history of diabetes, alcoholism, celiac disease, bipolar disorder	Omits food groups due to allergies and/or preferences Eats 1 meal/day Use mega-dose supplements Fad diet use	Wt gain/loss in past 3-6 months High Chol, Glu, Hcy Relevant score on ED assessment	Fatigue Headaches Irritability Trouble concentrating Oral lesions Nausea Diarrhea Poor appetite	Metabolic Syndrome Bulimia Gestational diabetes Alcohol dependence Medication with Nutrient-Drug interaction	Bariatric surgery Dementia Phenylketonuria Xerosis with ulceration Vit-A blindness Pernicious anemia
Examples of nutrients: Regular alcohol intake,		Pro, Carb Vit A,D, Fol, E, C Iron, CA	Low Fol, Vit B_{12}, C Iron, Ess. Fatty Acids Hi Trig,Chol, Hcy	B-Vit deficiency Low carb intake Insufficient or excess Calorie intake	Low Vit, Min intake MAOI Medication Lactose intolerance	Vit B_{12}, A Phenylalanine

Adapted with permission from Arroyave, G. Genetic and biologic variability in human nutrient requirements, *Amer J Clin Nutr* 1979; 32:486–500.

References

Alpers, DH, WF Stenson, and DM Bier. *Manual of Nutritional Therapeutics* 1995. Little Brown & Co. New York.

Arroyave, G. Genetic and biologic variability in human nutrient requirements, *Amer J Clin Nutr* 1979; 32:486–500.

Arterburn, LM, EB Hall, and H Oken. Distribution, interconversion, and dose response of omega-3 fatty acids in humans. *Amer J Clin Nutr* 2006.

Boullata, JB, and LM Hudson. Drug-nutrient interactions: A broad view with implications for practice. *J Acad Nutr Diet* 2012; 112(4):506–517.

Conklin, SM, SJ Manuck, JK Yao, et al. Serum omega-3 fatty acids are associated with depressive symptoms and neuroticism. *Psychosomatic Med* 2007; 69:932–934.

Diagnostic and Statistical Manual of Mental Disorders (DSM-IV).

Haller, J. Biokinetic parameters of vitamins A, B1, B2, B6, E, K and carotene in humans. *Nutritional Neuroscience*, HR Lieberman, RB Kanarek, and C Prasad, Eds. CRC Taylor and Francis, New York, 2005, p. 229.

Higdon, J. (2005), update Drake, V.J. (2009). Micronutrient Information Center.

Hightower, JM and D Moore. Mercury levels in high-end consumers of fish. *Environmental Health Perspectives* 2003; 111(4): 604–608.

http://www.nlm.nih.gov/medlineplus/ency/article/003645.htm

Ingenbleek Y., Ed. First International Congress on prealbumin in health and disease. *Clin Chem Lab Med* 2002; 40:1189–1369.

Kinsman, RA, and J Hood. Some behavioral effects of ascorbic acid deficiency. *Amer J Clin Nutr* 1971; 24:455–464.

Pagana, KD, and TJ Pagana. *Mosby's Diagnostic and Laboratory Test Reference*, 10th ed. 2011. Elsevier. St Louis, MO.

Linus Pauling Institute Oregon State University. Essential Fatty Acids. http://lpi.oregonstate.edu/infocenter/othernuts/omega3fa/ (Copyright 2003–2011).

Pronsky, ZM. *Food Medication Interactions*, 13th edition. 2004. Birchrunville, PA.

Sauberlich, HE. *Laboratory Tests for the Assessment of Nutritional Status*. 2nd ed. 1999. CRC Press, Boca Raton, FL.

Scriver, CR. Phenylketonuria: Paradigm for a Treatable Genetic Disease...? NIH Planning Committee on Consensus Development Conference on Phenylketonuria (PKU): Screening and Management, last update August 28, 2006. http://www.nichd.nih.gov/publications/pubs/pku/sub7.cfm.

Shlamovitz, GZ. http://www.medscape.org/viewarticle/704606

Simpson, MM, Goetz, RR, Devlin, MJ, Goetz, SA, and Walsh, B T. Weight gain and antipsychotic medication: Differences between antipsychotic-free and treatment periods. *J Clin Psychiatry*. 2001; 62(9):694–700.

Verrier, D and MI Greenberg. Care of patients who are worried about mercury poisoning from dental fillings. *J Amer Board Fam Med*. 2010; 23(6):797–798. www.medscape/com/viewarticle/733712

Walsh, WJ., HR Isaacson, F Rehman, and A Hall. Elevated blood copper/zinc ratios in assaultive young males. *Physiology & Behavior* 1997; 62(2):237–329.

Appendix B: Mental health concerns: Nutritional concerns matrix

Appendix B: Mental health concerns: Nutritional concerns matrix

Table B.1 Mental Health Concerns: Nutritional Concerns Matrix

Mental Health Concern	Nutritional Concern					
	Essential Fatty Acids, DHA, EPA, PUFA	Carbohydrate; Sugars	Amino Acids, Protein	Minerals	Vitamins	Other Concerns
Chapter 1. Addiction						
Alcohol	EFA and behavior			Cu, Zn	A, B_1, B_6, C, folate, K,	Supplements, cognition, co-morbidities Psychosis, tolerance, with alcohol Malnutrition
Chapter 2. Caffeine Food						
Chapter 3. Aggression, anger, hostility, violence	EFA, fish, low cholesterol			Cu, Zn, Se,	B_1	Supplements; chol-lowering drugs
Chapter 4. ASD	Fatty acids	Aspartame	Methionine, cysteine, glutathione gluten, casein		B_6, D, folate	Vitamin/mineral Supplements and enzyme Supplements
ADHD	Phospholipids and high-dose olive, flax, fish oils, omega-3 fatty acids	Sugar		Zn	Mg overdose	Common diets reported; diet patterns

Appendix B: Mental health concerns: Nutritional concerns matrix

Chapter 5. Genetics, nutrient-gene interaction				Vitamin influences w/SNPs; folate	Genetic stability, testing, biomarkers
Inherited disorders of metabolism	DHA, EPA bone density	Galactose	Phenylalanine; homocysteine BH4 MSUD	Cu, Se, Zn	Lipoic acid
Chapter 6. Intellect	EFA infants			Lead: spices, pottery, remedies; Fe, I, Se	B_{12}
Cognition	DHA, EPA, AA, Saturated/trans fats	Hypo-, hyperglycemia	Homocysteine	Boron, Cu, Zn	$B_1, B_{12}, B_6, C, D,$ folate, E, β carotene	Suppl (vitamin/mineral, fish oil), alcohol intake; Resveratrol; diabetes
Dementia	DHA			Cu, Fe, Zn, Hg	B_2, E,
Chapter 7. Mood disorders					
Depression	EPA, DHA, omega-3 fatty acids, cholesterol		Tryptophan		Biotin, B_2, B_3, B_{12}, folate, C, D, E	Vegetarian diets, anxiety, osteoporosis, eating patterns
Bipolar disorder	Omega-3 fatty acids, DHA		Choline	Mg	Folate, inositol	Vitamin/mineral suppl, weight

(Continued)

Table B.1 (Continued) Mental Health Concerns: Nutritional Concerns Matrix

Mental Health Concern	Essential Fatty Acids, DHA, EPA, PUFA	Carbohydrate; Sugars	Amino Acids, Protein	Minerals	Vitamins	Other Concerns
Suicide	Cholesterol, fatty acids, anorexia					Food insufficiency, leptin, arctic diet, serotonin
Chapter 8. Schizophrenia	AA, DHA, EPA, phospholipids		Gluten BH4	Cu, Mg, Zn	B_2, C, D, E, folate	Oxidative stress, tetrahydropiopterin comorbidities, multivitamin supplement, drug-nutrition interaction, weight management
Chapter 9. Starvation	EFA probable	Glucose	Protein, EAA likely	Any	Any	Calories, emotional and personality changes, experimental neurosis
Eating disorders	PUFA, EFA, cholesterol, phospholipids	Fiber		Mg, Fe	B_1, B_2, B_6, B_{12}, C, E, folate, K A—possible excess	Neuroimaging, genetics, nutritional monitoring

Appendix B: Mental health concerns: Nutritional concerns matrix 261

Craving	Carbohydrate, concentrated sugar		Chocolate, brain function
Dieting			Psychological effects, intuitive eating
Bariatric surgery		Ca, Fe	B_1, B_{12}	Psychology and morbid obesity, monitoring, stress following, managing the diet following, nutritional supplements Vitamin supplements
Chapter 10. Quality of life				
Well-being	DHA in healthy adults			
Stress	Ethane biomarker-lipid peroxidation	Hypoglycemia	Lysine	B_1, B_6, D	Anger, anxiety, GI symptoms, pain, oxidative stress
Chapter 11. Other				
Agoraphobia		Cu, Hg, Se, Mg, Mn		B_1, B_2, B_3, B_6, B_{12}, choline, folate	Supplement

Table B.2 Health Conditions or Health Histories that Warrant Evaluation for Nutritional Status

Condition/Health History	Potentially Relevant Nutritional Factors
History of gastric bypass surgery	Thiamin, iron, B_{12}, D, Ca
History of eating disorder	Essential fatty acids, protein, carbohydrate, B vitamins, iron, magnesium
History of regular alcohol intake >1 oz./day	Thiamin, folic acid, vitamin B_6, magnesium
Weight loss of 5% of body weight in past month or 10% in past six months, especially if unintentional	Protein, calories, many possible nutrients
Strict low carbohydrate diet	Thiamin, folic acid
Vegetarian diet (especially vegan diet)	Vitamin B_{12}, iron, protein
Hyperemesis gravidum during pregnancy	Thiamin plus other nutrients
Cigarette smoking	Vitamin C
Lactose intolerance	Vitamin D, calcium, riboflavin
Exclusion of fat from diet	Essential fatty acids, omega-3, omega-6, vitamin E, vitamin K
Omission of or infrequent vegetables or fruits	Vitamin C, vitamin A, folic acid
Irritable bowel syndrome, diarrhea several times a day over weeks	Many nutrients
Chronic dieting for weight loss, especially diets with an intake of <1500 calories/day	Dependent on which foods are chosen for dieting
Insufficient income for adequate food	All or a combination of many nutrients may result in general malnutrition
Physical or mental inability to prepare food	Dependent on which foods are consumed
Over 20% of calories from sweets, desserts, and refined or unenriched grains	Thiamin, chromium, magnesium
Head injury requiring IV glucose	Thiamin
Prescriptions that alter (1) appetite/intake, (2) nutrient absorption, (3) metabolism of nutrient, enzyme function, or drug, or (4) excretion of nutrient or drug	Evaluate on individual basis
Habits including eating only one meal/day	Dependent on which foods are consumed
Trouble adjusting visually to dim or dark lighting	Vitamin A, beta-carotene

Appendix B: Mental health concerns: Nutritional concerns matrix

Table B.3 Level I Evidence Reports

Definition of level I evidence: experimental, prospective, randomized, double-blind, placebo-controlled, clinical trials. Not all characteristics may be present in all reported studies.

Chapter 2
Moderate Level Supplements during Alcohol Rehabilitation
Chapter 3
Essential Fatty Acids, Anger, and Anxiety
Disciplinary Infractions in Prison and Nutrition Supplements
Juvenile Delinquency and Vitamin-Mineral Supplementation
Chapter 4
Research on Aspartame
High Dose Olive, Flax, or Fish Oil on Phospholipids
Sugar and Hyperactivity
Chapter 6
Vitamin-Mineral Supplementation and Intelligence in School Children
Supplements and Academic Performance in School Children
Homocysteine, B Vitamins, Brain Atrophy, and Cognitive Impairment
Folic Acid and B_{12} Supplements and Cognitive Decline
Resveratrol and Cognition
Copper and Cognition in Alzheimer's Disease
Chapter 7
No Effect of EPA and DHA on Depression
EPA Supplements along with Standard Drugs
Tryptophan and Chronic Insomnia
Ascorbic Acid, Depression, and Personality Changes
Chromium and Depression
Selenium, Depression, and Nursing Home Residents
Choline and Rapid-Cycling Bipolar Disorder
Chapter 8
Omega-3 Fatty Acids Decreased Progression to Psychosis
Chapter 10
Healthy Male Volunteers
Ethane as a Biomarker for Lipid Peroxidation
Vitamin D and Mental Well-Being
Chapter 11
Magnesium and Migraine Headache

Glossary

5-HIAA: 5-hydroxyindoleacetic acid, a breakdown product of serotonin. 5-HIAA test: N = 3–15 mg/24 h. Foods that can interfere with measuring and test results include plums, pineapple, bananas, eggplant, tomatoes, avocados, and walnuts. They should be omitted from the diet 3 days before the test.
5-HTP: 5-Hydroxytryptophan, a precursor to the neurotransmitter serotonin.
AA: An abbreviation for arachidonic acid, a conditionally essential, long-chain fatty acid that has 20 carbons and 4 double bonds.
Accommodation: The protein-sacrificing response to an inadequate intake; a physiologic compromise with adverse health consequences.
Adaptation: The adjustment of an organism to its environment, when, during starvation the brain switches from glucose-based (carbohydrate) fuel supply to ketone-based (fat) fuel supply.
ADHD: Attention deficit hyperactivity disorder.
AHRQ: Agency for Healthcare Research and Quality.
ALA: Alpha-linolenic acid.
Apoenzyme: The protein component of an enzyme which can be separated from a coenzyme (often a vitamin) but that requires the coenzyme to function.
Bariatric surgery: Surgery for treatment of excessive weight; often referred to as gastric bypass surgery.
BED: Binge eating disorder.
BH4: Tetrahydrobiopterin, used in the treatment for some patients with phenylketonuria; see also **Sapropterin**.
Bipolar disorder: A mood disorder characterized by vacillation between highs of euphoria and lows of depression.
BMI: Body mass index; a measure of weight relative to height.
Calorie, kcal: A measure of energy; either energy in foods, energy used in activity and body maintenance, or energy stored in the body for future use.
CAM: Complementary and alternative medicine.

Cheilitis: Inflammation of the lips or of a lip, with redness and the production of fissures radiating from the angles of the mouth and/or cracking of the lips; may be related to vitamin C or B-vitamin deficiency, mineral deficiency, sunburn, cosmetics, dermatitis, and other non-nutritional causes.

Cheilosis: A noninflammatory condition of the lips, characterized by fissuring. Cheilosis denotes relationship to the lips or to an edge.

CNS: Central nervous system.

Cochrane Review: The Cochrane Library is a collection of databases that contain high-quality, independent evidence to inform healthcare decision-making. To accomplish this, The Cochrane Collection is divided into Cochrane Review Groups, each of which concentrates on a specific healthcare area. Each group report represents the highest level of evidence to date on which to base clinical treatment decisions.

Cognition: Mental function, as contrasted with emotional function. Cognitive domains include executive function (abstraction, mental flexibility), memory (verbal, spatial, facial), and intellect (language, spatial).

Correlation: A condition when a phenomenon changes in conjunction with change in another phenomenon; the phenomena may change in the same direction ↑↑ (positive correlation) or in the opposite direction ↓↑ (negative correlation).

Cretin/cretinism: The congenital lack of thyroid secretion with arrested physical and mental development.

CRP: C-reactive protein, a metabolic indicator of the presence of inflammation.

CSF: Cerebral spinal fluid.

De novo mutations: A genetic mutation that is neither possessed nor transmitted by a parent.

DHA: Docosopentaneoic acid, a 22-carbon, omega-3 fatty acid that plays a role in membrane structure.

Dysthymia: A mild but chronic form of depression. Symptoms may be less intense than those of major depression, but can affect one's life seriously because it lasts for so long, perhaps years. Symptoms may include poor appetite or overeating, insomnia or excessive sleep, low energy or fatigue, low self-esteem, poor concentration, indecisiveness, and hopelessness.

Eating disorder: A condition characterized by abnormal, unrealistic perspective on weight, body image, eating habits, and food consumption. Eating disorders include anorexia nervosa, bulimia nervosa, and binge eating disorder.

EFA: Essential fatty acid, which includes linoleic, linolenic, and arachidonic acids.

Enzymopathy: Abnormalities in enzymes.

EPA: Eicosapentaenoic acid, a 22-carbon, omega-3 fatty acid that plays a role in membrane function.
EPC: Evidence-based practice centers.
Epigenetics: Having an effect on genetic transcription without a change in DNA sequence.
Epiphenomenalism: The doctrine that mental activities are simply epiphenomena of the neural processes of the brain and have no causal influence.
Epiphenomenon: An accessory, exceptional, or accidental occurrence in the course of an attack of any disease; an additional condition in the course of a disease not necessarily connected to the disease; a secondary phenomenon overlapping and resulting from another.
Essential fatty acids (EFA): Long chain fatty acids not made in the body including linoleic acid (O-6), linolenic acid (O-3), and arachidonic acid.
ETC: Electron transport chain.
Executive function: Refers to cognitive ability to think abstractly; mental flexibility.
Experimental neurosis: Neurotic symptoms produced by restriction of the diet and reversed by means of controlled nutritional rehabilitation.
Extrapyramidal symptoms: Include extreme restlessness, involuntary movements, and uncontrollable speech, such as tongue movements, lip smacking, movement of arms, legs, fingers, or eye blinking and may occur as a side effect of antipsychotic medications.
FAE: Fetal alcohol effect. FAE is a term less preferred than FAS, which refers to alcohol-related birth defects and neurodevelopmental disorders.
FAS: Fetal alcohol syndrome. FAS is a cluster of abnormalities including physical, behavioral, and cognitive effects associated with prenatal exposure to alcohol.
Fasting state: Often defined as the point in time when the liver glycogen stores are exhausted.
Fatigue: Has been described by three subscales of the Fatigue Symptom Checklist as being general, physical, or psychological fatigue.
fL: Femtoliter. The SI prefix "femto" represents a factor of 10^{-15}; One femtoliter = 10^{-15} liter. One U.S. ounce liquid is equal to 29,573,529,687,517.04 fL (29 trillion).
Folate/folic acid: A B-vitamin; folate is found in foods; functions as the tetrahydrofolate reduced form; folic acid is the synthetic oxidized form and must be converted to tetrahydrofolate by the enzyme dihydrofolate reductase (DHFR) to be used by the body.
Fluid intelligence: Involves reasoning, the capacity to solve complex problems, the ability to think abstractly, and the ability to learn.

fMRI: Functional magnetic resonance imaging.
Functional disease/disorder: A condition that affects function, but not structure.
Functional testing: Assessment of enzyme-dependent activity; necessary for detection of suboptimal levels of many nutrients and metabolites. A given nutrient may be present, but it may not be properly activated, appropriately localized, or have sufficient cofactors to function at a normal level of activity. The result will be a defect in the biochemical pathways that depend upon that nutrient for optimal function. A deficient or defective pathway may operate at a sub-optimal level for many months, or even years, before a clinical symptom becomes apparent.
GABA: Gamma-aminobutyric acid.
Gastric bypass surgery: A procedure for decreasing the volume of the stomach and absorption area for nutrients. In vertical banded gastroplasty, a small stomach pouch is constructed and the outlet from the stomach to the small intestine (duodenum) is restricted. In gastric bypass surgery, a small stomach pouch is constructed that has an outlet directly to the jejunum, rather than the usual outlet into the duodenum. Most nutrient absorption normally occurs in the duodenum.
g/dl: Grams per deciliter. A deciliter is a fraction of a liter; 1 liter is approximately 1 quart. A deciliter is 1/10 of a liter. Gram per deciliter refers to the weight per volume of a substance (e.g., 1 gm/dl of glucose indicates that there is 1 g of glucose in 1 dl of blood.)
Genotype: The internally coded, inheritable information carried by all living organisms. This stored information is used as a "blueprint" or set of instructions for building and maintaining a living creature. These are written in a coded language (the genetic code), and are passed from one generation to the next; the class to which that organism belongs is determined by the description of the actual physical material made up of DNA that was passed to the organism by its parents.
GLA: Gamma-linolenic acid, derived from omega-6 fatty acids.
Glossitis: Inflammation of the tongue.
Hcy: Homocysteine.
HDL: High-density lipoproteins, a fraction of cholesterol found in the body, commonly referred to as "good cholesterol."
High calorie malnutrition: Poor nutritional status and the associated compromise of metabolic systems resulting from the imbalance of nutrient need and supply. High calorie malnutrition may be the result of poor diet (an excess of empty calories). Increase of need may be due to disease, genetic alteration, or increased excretion. High calorie malnutrition frequently applies to an inadequate

intake of vitamins and minerals while calorie and protein needs are met.

Homeostasis; psychological homeostasis: Tendency to maintain stability or balance in normal bodily states; a state of psychological equilibrium obtained when tension or a drive has been reduced or eliminated; stability may be maintained even if it is detrimental.

HUFA: Refers to highly unsaturated fatty acids.

Hysteria: Current terms for hypochondriasis and hysteria include somatization disorder, conversion disorder, and dissociative disorder or histrionic personality. Part of the "neurotic triad."

Incidence: A statistical term defined as the number of new cases of disease over a period of time divided by the number of people at risk during that period of time (e.g., the number of new cases of polio in 2001 divided by the population of the United States in 2001). See also **Prevalence**.

Inositol: A six-carbon molecule present as a free molecule or as part of phospholipids. Unbound inositol is relatively higher in neural tissue than in plasma. Inositol is synthesized in the body and does not appear to be toxic in high dietary amounts unless inositol metabolism is impaired. Found in animal protein foods or in plant foods as phytates.

IU: International units, a measure of quantity or activity of some vitamins, especially fat-soluble vitamins with several molecular forms with different activity in the body (e.g., the alpha/α and gamma/γ forms of vitamin E/tocopherol).

LDL: Low-density lipoprotein, a fraction of cholesterol found in the body, commonly referred to as "bad cholesterol."

Linolenic acid: An omega-3 polyunsaturated fatty acid; alpha linolenic acid (ALA) has double bonds in the 9, 12, 15 position; gamma linolenic acid (GLA) has double bonds in the 6, 9, 12 positions.

LNAA: Large neutral amino acids, which include tryptophan, phenylalanine, tyrosine, lucine, isoleucine, valine, methionine, threonine, serine, and cysteine.

Macrocytosis: The enlargement of red blood cells.

Malnutrition: May refer to the state of being undernourished or overnourished; malnutrition is a disorder of nutrition. The AMA Council on Foods and Nutrition defined malnutrition as a state of impaired functional ability, of deficient structural integrity or development brought about by a discrepancy between the supply to the body tissues of essential nutrients and calories, and the biologic demand for them. (*Source*: Carl C, Pfeiffer, PhD, MD, *Mental and Elemental Nutrients*, p. 4. 1975, Keats Publishing Company, New Canaan, Connecticut.)

MAOI: Monoamine oxidase inhibitor, a class of anti-depressants.

mEq; milliequivalent: A measure of quantity of a substance. An mEq is the atomic weight (or molecular weight if the substance is a complex molecule) noted in mg (one one-thousandth [1/1000] of a gram, 10^{-3}).

Metabolic syndrome (also called syndrome X): A combination of medical disorders including the presence of high blood pressure, high LDL and low HDL in the blood, obesity, insulin resistance, and high glucose.

Metabolism: Refers to the sum of the physical and chemical processes by which simpler compounds are converted into living, organized substances.

Metabolite: A substance derived by metabolism.

Mentalization: Being mindful of what others are thinking and feeling as well as being mindful of your own thoughts and feelings.

Metaparadigm: An abstract statement of the domains of professional concern.

Mercury toxicity: Refers to the health hazards of mercury. Foods generally contain levels of mercury below 50 ng/g. Fish may contain 10–1500 ng/g or higher, depending on the environment of the fish. The acceptable limit for fish marketed in the United States and other countries is 0.4–1.0 mg/kg of mercury. This equals 400 ng/g. (John N. Hatchcock and Jeanne I. Rader in *Modern Nutrition in Health and Disease*, 2:1, 598, 1994.)

microgram: μg, mcg. Refers to a unit of weight; 1 millionth of a gram or one 1/1000 of a mg.

micromole: μmol. A measure of quantity/weight; refers to 1 millionth of a mole, which is the molecular weight in grams.

MMA: Methyl malonic acid, which is involved in vitamin B_{12} metabolism.

MMPI: Minnesota Multiphasic Personality Inventory.

MRI: Magnetic resonance imaging.

MTHFR: Methyltetrahydrofolate reductase, an enzyme involved in the metabolism of folate, which is a source of methyl groups; two polymorphisms and lowered enzyme activity of MTHFR demonstrates an association between the genetic variants and depression, schizophrenia, and bipolar disorder.

MUFA: Monounsaturated fatty acids.

Myopathy: A disease of the muscles.

N: May refer to the number of subjects in a research study (e.g., N = 24 males and N = 26 females), or may refer to normal, or expected, laboratory values for a healthy person (e.g., N = 75–100 mg/dl).

NAD: Nicotinamide adenine dinucleotide.

NADH: NAD plus an H+ ion.

NES: Night eating syndrome.

Neuropathy: A disease of the nerves.

Neurotic triad: A combination of hypochondriasis, hysteria, and depression; often derived from four scales of the MMPI psychological test. "Hysteria" is now described as somatization disorder, histrionic personality, or other terms.

Neuroticism: One of five domains of adult personality; six facets of neuroticism include anxiety, angry hostility, depression, self-consciousness, impulsivity, and vulnerability.

Neurotransmitters: Biochemicals in the body that participate in the transmission of signals along the central and peripheral nerves, including serotonin, epinephrine, dopamine, and norepinephrine. Amino acids from the diet that are converted to neurotransmitters include tryptophan, tyrosine, and choline. Tyrosine and L-phenylalanine are precursors to dopamine and norepinephrine. Tryptophan is the precursor to serotonin.

Neutraceutical: Any substance that may be considered a food or part of a food and provides medical or health benefits, including the prevention and treatment of disease.

ng: Nanogram, which is a measure of quantity equal to one-billionth of a gram, or one-one-thousandth of a microgram.

NHANES: National Health and Nutrition Examination Survey.

NIDDM: Non-insulin dependent diabetes mellitus (Type 2).

Nootropics: Cognition-enhancing drugs.

Normal: May refer to a usual, expected, or typical pattern state (e.g., a normal blood value is a value that falls within a range usually not associated with a health problem); in psychology, refers to the average in some quality such as intelligence or personality traits.

Nutrigenetics: Refers to the influence of the genes; gene alteration is the problem, foods and nutrients are the solution.

Nutrigenomics: Refers to the influence of nutrients (an environmental influence) on genes and gene expression.

Nutritional genomics: Refers to the scientific field of nutrient-gene interaction.

Nutritional injury: A compromise in metabolism related to the excess or deficiency of a nutrient, or dys-adaptation of a nutrient-related system in the body.

NQOL: Nutritional quality of life; may include "being able to eat what I want, when I want." The taste, pleasure, enjoyment, comfort, or socialization that we experience from food and mealtimes.

O-3: Omega-3 fatty acid.

O-6: Omega-6 fatty acid.

OCD: Obsessive-compulsive disorder.

Orthorexia: Obsession with a perfect diet or a fixation on righteous eating.

Oxidative stress: Refers to the effect of excess (greater than normal production) free radical molecules produced during ATP (energy) production in the mitochondria and their potential damage to lipid cell membranes, DNA, enzymes, neurotransmitters, and structural proteins.

p or P: A designation for statistical probability; indicating the probability of making an error in drawing a conclusion from data. A Type 1 error (false positive) occurs when the null hypothesis is rejected when it is actually true in the specified population. A Type II error (false negative) occurs when a null hypothesis is not rejected when it is actually false in the population.

Paresthesia: A diseased or unhealthy sensation.

pg: Picograms, a measure of quantity equal to one-trillionth of a gram.

Pharmacological/pharmaceutical: As in the "pharmacological/ pharmaceutical dose," of a nutrient refers to a dose higher than is expected to be effective or needed for nutritional functions alone (e.g., taking a vitamin at 20 times more than the Reference Daily Intake is not a neutraceutical dose, but a pharmacological dose. At such intakes, the vitamins may be expected to function differently. Niacin prescribed for lowering cholesterol levels is a pharmacological dose of a B-vitamin. Liver enzymes must be monitored in any individual taking the high dose to assure that the dose is safe.)

Phenotype: The outward, physical manifestation of an organism. These are the physical parts, the sum of the atoms, molecules, macromolecules, cells, structures, metabolism, energy utilization, tissues, organs, reflexes, and behaviors; anything that is part of the observable structure, function or behavior of a living organism; the class to which that organism belongs as determined by the description of the physical and behavioral characteristics of the organism.

Phospholipids: Refers to a fatty acid with a phosphorous-oxygen duo, plus possible other molecular combinations linked to it. Phospholipids provide a plausible biochemical explanation for interaction between genetic and environmental factors in psychiatric disorders. These compounds link the essential fatty acids from the environment (diet) and the genetic control of enzymes that control essential fatty acids in the body.

Polymorphism: A genetic substitution that occurs in more than 1% of a population.

PPM or ppm: Refers to parts per million, the amount of one substance dissolved in or mixed into another.

Prevalence: A statistic defined as the number of people who have a given disease at one point in time divided by the number of people at risk at that point in time (e.g., the number of people who are depressed in the year 2000 divided by the population of the

United States in the year 2000 would give you the prevalence of depression during that year). See also **Incidence**.

Pseudotumor cerebri: An enlargement that resembles a tumor possibly resulting from inflammation or fluid accumulation; cerebral edema and raised intracranial pressure with normal cerebrospinal fluid, headache, nausea, vomiting, and papilledema that is without neurological signs except occasional sixth-nerve palsy.

Psychiatry: A medical discipline that deals with illnesses, the signs and symptoms of which are manifest in disorders of emotion, thinking, and behavior. Psychiatry is concerned with disorders of mood, cognition, and behavior including depression, mania, cognitive disorders, mental retardation and organic brain syndromes, and personality and behavior disorders ranging from drug and alcohol abuse to neuroses and psychoses.

PTSD: Post-traumatic stress disorder.

PUFA: Polyunsaturated fatty acids.

r: A designation for statistical correlation.

RNA: Ribonucleic acid.

RR (Relative risk): A measure of how much a particular risk factor (say, cigarette smoking) influences the risk of a specified outcome (say, death by age 70). This specifies the amount of risk that the study indicates an individual might have in (1) contracting the studied condition or (2) having effective treatment from the treatment studied.

Sapropterin: Tetrahydrobiopterin (BH4), a co-factor for three enzymes used in the degradation of amino acid phenylalanine and in the biosynthesis of the neurotransmitters serotonin (5-hydroxytryptamine, 5-HT), melatonin, dopamine, norepinephrine (noradrenaline), and epinephrine (adrenaline).

Seborrheic dermatitis: Excessive secretion of sebum (sebum is a thick, semi-fluid substance composed of fat and debris from epithelial [skin] cells); pertaining to areas in which sebaceous glands are abundant.

Semi-starvation: A form of malnutrition in which food is eaten but in quantities not sufficient to meet bodily needs.

Serotonin syndrome: A drug reaction in which there is an accumulation of excessive amounts of serotonin; excess production or insufficient rate of breakdown of the neurotransmitter serotonin.

SFA: Saturated fatty acids.

Somato-psychic disorder: A nutritional-biochemical imbalance that creates an emotional effect (Abbey).

SSRI: Selective serotonin reuptake inhibitors.

Starvation: A form of malnutrition in which the intake amount of food and nutrients is negligible or minimal.

Stroop test: A test indicating attentional ability and executive function. In Stroop matching tasks, participants indicate whether the name of a color matches the word printed in color. The test may time how long it takes for an individual to say "green" when seeing the word "orange" displayed in the color green. The response time may be delayed for emotion-laden words, such as the word "fat" presented to an individual with an eating disorder.

Subthreshold disorders: Not meeting the diagnostic cutoffs in the DSMIV or ICD 10. (Shankman et al. 2009, Ch. 2) Subthreshold conditions as precursors for full syndrome disorders: a 15-year longitudinal study of multiple diagnostic classes. SA Shankman, PM Lewinsohn, DN Klein, JW Small, JR Seeley, SE Altman *Jour. of Child Psychol. and Psych.* 50 (12), 1485–1494.

Syndrome X: See **Metabolic syndrome**.

Tardive dyskinesia: Involuntary movements that sometimes develop from the use of antipsychotic medications.

TC: Transcobalamin or total cholesterol, depending on the context.

TCA: Tricarboxylic acid cycle, a cycle/flow of body metabolites that produces energy from food in the mitochondria.

Teratogen: Causing abnormal development in embryos.

Tertile: Division of a group into three equal parts for comparison (e.g., his score was in the upper tertile.)

Tetrahydrobiopterin (BH4): See **Sapropterin**.

TKA: Transketolase activity.

Tocopherol: Vitamin E; alpha-tocopherol is the most biologically active form and is a potent antioxidant. Gamma-tocopherol has anti-inflammatory properties and is more abundant in the U.S. diet.

TPN: Total parenteral nutrition, which refers to supplying nutrients via a tube into the blood stream, rather than into the gastrointestinal tract.

TPPE: Thiamin pyrophosphate effect on erythrocyte transketolase, which refers to the percentage of uptake with a known dose. (Interpretation of the stimulation assay: A change 0–15% is considered normal. A change of 16–24% is considered a marginal deficiency. A change >25% is considered a deficiency of thiamin. H.E. Sauberlich, p. 40).

Vitamin deficiency: An often co-occurring insufficient level of vitamins in the body to support energy production, normal metabolism, and growth. Unlike few other categories of disease, vitamin deficiency affects both the central nervous system and the peripheral nervous system.

Wernicke Korsakoff Psychosis (KP): The chronic form of WE characterized by severe short-term memory loss.

Wernicke-Korsakoff syndrome (WKS): The term used to refer to a combination of WE and KP as if it were a single entity.

Wernicke's Encephalopathy (WE): The acute phase of a relatively common and potentially lethal condition resulting from thiamine deficiency.
WKS: Wernicke-Korsakoff syndrome.
Xerosis: Abnormal dryness, as of the eye (xeropthalmia), skin (xeroderma), or mouth (xerostomia).

Terminology used in research

In a **case-control** study, patients who have developed a disease or condition are identified and their past exposure to suspected etiological factors is compared with that of controls or referents that do not have the disease or condition. The starting point for the follow-up may occur back in time (retrospective cohort) or at the present time (prospective cohort). In either situation, participants are followed to determine whether they develop the outcome of interest. For a case-control study, the outcome itself is the basis for selection into the study. Previous interventions or exposures are then evaluated for possible association with the outcome of interest.

A **case study** is an in-depth description of the factors related to a disease, disorder, or condition in a specific individual.

Cheilitis inflammation of the lips or lip, with redness and the production of fissures radiating from the angles of the mouth and/or cracking lips; may be related to vitamin C or B-vitamin deficiency, mineral deficiency, sunburn, cosmetics, or dermatitis. Causes may include infection with microorganisms such as *Candida albicans*, staphylococci; poor hygiene; drooling of saliva; over closure of the jaws in patients without teeth, or with ill-fitting dentures—also called perleche.

Clinical significance is usually based on the size of the effect observed, the quality of the study that yielded the data, and the probability that the effect is a true one. Clinical significance is not the same as statistical significance; a finding in a study may demonstrate a statistical difference in an attribute under review, but this may have no impact clinically.

Coherence examines whether the cause-and-effect interpretation for an association conflicts with what is known of the natural history and biology of the disease and the relevance for developing clinical recommendations.

A **cohort** is a subset of a population with a common feature, such as age, gender, or occupation. In a cohort study, a group is assembled and followed forward in time to evaluate an outcome of interest. Factors related to the development of disease are measured initially in a group of persons, known as a cohort. The group is followed over a period of time and the relationship of a factor to the disease is examined. The population may be divided into subgroups according to the level or presence of the factor initially and comparing the subsequent incidence of disease in each subgroup. For example, a study of the occurrence of heart attacks in a cohort of nurses who did or did not have high cholesterol at the beginning of an observation period.

Consistency is the extent to which diverse approaches (such as different study designs or populations) for studying a relationship or link between a factor and an outcome will yield similar conclusions.

Controls are a group of study subjects with whom a comparison is made in an epidemiologic study. In a case-control study, cases are persons who have the disease and controls are persons who do not have the disease.

A **crossover trial** is a research design in which subjects receive a number of treatments in sequence. Generally, this means that all subjects have an equal chance during the trial of experiencing both treatment and placebo dosages without direct knowledge, instead of either placebo *or* the treatment. Subjects may be transferred directly from one treatment to another or may have a washout period in-between test treatments. This type of trial can be randomized so that not all subjects get the alternative treatments in the same order.

An **epidemiological** study gathers data from large, defined populations.

Intention to treat gives a pragmatic estimate of the anticipated benefit from a change in treatment policy rather than of potential benefit in patients who receive treatment exactly as planned. In epidemiology, an intention to treat (ITT) analysis (sometimes also called intent to treat) is an analysis based on the initial treatment intent, not on the treatment eventually administered. ITT analysis is intended to avoid various misleading artifacts that can arise in intervention research. Full application of intention to treat can only be performed where there is complete outcome data for all randomized subjects (*BMJ* 319:670, 1999).

Meta-analysis is the process of using statistical methods to quantitatively combine the results of similar studies in a systematic review.

Null hypothesis refers to the assumption that there is no difference between two groups, individuals, conditions, or situations that research is addressing. If it is assumed there is no difference and this is shown to be statistically false, this indicates there is a difference. The difference may be due to a treatment or another factor being studied. For example, if a null hypothesis states there will be no difference in a health outcome between a group provided a nutrient and a group not provided the nutrient and the null hypothesis is shown to be false, then this is an indication the nutrient made a difference. This is not proof, and is not an indicator of a specific cause and effect, but is generally one step in the scientific process.

Number needed to treat (NNT) is an indicator of the effectiveness of a therapy. An NNT of one means that every person treated responds to treatment. An NNT of three means for every three people given the treatment, one is expected to respond. An NNT of 20 to 40 can still be considered clinically effective (Guyatt, G., J. Cairns, and D. Churchill.

Evidence-based medicine: A new approach to teaching the practice of medicine. *JAMA* 268 (17):2420–2425, 1992.)

Observational studies may be cohort and case-control studies; they may also be follow-up, incidence, longitudinal, or prospective studies. An observational study by its very nature "observes" what happens to individuals. Thus, to prevent selection bias, the comparison groups in an observation study are as similar as possible except for the factors under study. Establishment of similarity in nutritional status of research participants may be helpful in nutritional and mental health research.

Power refers to the probability of observing an effect in a sample if a specified effect size or greater exists in a population. If a sample is too small, an effect might not be evident, even though the effect actually exists.

Randomized clinical trials (RCT) require that allocation to treatment or control be randomized with the investigator "masked" (or "blinded") to the subsequently assigned treatment (allocation concealment). This helps to ensure comparability of study groups and minimizes selection bias. While not all controlled studies are randomized, all randomized trials are controlled.

Relative risk (RR) is a measure of how much a particular risk factor (say, cigarette smoking) influences the risk of a specified outcome (say, death by age 70). This specifies the amount of risk that the study indicates an individual might have in (1) contracting the studied condition or (2) having effective treatment from the treatment studied.

Significant refers to relevant or meaningful, or statistical significance, which refers to the probability that an outcome is unlikely to have occurred by chance. A frequent level of significance is $p = .05$, which means 5 times out of 100 the result would have been by chance. Significance is related to the statement of a null hypothesis.

Specificity refers to the proportion of truly non-diseased persons who are identified as such by the screening test; that is, the true-negative rate.

Statistical significance is the probability that an event or difference is real or occurred by chance alone. It does not indicate whether the difference is small or large, important or trivial. The level of statistical significance depends on the number of patients studied or observations made, as well as the magnitude of difference observed. Statistical significance observed in a clinical trial does not necessarily imply clinical significance.

Strength is the size of the estimated risk (of disease due to a factor) and its accompanying confidence intervals. Both of these concepts are directly related to grading the strength of a body of evidence.

Surveys are a quantitative method generally cross-sectional or longitudinal, using questionnaires or structured interviews with the intent of generalizing from the sample to the general population.

Systematic review is an organized method of locating, assembling, and evaluating a body of literature on a particular topic using a set of specific predefined criteria. A systematic review may be purely narrative or may also include a quantitative pooling of data, referred to as a meta-analysis. In a systematic analysis, only those trials which meet a number of pre-set conditions in relation to research design (e.g., sample size, randomization) are included in the final meta-analysis.

Temporality refers to the relationship of time and events such as exposure to a risk factor and the development of disease.

Washout period is the stage in a crossover trial where treatment is withdrawn before a second treatment is given. This is usually necessary to counteract the possibility that the first substance can continue to affect the subject for some time after it is withdrawn.

Index

A

A, vitamin
   alcohol, interaction between, 18
   alcohol-caused deficiency, case study, 19
   autism, relationship between, 51
Acetylcholine, 31
Addiction
   abuse, *versus*, 13
   alcohol; *see* Alcohol addiction
   caffeine; *see* Caffeine addiction
   defining, 13
   food; *see* Food addiction
   overview, 13
Addiction Severity Index (ASI), 22
Agency for Healthcare Research and Quality (AHRQ), 85
Aggression, anger, hostility, violence
   body mass index, relationship between, 37
   calorie intake, relationship between, 37
   fatty acids, relationship between, 32
   juvenile delinquency studies, 36
   labile moods, 37
   lipids, relationship between, 37
   metabolic syndrome, 37
   minerals, relationship between, 34
   neurotransmitters, relationship between, 31
   nutritional supplement studies, 35–36
   review of past literature for links to nutrition, 34–35
   thiamin, relationship between, 33
   waist-hip ratio, relationship between, 37
Agoraphobia, 197–199
Albumin, 100, 101
Alcohol addiction
   anxiety related to, 22
   bone loss related to, 20
   cognitive decline related to, 20
   congeners, 22
   depression related to, 22
   diabetes, complications with, 22
   eating disorders comorbidity, 21
   genetics, role of, 63–64
   hangovers, 22
   liver, role of, 16
   mental well-being effects, 188
   metabolism by genetic groups, 64
   metabolism of alcohol, 15–16
   minerals, interaction between, 20
   nutrient intake, role in, 16
   paternal consumption, effect on fetus, 17
   pregnancy, during, 16–17
   psychiatric comorbidity, 21
   screening for, 21
   toxicity, 18
   vitamin supplements taken during rehabilitation, 19–20
   vitamins, interaction between, 16–17, 18, 19–20
   withdrawal from, 22
Alcohol Use Disorders Identification Test (AUDIT), 21
Alcohol-containing caffeine drinks, 24–25
Aldehyde dehydrogenase (ALDH), 63
Alpha-linoleic acid (ALA), 31
Alzheimer's disease, 87. *See also* Dementia
   copper, relationship between, 91
   niacin, relationship between, 89
   vitamin C, relationship between, 90
   vitamin E, relationship between, 89–90
Ames, Bruce, 61
Amino acids. *See also specific amino acids*
   anxiety levels and, 191
   balanced, 224

depression, role during, 103
recommended intake levels, 252
schizophrenia, role in, 141
Anemia, 225
Anger. *See* Aggression, anger, hostility, violence
Anorexia nervosa (AN), 154. *See also* Eating disorders
  alcoholism, comorbid with, 21
  BMI and death, 164, 220
  brain neuroimaging, 161
  cholesterol, 162
  magnesium levels, 164
  pellagra in, 161
  scurvy, 163–164
  suicide, 162–163
Antabuse, 22
Anticonvulsants, 106
Antidepressants, 101, 172
Antipsychotic medications
  minerals, interactions with, 143
  omega-3 fatty acids, antipsychotic properties of, 140
  weight gain associated with, 148–149
Anxiety
  adolescents, 116–117
  alcohol use, related to, 22
  beriberi, links to, 33
  depression, comorbidity with, 116–117
  essential fatty acids, relationship between, 32
Arachidonic acid (AA), 31
Ascorbic acid (vitamin C)
  alcohol-caused deficiency, case study, 19
  Alzheimer's disease, relationship between, 90
  circulating plasma, role in, 89
  deficiency, 62, 104, 105–106
  deficiency and mood disorders, 104
  depression, relationship between, 105
  high doses of, 4
  mood, impact on, 106
  plasma levels, schizophrenic patients, 105
  supplement levels, 144
Aspartame
  autism spectrum disorders (ASDs) studies, 50–51
Attention deficit hyperactivity disorder (ADHD)
  bipolar disorder, comorbid with, 119
  brain imaging studies, 53
  dietary patterns and, 56

Feingold diet, 55
lipids, effects of high doses of, 53
nicotine as self-medication, 56
nutritional adequacy, 57
nutritional supplement treatment, 55, 57
omega-3 treatment, 53–54
overview, 41, 53
prevalence, 53
sugar, relationship between, 54–55
vitamin D, relationship between, 55
zinc, relationship between, 55
Autism spectrum disorders (ASDs)
  aspartame, relationship between, 50–51
  carbohydrates, relationship between, 45
  developmental disabilities, 42
  diagnosis, 41
  dietary interventions, 51, 52
  digestive enzyme supplements, 51
  digestive issues, 43–44, 47
  environmental factors, 44, 45
  genetic factors, 44–45
  history of, 41
  metabolism alterations, 42
  micronutrient management, 52
  minerals, relationship between, 48–49
  nutrients, relationship between, 42–43
  nutritional interventions, 43
  overview, 41
  retardation, *versus*, 45
  risk factors, 45
  supplements, relationship between, 49, 50
  symptoms, 41–42
  treatments, 43
  vaccine controversy, 45
Ayurvedic medications, 79

B

B_1
  autism, relationship between, 50, 51
  deficiency, 33
B_{12}
  autism, relationship between, 50, 51
  cognitive decline relationship between, 84
  cognitive function, relationship between, 85
  cognitive impairment, relationship between, 84
  deficiency, 62, 76–77, 107, 108, 200–201
  depression, relationship between, 107, 108

infant development, importance to, 76
pernicious anemia treatment, 226
post-bariatric surgery, 175
toxicity, 226
vegetarianism as cause of deficiency in, 200
B_2, autism, relationship between, 48
B_6
  alcohol, interaction between, 18
  autism, relationship between, 48, 49, 50, 51
  bereavement stress, relationship between, 191–192
  cognitive function, relationship between, 85
Baltimore Memory Study, 92
Bariatric/gastric bypass surgery, 154
  anxiety, post-surgery, 171
  B_{12} levels, 174
  calcium levels, 176
  cognitive behavioral therapy (CBT) following, 173
  comorbid issues, 169
  depression, post-surgery, 171
  dietary management following, 177
  eating disorders and, 169
  fat malabsorption, 179
  food avoidance post-surgery, 171
  health status, impact on, 172
  long-term effects, 170
  maltreatment, prior, 169
  nutritional monitoring following, 176–177
  obesity, psychological effects of, 168
  postoperative nutrition, 174
  preoperative nutrition, 174
  psychiatric diagnoses prior to, 168
  psychological support, 173
  psychopathology pre- and post-surgery, 169–170
  quality of life post-surgery, 171
  reappearance of symptoms years later, 172
  screening, 168
  stressful consequences of, 177, 179
  supplements, additional (more than multivitamins), 176
  supplements, multivitamin, 176
  thiamin levels, 175–176
  types of, 167, 168
  unfavorable outcomes, factors influencing, 173
Beck Depression Inventory-II, 168, 170
Belief, biology of, 210–211
Bereavement stress, 191–192
Beriberi, 33

Binge-eating, 162. *See also* Bulimia nervosa
  bariatric surgery, before and after, 170
Biofeedback, 209–210
Biological psychiatry, origins of, 2
Biotin deficiency, 106, 107
Bipolar disorder
  brain imaging studies, 120
  choline supplementation, 121, 122
  comorbidities, 119–120
  depressed phase, 119
  DHA supplements, 121
  heritability, 119
  insitol, relationship between, 123–124
  lithium treatment, 119
  micronutrient supplements case study, 122, 123
  nutrition-related behaviors, 120
  omega-3s, relationship between, 121
  overview, 118–119
  overweight status, 120
  suicide rates, 119
  symptoms, 118–119
Bland, Jeffrey, 7
Body odor,
  genetically caused, 66, 71
  genetically caused, 66, 71
Bone mineral density, 68
Boron, 85
Brain. *See also* Cognition; Intellect
  atrophy of, 84
  cholesterol's impact on, 102
  food cravings, role in, 166
  meal contents, relationship between, 6
  nutrition for, 5–6
  postmortem fatty acid studies of, suicide autopsies, 127
Brain Bio Center, Princeton, NJ, 5
Brain Bio Centre, Richmond, London, 5
Brozek, Josef, 154
Bulimia nervosa (BN). *See also* Eating disorders
  alcoholism, comorbid with, 21
  body weight loss, 159
Bush, George W., 10

C

C, vitamin. *See* Ascorbic acid (vitamin C)
Caffeine addiction
  absorption, 23
  alcohol-containing caffeine drinks, 24–25
  effects of, 23

Journal of Caffeine Research, 25–26
noncoffee forms of, 24
overview, 22, 23
paranoia related to, 23–24
psychosis, case study, 23–24
tolerance, 24
Calcium
autism, relationship between, 51
post-bariatric surgery, 176
Caloric intake, 15
schizophrenic patients, 146–147
Carbohydrates
cravings for, 165
hypoglycemia and stress, 191
mixed meals, effect on mood, 104
recommended intake levels, 252, 254
restricted diets, 207
Catecholamine, 159
Celiac disease, 141
Chelating therapies
copper toxicity, 90–91
Cholesterol
brain, impact on, 102
depression, relationship between, 102, 103
statin drugs, 33
suicide studies, 126–127, 162–163
violent crime, relationship between, 32–33, 38
Choline, 8
rapid-cycling bipolar disorder and, 121, 122
Chromium, 110–111, 227
Chronic fatigue syndrome, 208
Cochrane review, 113–114
Coenzyme Q, 49, 199
Cognition
beta-carotene, relationship between, 86
copper, relationship between, 85–86
defining, 75
executive domain, 137
fish oil supplements, 82
impairment, 75–76, 84
intellectual domain, 137
lipids, role of, 82
memory domain, 137
resveratol, relationship between, 86–87
vitamin C, relationship between, 86
zinc, relationship between, 86
Copper
-to-zinc plasma levels as schizophrenia marker, 143
alcohol withdrawal and, 20
cognition, relationship between, 91

dementia, relationship between, 90–91
elevated, 86
free plasma *versus* bound, 85–86
intellect, relationship between, 85–86
levels of, schizophrenic males, 142
liver, accumulation in, 90
plasma levels, 68
ratios to zinc in assaultive males, 34
toxicity, 86
Wilson's disease, relationship between, 201
Craving, food. *See* Food craving
Crohn's disease, 192
Cytidine, 8

D

D, vitamin
alcohol, interaction between, 18
attention deficit hyperactivity disorder (ADHD), relationship between, 55
autism, relationship between, 47–48
chronic pain, connection to, 193
cognition, role in, 84–85
deficiency, 193
DNA repair, role in, 45
mental illness, links to, 142
mental well-being, connection to, 193
mental well-being, link to, 187
De novo mutations, 45
Dementia
copper, relationship between, 90–91
defining, 75
depression, comorbid with, 88
iron, relationship between, 91–92
mercury toxicity, relationship between, 92
prevention, 87–88
Deoxyhemoglobin, 87
Depression. *See also* Major depressive disorder
alcohol withdrawal, related to, 22
anxiety, comorbidity with, 116–117
ascorbic acid, relationship between, 104, 105
atypical, 97
autism spectrum disorders (ASDs), comorbid with, 42
B_{12}, relationship between, 107, 108
body mass index of patients with, 99
carbohydrate craving, relationship between, 8
cholesterol, relationship between, 102, 103

chromium, relationship between, 110–111
costs of treating, 6
dementia, comorbid with, 88
docosahexaenoic acid (DHA),
    relationship between; see under
    Docosahexaenoic acid (DHA)
eating patterns of patients with, 118
eicosapentaenoic acid (EPA) levels in
    patients with, 100
eicosapentaenoic acid (EPA)
    treatment, 100
essential fatty acids (EFAs) levels in
    patients with, 101
folate, relationship between, 107
folic acid, relationship between, 108, 116
immune-inflammatory response
    during, 103
niacin, relationship between, 109
omega-3s, relationship between; see
    under Omega-3 fatty acids
osteoporosis, comorbidity with, 117
prevalence, 6
riboflavin, relationship between, 109
symptoms, 97
tocopherol, relationship between, 110
vitamin D, relationship between, 109
vitamin E, relationship between, 110
Deprivation, food, 164–165
DHA. See Docosahexaenoic acid (DHA)
Diabetes
    alcohol addiction, complications with,
        22, 88
    cognitive impairment in, 87
Diacylglycerol kinase eta (DGKH) gene, 120
Dietary Guidelines, U.S., 15
Dietary Reference Intake (DRI), 19
Dieting/fasting
    Health at Every Size (HAES), 167
    physical effects of, 166
    psychological effects of, 166–167
Dilutional hyponatremia, 146
Docosahexaenoic acid (DHA), 31
    bipolar disorder, supplements taken for,
        121
    cognitive function, role in, 82, 83, 88
    depression, role in, 98, 99, 100
    supplements, 68, 121
    violence behavior, relationship
        between, 32
Docosapentaenoic acid (DPA), 32
Dopamine, 6, 31
Down syndrome, 42
Dysthymia, 128

E

E, vitamin
    Alzheimer's disease, relationship
        between, 89–90
    autism, relationship between, 50
    deficiency, 62
    depression, relationship between, 110
Eating Disorder Examination (EDE), 170
Eating Disorder Inventory-II (EDI-II), 171
Eating disorders. *See also* Anorexia nervosa
        (AN); Bulimia nervosa (BN)
    adaptations to starvation, 159
    alcoholism, comorbid with, 21
    enzyme functions, 161
    laxative use, 160
    nutritional consequences, 160
    plasma observations, 161
    twin studies, 162
    vitamin depletion, 160
Efamol, 55
Eicosapentaenoic acid (EPA), 82
    -DHA treatment in schizophrenics,
        139–140
    levels, in depressive patients, 100
    use, risks of, 220
Elderly patients
    dementia; see Dementia
    selenium supplements, 112
    vitamin D status of, 109
Electrolytes
    mood, impact on, 111
English Longitudinal Study of Ageing, 20
Essential fatty acids (EFAs), 75–76
    levels of in depressive patients, 101
    levels of in schizophrenic patients, 139, 140
    precautions of use, 222
    recommended intake levels, 252
    treatment of schizophrenic patients,
        139, 140
Ethyl alcohol, 15

F

Fasting, 153, 166–167
Fatty acids, 8, 17
    anger and aggressive behavior,
        relationship between, 31–32
    autism spectrum disorders (ASDs),
        relationship between, 46
    infant development, importance to, 75–76
    plasma lipids, relationship between, 46
Fatty liver, 16

Feingold diet, 55
Fernstrom, John, 6
Fetal alcohol effects, 16–17
Fetal alcohol syndrome, 16–17
Fibromylagia, 208
Fight of flight response, 189–190
Fish oil supplements, 82
Fish, mercury in, 203–204, 228
5-hydroxytryptophan (5-HTP), 114
  supplements, 115–116
Folate
  deficiency, 142
  negative schizophrenic symptoms and, 141–142
Folic acid
  alcohol, interaction between, 18
  cognitive decline relationship between, 84
  cognitive function, relationship between, 85
  depression, fortification treatment, 116
  depression, interaction between, 108
  genetic stability, relationship between, 62
  genetic testing, 64
  self-induced deficiency, study, 3
Food addiction
  assessment tools, 15
  cravings, 14
  food-as-drug model, 14
  genetics of, 13
  malnutrition, role of, 14–15
  neurochemistry, 13
  overview, 13
Food craving
  brain function's role in, 166
  calorie restriction, relationship between, 165–166
  carbohydrates, 165
  chocolate, 165
  overview, 164–165
Free radicals, 89
Functional gastrointestinal disorders (FGID), 192
Functional medicine, 7, 8
Functional Medicine, Institute for, 7
Functional MRI (fMRI), 83

G

Galactose, 46
Galactosemia, 69
Gamma-linolenic acid (GLA), 31
Gastrointestinal Quality of Life Index, 174
Gastrointestinal symptoms, stress-related, 192
General adaptation syndrome (GAS), 189–190
Genome, human
  aberrations, 61
  sequencing of, 9, 61
Genomic variability, 7
Gliadin, 141
Glipizide, 22
Glucotrol, 22
Gluten sensitivity, 141, 208
Gluten-free diets, autism spectrum disorders, 43
Guthrie screening test, 66

H

Haller, Jung, 109
Haloperidol, 143, 148
Hamilton Depression Rating Scale Score, 100
Hartnup disease, 71
Health at Every Size (HAES), 167
Health-related quality of life (HRQOL) assessments, 185
Healthy Eating Index, 118
Hedaya, Robert J., 8
Henshel. Austin, 154
Highly unsaturated fatty acids (HUFA), 46
Histamine, 31
Hoffer, Abram, 5
Holford, Patrick, 5
Homeostasis, psychological, 210
Homocysteine, 84, 108, 188, 225
Homocystinuria, 70, 71
Horrobin, David, 8
Hostility. See Aggression, anger, hostility, violence
Human genome. See Genome, human
Human Genome Project, 61
Hyperglycemia, 83, 225
Hypericum, 113
Hypoglycemia, 83, 191, 253

I

Impact of Weight on Quality of Life (IWQOL), 185
Insomnia
  chronic, 103
  defining, 6
  tryptophan treatment, 103, 224

Index

Intellect
  defining, 75
  fears of losing mental capacity, 75
  lipids, role of, 75–76
  overview, 75
  supplement studies, 81–82
Intuitive eating, 167
Iodine, 80
Iron
  deficiency, 62, 79–80
  dementia, relationship between, 91–92
  uptake, 79–80
Irritable bowel syndrome (IBS), 192, 208
Isoleucine, 103

K

K, vitamin, supplementation and alcohol dependence, 19
Kanner, Leo, 41
Keys, Ancel, 154
Kitchen spirituality, 154
Kwashiorkor, 153

L

L-tryptophan, 114
Lead
  Ayurvedic medications, in, 79
  behavioral signatures of exposure, 78
  exposure levels, 78
  pottery from Mexico, in, 79
  spices from India, in, 78–79
  symptoms of toxicity, 77
  toxicity, adults, 77
  toxicity, children, 77–78
Leptin, 127
Leucine, 70, 103
Levels of Evidence, defined, xxviii
Lipids
  cognition, role in, 82
Lithium
  bipolar disorder treatment, 119
  insitol interaction, 123–124
  magnesium binding sites, 122
  weight gain associated with, 125
Liver, role in metabolism of alcohol; see under Alcohol addiction
Low-density lipoprotein (LDL), 222

M

Macrobiotic diets, 77
Magnesium, 227
  deficiency, 112–113, 143
  excess, 113
  migraine treatment, 206
  overdose, 48
Major depressive disorder, 97
  elevated lipid levels in, 100, 101
Malnutrition
  definition, 153
  subclinical levels, eating disorder patients, 160
Manganese, 206
Maple syrup urine disease (MSUD), 70
Marasmus, 153
Marasmus-kwashiorkor mixed, 153
Mental energy, 186, 187
Mental Health Parity and Addiction Act, 10–11
Mental health professionals,
  need for nutritional education, 216
Mercury toxicity, 92, 201, 203–204, 227
Messenger RNA, 158
Methyltetrahydrofolate reductase, 107
Metabolic disorders, inherited, 64–65. *See also* Phenylketonuria (PKU)
Metabolic syndrome, 37, 87, 120
Methionine malabsorption, 72
Methionine synthase, 45
Methylation, 44
Methylmalonic acid (MMA), 225
  deficiency related to diet, 77
Micronucleus assays, 63
Migraines, 199, 206
Mini-Mental State Examination (MMSE), 20, 82
Minnesota Multiphasic Psychological Inventory (MMPI), 155
Mitochondrial decay, 62
Mood disorders. *See also specific mood disorders*
  bipolar depression; *see* Bipolar disorder
  mixed meals and impact on mood, 104
  overview, 97
  prevalence, 97
MTHFR, *See* Methyltetrahydrofolate reductase
Multitasking, 188
Multivitamins, 187–188
Myelin sheath, 102, 120

N

N-acetyl-cysteine, 49
Nardil, 114
National Health and Nutrition Examination Survey (HANES), 109
National Health and Nutrition Examination Survey (NHANES), 15
Neurosis, experimental, 156, 260
Neurotransmitters. *See also specific transmitters*
  definition, 31
  role of, 31
New Freedom Commission on Mental Health, 10
Niacin
  absorption capacity, 109
  alcohol, interaction between, 17, 18
  Alzheimer's disease studies, 89
  deficiency, 62, 109
  migraine treatment, 199
  pellagra; *see* Pellagra
  pigmentation, effect on, 18
  skin flush tests for schizophrenia, 138
Nicotine
  attention deficit hyperactivity disorder (ADHD), as self-medication for, 56
Norepinephrine, 6, 31
Nutraceuticals, 9, 113. *See also specific nutraceuticals*
Nutrient-gene interaction, 61
Nutrigenetics, definition, 9
Nutrition
  genotype to the phenotype translation, role in, 7
  psychiatry, role in, 6–7
  traditional diets, 118
  vegan/vegetarian diets; *see* Vegan/vegetarian diets
  Western diets, 118
Nutritional assessment, 216–217
Nutritional biochemistry, 8–9
Nutritional quality of life (NQOL), 185–186
Nutritional status, assessment for, 217–220

O

Obesity
  malnutrition, as category of, 153
  psychiatric disorder, classification as, 125
  psychological aspects of, 168
Obesity Related Well-Being (ORWELL), 185
Obsessive-compulsive disorder (OCD), 207
Olanzapine, 148
Omega-3 fatty acids, 9, 31
  antipsychotic properties, 140, 221
  attention deficit hyperactivity disorder (ADHD) treatment, 53–54
  bipolar disorder studies, 121
  cognition in midlife, relationship between, 83
  depression, relationship between, 98
  depressive symptoms, relationship between, 99
  hostility in young adults, relationship between, 32
Omega-6 fatty acids, 31
  hostility in young adults, relationship between, 32
Orthomolecular medicine
  acceptance of, 5
  history of, 3–4
  overview, 3–4
  Pauling's studies, 4
  pioneers in, 5
Orthorexia, 154
Osteoporosis, 117
Oxidative stress, 190–191

P

Panic disorder, 119
Parkinson's disease, 115
Parnate, 114
Pauling, Linus, 3, 4
Paxil, 114
Pellagra, 161
Peripheral neuropathy, 18
Pernicious anemia, 226
Phenylalanine, 65–66, 103
Phenylketonuria (PKU), 65–68
Phospholipids, 141
Pica, 42
Polydipsia, psychogenic, 146
Polymorphisms, 62–63
Protein, recommended intake level, 252
Prozac, 114
Psoriasis, 124
Psychiatric Disorders Screening Questionnaire, 21
Psychogenic polydipsia, 146
PsychoNutriologica person, defining, 1

Index

Psychopathic Deviation scale, 156
Pyridoxine, 144–145, 160, 191 *See also* Vitamin B6

Q

Quality of life, overview, 185
Quality of life scores, 112
Questionnaire on Eating and Weight Patterns, 168

R

Ratios
    0–6 : 0–3 ratio, 101
    Copper and other minerals 20, 34,143, 149
    Zinc, 34
Red Bull, 24
Resveratol, 86–87
Riboflavin, 109, 199
Risperidone, 143, 148
Rotterdam study, 101
Roy-Byrne, Peter, 6–7, 106

S

S-adenosyl-methionine (SAM), 113, 124
Satcher, David, 10
Schizophrenia
    caloric needs, 146–147
    copper nutrition in schizophrenic males, 142
    diet history of patients with, 147
    diet pattern of patients with, 147
    EPA/DHA treatment, 139–140
    folate levels, 141–142
    magnesium deficiency, 143
    megavitamin therapy, 144–146
    minerals and antipsychotic medications interactions, 143
    niacin skin flush test for, 138
    nutrition-related comorbidities, 138
    omega-3 treatment, 221
    overview, 137
    oxidative stress, 139
    symptoms, 137
    vitamin C studies, 143–144
    vitamin D levels, 142
    vitamin studies, 4
    water intoxication, 146
    weight gain, medication-related, 148
    zinc nutrition in schizophrenic males, 142
Scleroderma, 115
Scurvy, 105, 163–164
Seasonal affective disorder (SAD), 126
Selegiline, 114
Selenium, 228
    deficiency, 80
    environmental exposure, 34
    mood, relationship between, 112
    quality of life scores, relationship between, 112
    quality of life, relationship between, 205–206
    supplements given in nursing home, case study, 112–113
    toxicity, 204
    uptake, 204–205
Selye, Hans, 189, 208
Serotonin, 31
    depression, role in, 113
    syndrome, 114–115
Single-nucleotide polymorphisms (SNPs), 61
Society of Biological Psychiatry, 2
Somatoform disorders, 207, 208
St. John's wort, 113
Stages of Nutritional Inquiry, xxvii, 109, 237, 255
Starvation
    behavioral changes wrought by, 157–158
    case study, conscientious objectors, 154–158
    case study, self-induced, 105–106
    definition, 153
    emotional changes wrought by, 154–155
    mental changes during, 154–155
    messenger RNA, reduction in, 158
    personality changes wrought by, 154–155
    psychological consequences of, 1–2
Statin drugs, 33
Stress
    bereavement stress, 191–192
    definition, 189
    gastrointestinal symptoms, relationship between, 192
    general adaptation syndrome (GAS); *see* General adaptation syndrome (GAS)
    impact of, 189
    mood, impact on, 190
    oxidative, 190–191
    persistent, 189
    physical response to, 190

Sugar
  ADHD, 54
  Intake, 54
  WHO, 54, 185
  IOM, 54, 25
Suicide
  bipolar disorder patients, in, 119
  cholesterol studies, 126–127, 162–163
  diet and mental health in the arctic, related to suicide, 126
  food insufficiency, relationship to, 128
  leptin studies, 127
  overview, 125–126
  postmortem fatty acids in brain following, 127
  prevalence, 125
  risk factors, 126
Supplements, multivitamin. *See* Multivitamins
Surgeon General's Report on Mental Health, 10
Syndrome X, 87
Systems theory, 8

T

Temporomandibular joint disorder, 208
Thiamin
  absorption, 226
  aggression, links to, 33
  alcohol, interaction between, 18
  deficiency, 18, 19
  mood, impact on, 187
  neuropathy associated with deficiency, 19
  post-bariatric surgery, 175–176
Three-Factor Eating Questionnaire (TFEQ), 170
Tocopherol. *See* E, vitamin
Trimethylaminuira, 71
Tryptophan, 31, 89, 103
  cautions to use, 114
  chronic insomnia treatment, 103, 224
  depression, impact on, 103–104
  metabolism, impact on, 103–104
  pellagra; *see* Pellagra
Tuberous sclerosis, 42
Tyrosine, 6, 103

U

Uridine, 8

V

Valine, 103
Vegan/vegetarian diets
  B_{12} deficiency, linked to, 200
  cognitive performance, relationship between, 76–77
  mood states, relationship between, 115
Violence. *See* Aggression, anger, hostility, violence
Vitamin-dependent enzymes, 62–63
Vitamins. *See also specific vitamins (by letter or name)*
  alcohol, interaction between, 16–17
  genetic stability, relationship between, 61–62
  supplements, relationship to intelligence, 81

W

Wakefield, Andrew, 45
Weight and Lifestyle Inventory, 168
Well-being, definition of, 186
Wernicke–Korsakoff syndrome (WKS), 18
Wernicke's encephalopathy (WE), 18, 226
Williams, Roger, 8
Wilson's disease, 91, 201
Wurtman, Judith, 8
Wurtman, Richard, 6, 8

Y

Yale Food Addiction Scale, 15

Z

Zinc
  alcohol withdrawal and, 20
  attention deficit hyperactivity disorder (ADHD), relationship between, 55
  deficiency, 62, 80–81
  depression, link between, 113
  gray matter, abundance in, 80
  levels of, schizophrenic males, 142
Zoloft, 114